The Purpose Driven Design Framework for Modern Enterprises

A Systemic Approach to Enterprise Architecture Using the Purpose Driven Design Framework

Andre Milchman

Apress®

The Purpose Driven Design Framework for Modern Enterprises: A Systemic Approach to Enterprise Architecture Using the Purpose Driven Design Framework

Andre Milchman
Toronto, ON, Canada

ISBN-13 (pbk): 979-8-8688-2222-3 ISBN-13 (electronic): 979-8-8688-2223-0
https://doi.org/10.1007/979-8-8688-2223-0

Managing Director, Apress Media LLC: Welmoed Spahr
Acquisitions Editor: Aditee Mirashi
Editorial Assistant: Gryffin Winkler

Cover designed by eStudioCalamar

Cover image designed by Freepik (www.freepik.com)

Distributed to the book trade worldwide by Springer Science+Business Media New York, 1 New York Plaza, New York, NY 10004. Phone 1-800-SPRINGER, fax (201) 348-4505, e-mail orders-ny@springer-sbm.com, or visit www.springeronline.com. Apress Media, LLC is a Delaware LLC and the sole member (owner) is Springer Science + Business Media Finance Inc (SSBM Finance Inc). SSBM Finance Inc is a **Delaware** corporation.

For information on translations, please e-mail booktranslations@springernature.com; for reprint, paperback, or audio rights, please e-mail bookpermissions@springernature.com.

Apress titles may be purchased in bulk for academic, corporate, or promotional use. eBook versions and licenses are also available for most titles. For more information, reference our Print and eBook Bulk Sales web page at http://www.apress.com/bulk-sales.

Any source code or other supplementary material referenced by the author in this book is available to readers on GitHub. For more detailed information, please visit https://www.apress.com/gp/services/source-code.

If disposing of this product, please recycle the paper

To Daniel, Kaylee, Nathan, and Sean

Table of Contents

About the Author

Andre Milchman is a hands-on enterprise and software architect with a professional background that bridges civil engineering and computer science. Beginning his career as a construction engineer in the dynamic oil and gas industry, Andre transitioned in the 1990s into computer science, earning a degree and later becoming a Sun Certified Enterprise Architect. His expertise spans systems thinking, sociotechnical systems architecture, advanced analytics, and artificial intelligence. Andre's passion for purposeful systems is reflected in his extensive body of work, which includes *Enterprise Architecture Reimagined, The Triadic Evolution*, and *AI Safety and Security*. He has also co-authored works with Noah Fang, such as *Prescriptive Analytics* and *The Transformable Enterprise*, focusing on aligning technological advancements with strategic organizational design. Currently based in Toronto, Canada, Andre adapted the principles of Purpose Driven Design, originally developed as part of Unit Oriented Architecture, into the Purpose Driven Design Framework, a practical system design guide for a broad audience. This includes solo entrepreneurs, creatives, executives, managers, and professionals interested in designing and constructing purposeful systems that align goals with meaningful outcomes.

About the Technical Reviewer

Uttam Kumar is an Engineering Manager with 17+ years of experience in retail technology, specializing in POS and payment system integration. He currently leads POS modernization and global payment solutions at American Eagle Outfitters, where he has driven innovations in omnichannel commerce and customer experience.

A recognized thought leader, Uttam actively contributes to industry research, including data-driven retail decision-making and human-AI collaboration in POS systems. He also advises on emerging retail technologies and frequently shares insights through publications and professional networks.

Acknowledgments

Introduction

We no longer build systems—we build worlds. Layer by layer, line by line, humanity is constructing realities composed not only of matter, but of data, code, and belief. The physical world expands through infrastructure and machines. The digital world accelerates through algorithms and networks. The social world persists through norms, narratives, and governance. Yet these three worlds—though inseparable—are not being architected together. They collide, drift, and entangle without shared structure. The result is not progress but accumulation. Not coherence but complexity.

Throughout history, humanity has reached for grand structures—civilizations, institutions, digital platforms. But ambition alone does not ensure coherence. Without shared foundations, even the most sophisticated endeavors—like the *Tower of Babel*—falter. The Tower of Babel is not just a story; it is a recurring pattern, a structural condition that haunts the architectures of our most complex modern systems. Its lesson is not about linguistic diversity or divine punishment. It is about the consequences of building without architecture.

Babel Is Not a Myth

The Tower of Babel is often misread as a tale of hubris punished by divine intervention. In its popular form, it recounts how a unified people sought to build a tower that reached the heavens, only to have their language confounded and their efforts scattered. But architecturally, Babel is not a myth—it is a diagnostic pattern. It reveals what happens when complex systems scale without internal differentiation, when coherence is assumed rather than structured, and when coordination is overwhelmed by uniform ambition.

In today's terms, the Tower is not a ruin of ancient stone—it is a modern enterprise rising into the digital heavens, built on clouds yet burdened by unmanaged complexity. It monumentalizes everything. Data lakes become oceans. Strategies become doctrines. Infrastructure becomes bureaucracy. These enterprises ascend ever higher, technocratic by design—structured for hierarchical subordination, governed by abstraction, and steered through centralized control. What appears organized from above often dissolves into confusion below, as purpose is diluted and decision-making becomes detached from consequence.

Their languages are financial, managerial, and technical—optimized for reporting, but opaque to most who participate in or are affected by them. Communication scales; coordination does not. The result is legibility without comprehension, integration without coherence, growth without evolution.

The fragmentation of language in Babel was not a punishment—it was a structural correction. A system that spoke only one language could not sustain differentiation. Diversity was not introduced to destroy the tower, but to prevent its collapse under the weight of its own univocality. In architectural terms, the event did not mark the end of building—it marked the beginning of modularity. The scattered groups did not stop constructing; they began doing so with local languages, bounded contexts, and differentiated purposes.

In this interpretation, Babel is not a warning against ambition. It is a case study in the failure of unstructured scale. It shows what happens when systems are built without a logic of composition, when roles are undefined, when coordination is implicit, and when purpose is divorced from structure. The lesson is not to build smaller—but to build differently. Not to unify by control—but to compose by purpose. That is the foundation of Unit Oriented Architecture and Purpose Driven Design.

Clouds, Monuments, and the Myth of One Language

The Babel pattern did not vanish. It evolved. What once was a singular tower built of stone is now distributed across cloud regions and enterprise roadmaps—scaled, virtualized, and institutionalized. But the structural flaws remain. The modern enterprise continues to reproduce the same conditions that led to Babel's collapse: over-centralization disguised as integration, monumentality mistaken for progress, and communicative sameness mistaken for coherence. These flaws reveal themselves in three dominant forms.

a. ***Cloud-Based Ascent***

   In the modern enterprise, elevation is strategic. Systems are lifted into the cloud not merely to modernize but to transcend. Migration is framed as progress, elevation as inevitability.

The cloud—geographically nowhere, architecturally opaque, and administratively distributed—has become the dominant terrain of contemporary enterprise design. These organizations now operate closer to the heavens than ever before. And while altitude suggests advantage, it rarely guarantees structure.

Proximity to the cloud implies abstraction but also distance: distance from those who govern the systems, distance from those who rely on them, and distance from the foundational clarity that once defined the enterprise's ground-level purpose. In seeking elevation, enterprises lose gravity—platforms float, responsibilities diffuse, and no one quite recalls who laid the first stone.

b. ***Monumentalism of the Platform Era***

At these altitudes, every platform aspires to permanence. Data lakes are scaled without constraint. Integration programs accumulate mandates. Enterprise-wide strategies solidify into doctrine. Scale becomes self-justifying. What begins as an initiative to improve interoperability becomes a monument to centralized ambition. These structures are not composed; they are imposed. They stand tall but are rarely designed to evolve.

Monumentalism resists feedback. Its architectures are meant to be observed, not questioned. The deeper the investment, the harder it becomes to adapt. Every layer added to improve reach or coherence instead amplifies opacity and slows decision-making. Purpose is no longer articulated by the system. It is assumed by those tasked with maintaining its continuity.

c. ***The Technocratic Monolingualism***

Beneath these structures lies the myth of one language. To manage complexity, enterprises reduce communication to a single dominant logic—typically technical, financial, or procedural. These languages are optimized for standardization, not interpretation. What cannot be expressed in the dominant syntax becomes noise. Context collapses into compliance. Difference is treated as inefficiency.

This technocratic compression excludes more than it coordinates. Legal frameworks, ethical reasoning, social knowledge, and human judgment are all present—but unrecognized. The system appears unified but is internally fractured. Communication flows outward; understanding does not return. Coordination is replaced by interface tokens, performance metrics, and synthetic dashboards.

The Tower of Babel did not fall because humans aimed too high. It fell because the structure could not carry differentiated purpose. In architectural terms, the collapse was not a failure of ambition—it was a failure of design. The fragmentation of language was a correction: a reintroduction of modularity, of contextual coherence, of structural realism.

Purpose Driven Design begins where this correction left off. It does not eliminate difference—it structures it by giving it meaning. While the purpose of each sociotechnical entity already exists in implicit form, Purpose Driven Design makes it visible and explicit, showing that Purpose, Value, Function, and the other architectural constructs are not abstractions but meaningful structural expressions that allow difference to coexist coherently. Coherence is not achieved by eliminating multiplicity, but by designing for it. Purpose-bound units, each speaking their own functional language, coordinated through interfaces and governed by architecture, form a system capable of evolving without collapsing.

When Collapse Was a Correction

The collapse of the Tower of Babel is often interpreted as a divine rebuke—a cautionary tale about the dangers of collective arrogance. But structurally, it reveals something more fundamental: systems that scale without internal differentiation eventually fail—not because they are punished, but because they are unsustainable. The collapse was not an interruption of progress; it was a correction of trajectory.

Uniformity offers the illusion of control. A single purpose, a single language, a single logic—these simplify governance in the short term but erode adaptability in the long term. As systems grow, so do their internal contradictions: overcentralized decisions, illegible operations, mismatched incentives, and silent failure modes. The higher the system rises without decomposing, the more brittle it becomes. Collapse, in such cases, is not a malfunction. It is the system's only remaining form of self-regulation.

This pattern repeats in modern organizations and infrastructures. Strategic programs span so many domains that no single entity can govern them meaningfully. Platforms accumulate functionality until they become immobile. Compliance mechanisms grow dense while interpretability thins. By the time failure occurs, the root cause is not always visible—but it is almost always structural. What fails is not the original intent. It is the architecture that was never equipped to support differentiated purpose.

The collapse of Babel made visible what the structure concealed: that a singular, undifferentiated purpose cannot sustain complexity. What followed was not the end of building, but the beginning of purpose-specific composition—where differentiated structures became necessary for sustainable scale. A mono-purpose system gave way to a multi-purpose world. This transformation is foundational: it marks the shift from the logic of uniformity to the logic of composition.

Unit Oriented Enterprise Architecture, described in my previous book *Unit Oriented Enterprise Architecture: Constructing Large Sociotechnical Systems in the Age of AI*, introduced the idea that enterprises should be composed of bounded, purpose-oriented sociotechnical units—each accountable, interoperable, and capable of evolving independently within a coherent system. This architectural style decomposes the Tower, replacing the monolith with a network of structurally bounded, autonomous sociotechnical units. That work focused on external composition: how enterprises should be broken down into accountable, evolution-ready components. *Purpose Driven Design*, the present volume, focuses on internal construction: how each unit should be designed to retain coherence, express purpose, and operate intelligibly within the larger system.

Collapse, in this framework, is not something to fear. It is something to anticipate and structurally prevent—by designing units that carry purpose, express value, and coordinate through architecture, not control.

From Alienation to Architecture: Belonging As Structural Inclusion

Large systems tend to alienate. As they scale, the clarity of individual placement erodes. Roles become abstract, decision rights become ambiguous, and work becomes decoupled from outcomes. Participants can no longer see where they fit, why their efforts matter, or how they relate to the larger whole. This is not merely a cultural failure. It is an architectural one. Belonging is not an emotional state to be cultivated after the fact—it is a structural feature that must be designed into the system itself.

In Purpose Driven Design, every unit must belong in two senses. First, it must belong *systemically*—as a clearly positioned part of a larger sociotechnical architecture. It must be structurally defined, internally coherent, and externally interoperable. Second, it must belong *environmentally*—in relation to the broader world of entities it serves and responds to. This includes customers, partners, regulators, investors, tax authorities, and the general public. A unit that cannot relate to its environment cannot generate value; a unit that is not structurally placed within a system cannot sustain meaning.

This dual belonging is enabled by design. A *purposeful* unit possesses desire, intention, and the capacity to serve real needs in its environment. It is not simply performing tasks—it is pursuing structured value on behalf of others. A *fully accountable* unit maintains internal control over its own structure, processes, and outcomes. It does not delegate responsibility beyond its boundaries, nor obscure it within them. An *appreciative* unit recognizes its own interdependence—it sees the system it belongs to, not merely the outputs it produces. It designs with awareness of upstream dependencies, downstream consequences, and adjacent impact. A *legitimate* unit is one that earns trust not through claims but through structure: it can be understood, evaluated, and aligned with a larger societal trajectory. Its existence contributes to the sustained adaptability and directionality of the world it inhabits. These four qualities—purposeful, accountable, appreciative, and legitimate—form the structural basis for architectural belonging.

Decomposing the enterprise into such units is not just a matter of efficiency or manageability. It is a structural response to alienation—an architectural method for restoring placement, orientation, and accountability. In this model, units are not fragments. They are integrated entities: positioned within a system and oriented toward an environment. Belonging is not asserted—it is composed.

Decomposition, Not Demolition: From UOEA to Purpose Driven Design

The Tower of Babel did not need to be destroyed—it needed to be decomposed. Its collapse was not a verdict against building, but a revelation that the structure could not bear the scale and uniformity it had imposed. In architectural terms, what failed was not the ambition but the absence of modularity, intelligibility, and differentiated purpose. That lesson remains urgent today. As modern enterprises grow in size and complexity, the answer is not demolition but architectural decomposition.

Unit Oriented Enterprise Architecture addressed this challenge directly. It proposed a structural alternative to overcentralized, fragmented, and opaque enterprise systems by decomposing them into *networks of bounded, autonomous, sociotechnical units.* These units are not technical modules or business domains—they are accountable architectural components, each responsible for delivering value, maintaining coherence, and coordinating with others. UOEA reframed the enterprise not as a singular structure but as a system of systems.

Purpose Driven Design, the present volume, continues this architectural trajectory. If UOEA answered the question *"What should the enterprise be made of?"*, PDD addresses the next question: *"How should each of those components be designed?"* Where UOEA decomposes, PDD composes. It defines the internal architecture of each unit—ensuring that purpose, value, process, accountability, and adaptability are structurally embedded.

This is not a pivot from one framework to another. It is a continuation of the same architectural logic—moving from enterprise-level organization to unit-level composition. Together, UOEA and PDD form a two-tiered architecture: *the system of units*, and *the architecture of the unit*. Decomposition is not an end. It is the structural precondition for purposeful composition.

The Structural Grammar of Purpose

If decomposition identifies the units, composition defines their architecture. Purpose Driven Design provides a structural grammar for composing each unit—not stylistically but architectonically. This grammar is not a collection of best practices or functional checklists. It is a set of foundational principles that make a unit structurally coherent, externally legible, internally governable, and evolution-ready. Each principle addresses a necessary architectural function—together they form the blueprint for creating purposeful sociotechnical units.

The grammar consists of nine foundational principles:

- **Purpose As Architectonic Driver**

  Every unit must begin with Purpose—not as aspiration but as structural orientation. Purpose answers why the unit exists and whom it serves, defining its structural relevance within the system and the environment.

- **Value As Justification**

  Units must manifest Purpose through externally observable Value. Value is not claimed—it is realized through meaningful service to customers, regulators, partners, or other stakeholders.

- **Function As Interface**

  Function defines how a unit interacts with its environment and with other units. It includes both technical and social interfaces and determines how value is exchanged, monitored, and governed.

- **Process As Operational Logic**

  Process articulates how value is produced. It includes activities, decision logic, sequences, and rules—designed to be repeatable, transparent, and aligned to Purpose.

- **Structure As Agent Configuration**

  Structure defines the arrangement of agents—human and artificial—and the roles they perform. It establishes clarity in accountability, capability, and escalation.

- **Order As Constraint Mechanism**

  Order governs behavior through rules, escalation paths, policies, and commitments. It makes coordination legible and enables resolution when ambiguity or conflict arises.

- **Culture As Work Climate**

  Culture shapes behavior through norms, values, and judgment patterns. It is not decorative—it governs how decisions are made when procedures are insufficient.

- **Memory As Structural Continuity**

  Memory ensures that the unit learns, evolves, and remains traceable. It encompasses records, histories, rationales, and decision paths that can be interpreted over time.

- **Storage As Readiness**

  Storage includes the preservation of resources, deliverables, and assets necessary for functioning and adaptation. It ensures that value is not only produced but retained and made available.

This grammar is not sequential but compositional. The nine principles do not form a life cycle or a checklist—they form a *structural system*, each reinforcing the others. A unit without Process may have Purpose but cannot deliver. A unit without Memory may perform well now but cannot evolve responsibly. Coherence is achieved when all nine principles are addressed as architectural concerns, not procedural add-ons.

Purpose Driven Design offers this grammar as a way to build sociotechnical units that are not merely functional but structurally intelligible—capable of sustained contribution, principled coordination, and adaptive evolution within a complex system.

Evolving Without Collapsing: Why Architecture Now

Enterprises are no longer isolated machines. They are nodes in global, interdependent sociotechnical systems—linked to markets, regulations, digital platforms, and societal expectations. Their internal structures produce external consequences. Their decisions ripple across ecosystems. In this environment, the ability to evolve without collapsing is not a competitive advantage; it is a structural necessity.

The scale, speed, and abstraction of modern systems have outpaced the structures meant to contain them. Cloud platforms scale compute before coordination. AI agents make decisions before accountability is defined. Strategic programs launch before structural readiness exists. The result is an architectural crisis: systems accumulate mass faster than meaning. The problem is not complexity itself but the lack of structure to absorb, orient, and evolve with it.

Architecture, in this context, is not an aesthetic discipline or a planning function. It is a means of survival. Without it, systems revert to reactive adaptation: patching symptoms, institutionalizing exceptions, or centralizing control. These responses exhaust capacity without resolving structural misalignment. Sustainable evolution requires an architecture that can change—intelligently, intentionally, and without systemic breakdown.

Purpose Driven Design provides that architecture at the level of the unit. It is not a blueprint for stability but a grammar for transformation. When units are structurally coherent, purposive, and accountable, they can evolve independently while remaining interoperable. They do not require orchestration through hierarchy or constant reinforcement through culture. Their structure carries their logic. Their design embeds their role in the larger system.

This is why architecture now. Not because change is new—but because the conditions of change have become unmanageable without structural alignment. Purpose Driven Design is not a framework for better projects. It is a framework for building the kinds of components a volatile, intelligent, interdependent world can live with—at scale, under pressure, and over time. The goal is not to resist collapse through force. It is to evolve without collapsing—by design.

To support that evolution, this book does not aim to prescribe every solution but to provide the conceptual structure from which enduring ones can be built.

This volume should therefore be read as a foundation rather than a manual. It establishes the architectural grammar of Purpose Driven Design—the principles, constructs, and relationships from which further specialization can evolve. The book's aim is to clarify structure and meaning, not to prescribe every implementation detail. Future practitioners and authors may extend these ideas with case studies, diagrams, and code; this work provides the conceptual ground on which such elaborations can stand.

CHAPTER 1

From Towers to Fleets: The Architectural Shift Behind Purpose Driven Design

The collapse of the *Tower of Babel* marked more than the failure of a singular structure—it marked the end of an architectural era. What followed was not a rejection of ambition, but a redirection of it. That alternative is captured in the *Fleet* metaphor, which inspired the Unit Oriented Enterprise Architecture framework. The Fleet stands in structural opposition to the Tower: not a singular monument, but a system—composed of structurally distinct, navigationally autonomous, and interoperable units. Where the Tower sought uniformity through elevation and centralized control, the Fleet organizes outward through purposeful differentiation. It does not ascend; it aligns. Before we can design these units, we must first understand the architectural shift that made them not only possible—but necessary.

Unit Oriented Enterprise Architecture, introduced in *Unit Oriented Enterprise Architecture: Constructing Large Sociotechnical Systems in the Age of AI*, established the structural foundation for what followed. It reframed the enterprise not as a monolith but as a network of structurally bounded, purpose-oriented sociotechnical units—each designed to hold its own roles, rules, and responsibilities. That shift from singular tower to coordinated fleet made possible a new form of architectural viability: one in which structure enables adaptability, not merely control. Purpose Driven Design begins where that foundation ends—by defining how each unit must now be internally composed.

A. Milchman, *The Purpose Driven Design Framework for Modern Enterprises*,
https://doi.org/10.1007/979-8-8688-2223-0_1

KEY CONCEPTS OF UNIT ORIENTED ENTERPRISE ARCHITECTURE (UOEA)

Unit Oriented Enterprise Architecture reimagines the modern enterprise as a fleet of bounded, purpose-oriented sociotechnical units—each integrating human (crew) and digital (craft) elements within a coherent, accountable structure.

Core Ideas

Fleet Metaphor: Enterprises are designed as networks of autonomous yet coordinated units, much like ships in a fleet—each distinct in function but aligned toward a shared mission.

Dual-Level Design: Architecture operates at two levels: the macro level (a flat network of units) and the micro level (internal alignment of social and technological components within each unit).

Purpose-Driven Wholeness: Each unit has a defined purpose guiding its actions and relationships with other units and stakeholders.

Core Principles: Harmonious integration of social and digital realms, coherence in purpose implementation, and a flat, system-of-systems structure.

Benefits: Enhanced clarity, accountability, and coordination; scalable, adaptable structures; empowerment of humans through control over digital resources; and alignment between business and technology.

UOEA thus provides the structural foundation for Purpose Driven Design—moving from the external composition of enterprises to the internal design of each purpose-driven unit.

1.1. Beyond the Collapse: From Monument to Motion

The collapse of the Tower did not eliminate the need to build—it exposed the limits of building upward alone. The Tower embodied an architectural logic rooted in singular ambition: centralized coordination, rigid hierarchy, universal abstraction. It reached toward the heavens but failed to evolve outward. Its scale increased, but its adaptability deteriorated. Its language expanded, but its meaning fractured. When it fell, the cause was not external disruption—it was internal contradiction.

In the wake of this collapse, the challenge was not whether to build again but how to build differently. Structure could no longer mean vertical accumulation. Architecture could no longer assume universality. The solution was not to abandon ambition, but to rearticulate it across differentiated, navigable components. Thus emerged the architectural shift from monument to motion—from Tower to Fleet.

The Fleet is not a metaphor for decentralization alone. It is a metaphor for *coordinated autonomy*: a system in which each component is bounded, accountable, and capable of purpose driven action within a larger navigational logic. Unlike the Tower, the Fleet does not collapse under its own scale. Its ships move independently yet remain in formation, able to absorb change without structural failure. It trades uniformity for composability.

This shift is not merely aesthetic. It demands a redefinition of architectural primitives. The Tower relied on central control and architectural totality. The Fleet depends on intelligible units—each with internal integrity and interoperable form. It is not the absence of architecture; it is architecture restructured.

What follows in this chapter is a detailed exploration of the context and this transition. We will examine why the Tower's collapse was inevitable, what structural assumptions it encoded, and how the Fleet introduces a viable architectural alternative. From this foundation, *Purpose Driven Design* will then proceed inward—into the units themselves.

1.2. The Two Triads: Evolution and Structure

To understand what must be architected, we must first understand how we evolve—and what must be held together for that evolution to remain viable. In Purpose Driven Design, this is expressed through **two triads**: the first describes the forces shaping our evolution; the second describes the domains that every sociotechnical system must structurally align.

1.2.1. Triadic Evolution: How Humans Evolve

Human systems evolve through three intertwined trajectories:

- *Evolution by Natural Selection*: The biological dimension—emergent, environment-driven, and non-teleological. It shapes the body, the senses, and the brain through inheritance and survival.

- *Evolution by Organization*: The social dimension—driven by coordination, institutional memory, and the creation of roles, rules, and meaning. It enables cooperation, governance, education, and cultural reproduction.
- *Evolution by Extension*: The technological dimension—shaped by tools, language, computation, and system design. It externalizes memory, amplifies capability, and increasingly modifies the conditions of natural and social evolution itself.

These three modes are not sequential or exclusive. They operate concurrently, with growing interdependence. What once evolved biologically is now reconfigured organizationally—or directly extended through artificial systems. As extensions accelerate, *architecture becomes an evolutionary variable*: it defines what can be coordinated, delegated, stored, or changed without collapse.

1.2.2. The Social–Physical–Digital Triad: What Architecture Must Hold Together

Sociotechnical systems are not constructed in abstract space. They are built across three interdependent domains of human activity:

- *Social*: The domain of relating. Here, humans *associate, organize, empathize, communicate, and coordinate*. It includes trust, legitimacy, rules, belonging, and the human dimension of judgment and purpose.
- *Physical*: The domain of making. This includes *processing, transporting, storing, and exchanging matter*—through machines, labor, energy, infrastructure, and material workflows. It grounds systems in space, cost, and constraint.
- *Digital*: The domain of computing. In this domain, we *process, store, transport, and exchange representations*. Algorithms, data structures, software systems, and models operate here—detached from material form but dependent on physical substrates.

These domains overlap in every real-world system. A healthcare platform, a logistics operation, or a tax authority is never purely digital, purely physical, or purely social. Yet many architectures treat them as separable or subordinate—building digital abstractions that ignore social implications or optimizing physical flows without coordinating meaning or governance.

Legible, resilient, and ethical architecture requires *alignment across all three domains*. Misalignment leads to systems that are efficient but illegible, connected but incoherent, or scalable but socially unstable. Every unit must be composed with an awareness of its social legitimacy, physical constraints, and digital capabilities—not as checklists, but as structurally integrated concerns.

Together, these two triads define the context for Unit Oriented Architecture and Purpose Driven Design. Architecture is not only about how systems are built. It is about how human evolution is supported, guided, and made accountable through structure. What we build must reflect what we are becoming—and what we intend to remain.

1.3. Why the Tower Failed: Misalignment, Monolog, Monolith

The Tower did not fail because it lacked ambition. It failed because it lacked structural coherence. Its singular form was not matched by differentiated function. Its expansive language promised universal coordination, yet silenced context and constraint. Its scale concealed the absence of adaptability. The Tower was not defeated by external adversity; it collapsed under the weight of its own architectural logic.

Three core characteristics defined the Tower: *misalignment*, *monolog*, and *monolith*. Each contributed not only to its instability but to the pattern of systemic failure that continues to manifest in modern sociotechnical systems.

Misalignment begins subtly—when purpose becomes diffuse, functions drift, and structures persist out of inertia rather than necessity. In Tower-style systems, functions often expand without recalibration. Capabilities are added, not integrated. Accountability becomes distributed so thinly that no decision can be traced, and no outcome owned. The system grows, but its logic fragments. Eventually, direction is lost—not because no one is steering, but because no one knows where the structure is meant to go.

Monolog refers to the unidirectional flow of language, authority, and logic. In the Tower, one architectural vision dominates: managerial, financial, or technical—much like when a technology company allows its engineering logic to overrule ethical, social, or human considerations, translating every discussion into efficiency metrics and release cycles. Alternate perspectives are not coordinated—they are suppressed. Even when feedback loops exist, they are decorative, not structural. This breeds opacity. Systems begin to communicate more, but understand less. The language of decision-making becomes increasingly abstract—fluent in performance indicators, transformation roadmaps, and optimization targets—yet detached from operational reality.

Monolith is the outcome: a structure that is large, rigid, and internally homogeneous. Its uniformity is mistaken for order. It treats coherence as something imposed from above, not composed from within. The monolith resists variation. It cannot tolerate local adaptation without compromising global integrity. As a result, it either centralizes every exception or breaks under exceptions it cannot absorb.

These characteristics are not historical relics. They remain prevalent in modern organizational designs that confuse integration with sameness, alignment with subordination, and control with comprehension. Platforms are scaled without differentiated interfaces. Data is consolidated without purposeful segmentation. Transformation programs are launched without decomposable structures. Misalignment persists, not in the absence of planning but in the absence of architectural reasoning.

The Tower failed because its structure could not evolve. It optimized for scale, not structure—for uniform control, not adaptive coordination. Its collapse was not a pause in progress, but a structural correction. What followed was not silence—but the search for a new architectural form. The Fleet emerged not as a retreat from ambition, but as a reconstitution of it.

1.4. The Fleet As Architectural Response

The collapse of the Tower demanded not only a critique of its form but the proposal of a viable architectural alternative. That alternative is the *Fleet*—a metaphor not of fragmentation, but of *coordinated differentiation*. If the Tower symbolized centralization, uniformity, and vertical ambition, the Fleet represents distributed coherence, modular accountability, and purposeful directionality.

The Fleet is not a singular structure. It is a system of systems—a network composed of distinct yet interoperable components. Its power lies in its *structural grammar*: every fleet, regardless of its domain, maintains a composable architecture that accommodates variation without sacrificing coordination.

1.4.1. The Architectural Composition of Sociotechnical Formations

The Tower sought unity through elevation. The Fleet, by contrast, achieves coherence through composition. But even within the Fleet metaphor, architectural differentiation matters. Not every structure is a fleet. Some are flotillas—smaller, more agile formations. Others are armadas—large, federated constructs coordinating multiple fleets across geographies or domains. Purpose Driven Design operates at the level of the unit, but every unit exists within a broader structural form. Understanding these nested formations is essential for architectural precision.

The Armada

The *Armada* represents the highest-order structural formation: a federation of fleets, each potentially representing a region, business group, or legal entity. Armadas are characteristic of large, complex, multinational enterprises or distributed institutional structures. Each Armada includes a single *flagship*, responsible for strategic navigation, regulatory posture, cross-fleet coordination, and systemic legitimacy. The flagship does not control the Armada—it integrates it. Armadas must balance global coherence with local viability. The architecture of an Armada demands explicit structural agreements between fleets, without flattening their individual contexts.

The Fleet

The *Fleet* is the canonical enterprise form: a structurally coherent system composed of multiple flotillas and stand-alone units. A fleet may span several functional domains, jurisdictions, or products, but it is governed through shared architectural principles, clear coordination boundaries, and purpose-aligned composition. A fleet includes a *command ship*, responsible for synchronizing flotilla activity and interfacing with the flagship above or with external entities. While the Armada governs strategic federation, the Fleet governs operational orchestration. It is at the fleet level that structure, policy, and direction converge.

The Flotilla

The *Flotilla* is a cluster of units organized around a shared operational mission, functional theme, or localized scope. Flotillas are typical of business units, service areas, or product groups within a fleet. Each flotilla includes its own *command ship*, tasked with localized decision-making, coordination, and escalation. What distinguishes a flotilla is not size, but *semantic coherence*—its units are structurally aligned and collectively accountable for a defined area of purpose. A flotilla enables agility without fragmentation and supports learning, prioritization, and adaptation at a manageable scale.

The Unit (Ship)

The *Unit*—or ship—is the foundational architectural element. It is a bounded sociotechnical system designed using the grammar of Purpose Driven Design. A unit is not an implementation detail. It is a complete architectural component: *purposeful, accountable, coherent, interoperable, and evolution-ready*. Each unit must carry its own structure, retain its own memory, enact its own function, and interface legibly with its environment. Units may exist independently or within flotillas, but their viability does not depend on constant external governance. They are the vessels that make the Fleet navigable.

Across all levels, these formations share a common design principle: *flat composition within clear architectural boundaries*. The Fleet metaphor is powerful not because it evokes machinery or scale but because it expresses *structural clarity, role differentiation, and compositional coherence*. Armadas are not Towers in disguise. Their viability depends not only on the soundness of each fleet, the navigability of each flotilla, and the structural integrity of each unit—but also on the differentiated reach each formation enables: global, regional, thematic, and local.

In *Purpose Driven Design*, this structure is not decorative. It is foundational. Every chapter that follows will focus on how to architect the unit—the vessel upon which all higher formations ultimately depend.

Regardless of how the Fleet is composed, it preserves *modularity, traceability, and a networked form*.

1.4.2. Integration and Differentiation

Every well-structured Fleet exhibits two zones:

- *An integrative core*: Composed of the flagship and/or command ships—which is responsible for *systemic coherence, inter-fleet coordination*, and alignment with environmental conditions and long-range navigation.
- *An operational periphery*: Comprising vessels focused on execution, exploration, logistics, and other functional missions. These ships do not coordinate the Fleet but fulfill roles in alignment with its navigational logic.

The core ensures that the periphery remains *composable*. The periphery ensures that the core remains *relevant*.

This separation of roles does not weaken unity—it preserves viability. A Tower centralizes everything, mistaking concentration for coherence. A Fleet distributes responsibility *architecturally*: integrative ships manage systemic concerns (navigation, legality, provisioning, mission continuity), while operational ships execute defined processes within their own boundaries.

1.4.3. Domain-Agnostic Design

The Fleet metaphor transcends medium. It is not confined to oceans.

- *Space Fleets* coordinate satellites, drones, or exploratory missions.
- *Air Fleets* navigate layered airspaces with differentiated altitude, range, and mission scope.
- *Ground Fleets*—logistics networks, supply chains, mobile units—structure execution across varied geographies.
- *Underwater Fleets* operate with silent coordination, long-memory systems, and constrained communication.

Each fleet—whether imagined in space, air, ground, or sea—shares structural characteristics: boundary clarity, interface intelligibility, role accountability, and composable alignment. These variations are not architectural prescriptions, but *sources of metaphorical insight*. They offer designers and architects a broader conceptual palette for imagining structure, coordination, and adaptation across diverse contexts.

1.4.4. Purpose Specific Composability

A Fleet is never just a collection of vessels. It is a *purpose structured system*:

- A *Treasury Fleet* focuses on secure transfer, accounting, and value integrity. Its operational ships may include financial processors, regulatory interfaces, and audit vessels.
- An *Exploration Fleet* prioritizes adaptability and reach. Its vessels are optimized for sensing, experimentation, and mapping.
- A *Pirate Fleet*, though illegitimate, still follows a form: opportunistic, fast-moving, loosely structured around extractive operations.

Purpose defines structure. No one structure suits all purposes—but all structurally coherent fleets share *internal grammar*: orientation, function, process, structure, control, and memory. These examples—whether treasury, exploration, or even pirate fleets—are not taxonomies, but *narrative scaffolds*. Architects and designers are encouraged to draw from a wider repertoire of missions, both real and speculative: defense, diplomacy, colonization, humanitarian aid, logistics, surveillance, and beyond. Each type offers distinct structural implications, trade-offs, and coordination patterns. These variations are not prescriptions for enterprise form, but *conceptual instruments*—models for clarifying what a system must do, how it must operate, and why it must hold together.

1.4.5. The Fleet As Precondition for Unit Design

Unit Oriented Enterprise Architecture proposed the Fleet as an answer to the Tower—not merely as metaphor, but as structural architecture. In this framing, *the unit is the ship*. It is bounded, purpose driven, and situated within a navigable, interoperable system.

But designing a fleet is only half the challenge. To sustain viability, *each ship must be structurally sound*—coherent internally, responsive to its environment, and legible to other ships. That is the work of Purpose Driven Design: to define the grammar that allows each unit to hold structure, carry function, and sustain purpose under conditions of uncertainty.

The Fleet corrects the Tower. But without structurally designed ships, even fleets fail. The next section begins the transition inward—from systems of coordination to the structures that make coordinated systems possible.

The richness of the Fleet metaphor is not accidental. It is architecturally useful precisely because it is multidimensional—spanning physical domains, mission types, and structural roles. For designers and architects, this expressiveness is not decorative; it is functional. It supports imagination without sacrificing precision. It invites one to think structurally and compositionally at once—to construct systems that are not only operable and accountable, but also navigable, reconfigurable, and narratively coherent. The Fleet does not just correct the Tower. It enables the design of viable, evolvable futures.

1.5. Seeing Structure Again

The Tower collapsed not only because of how it was built—but also because no one could see how it was built. Structure had become implicit, submerged beneath abstraction, hidden behind performance indicators, and buried under process documentation. Coordination was assumed rather than designed. Interfaces were emergent rather than intentional. What passed for architecture was often just scale.

The Fleet reintroduces *architectural visibility*. It forces structure back into focus—not as a control layer, but as a design surface. Each vessel must be distinct, comprehensible, and aligned. Each interaction must be mediated by intelligible boundaries and accountable roles. Structure becomes observable again, not in the form of hierarchy charts or compliance matrices but through clearly defined *units of purpose.*

Yet for many modern organizations, structural invisibility remains the default condition. Consider systems where no one knows which team owns a workflow, or where a data pipeline has no clear steward, or where decisions must pass through six committees with no architectural logic for escalation. These are not isolated inefficiencies; they are symptoms of *unstructured complexity*—systems that are dense in effort but poor in design.

Structural invisibility is reinforced by tooling. Contemporary platforms encourage layering without composition. Dashboards proliferate, integrations multiply, and automations accumulate—yet the underlying structure remains unclear. What connects to what? Who governs whom? How do changes in one subsystem affect another? Answers are elusive, not because they are complex, but because they were never structurally defined.

To design viable systems, we must first *recover structure as a primary concern*. This means distinguishing between workflows and architectures, between connectivity and coordination, between visibility as transparency and visibility as comprehension. A structure is not visible because it is documented; it is visible because it is designed to be *legible, inspectable, and governable.*

In a Fleet, this visibility is not optional. Each ship must be *structurally coherent and contextually intelligible.* No command ship can coordinate what it cannot see. No operational ship can operate sustainably without knowing where it fits. Visibility is not the opposite of complexity—it is what makes complexity tractable.

Purpose Driven Design does not begin with analytics or measurement. It begins with *structure that can be seen*, understood, and evaluated. It assumes that clarity is not a luxury—it is a precondition for architectural viability. Only when structure becomes visible again can the enterprise evolve without collapse.

1.6. The Logic of Decomposition

After the collapse of the Tower, it became clear that structural visibility alone was insufficient. Seeing complexity was necessary—but not yet sufficient—for managing it. The next architectural move was decomposition: not fragmentation but the *purposeful partitioning of structure into intelligible, bounded components.*

In Unit Oriented Enterprise Architecture (UOEA), decomposition is not a technical maneuver. It is a structural strategy that begins with the recognition that no single system can remain coherent at scale unless it is *differentiated by design*. The aim is not to break the enterprise into pieces, but to *compose it through autonomous units* that are accountable, governable, and interoperable. Each unit is bounded, purposeful, and structurally complete—not a team, not a microservice, not a platform slice but a *sociotechnical system* capable of sustaining value on its own terms while aligning with broader enterprise intent.

This type of decomposition differs from conventional modularization. It is not based on technical interfaces alone or on superficial business functions. It is based on the identification of distinct *domains of purpose* that require differentiated coordination, differentiated risk tolerance, and differentiated architectural expression. In other words, a unit is created not because it fits neatly into an org chart or API map but because it *serves a specific function that cannot be meaningfully absorbed elsewhere* without distortion or opacity.

To decompose well requires more than dividing responsibilities. It requires designing each unit so that it can be governed from within. A unit must hold *its own structure, memory, processes, and interfaces*. If it cannot account for its actions, preserve its knowledge, coordinate its roles, and communicate with others intelligibly, then it is not a unit—it is a fragment.

The strength of decomposition lies in what it prevents. It prevents cross-cutting dependencies from becoming entanglements. It prevents central platforms from being bottlenecks. It prevents one-size-fits-all solutions from overrunning context. And it prevents systems from scaling beyond the limits of comprehension.

The Fleet offers a metaphor for how decomposition can be structured without disintegration. Its coherence does not depend on uniformity, but on *clarity of structure and fitness of purpose*. Each vessel is designed to function autonomously but not independently. It knows its mission, carries its own means of coordination, and integrates through interfaces—not through compliance with a single command logic.

The logic of decomposition is ultimately an ethical stance. It refuses to centralize what should be localized. It avoids abstracting out the very constraints that make systems intelligible and accountable. It treats scale not as a goal but as a condition that requires structure.

Purpose Driven Design picks up from this logic. Having decomposed the Tower into a system of vessels, the challenge now becomes: *How should each vessel be built? What internal architecture must a unit possess to fulfill its purpose, maintain its coherence, and evolve without collapse?* Decomposition is not the end. It is the precondition for compositional design.

1.7. Composing the Unit: Craft, Crew, and Cohesion

Decomposition makes the system navigable. But decomposition alone does not make the parts viable. A fleet may be well-structured on paper, but if its vessels are hollow, misaligned, or unsustainable, coordination becomes a burden, and coherence breaks down. To make decomposition meaningful, each part must be designed with architectural integrity. This is the task of unit composition—*the integration of craft, crew, and cohesion into a purposeful whole.*

In *Purpose Driven Design,* the unit is the primary building block of sociotechnical architecture. A unit is not a team or a function. It is a *bounded, accountable, purpose-oriented sociotechnical system,* composed to produce value for defined entities in its environment. It includes both human agents and technical components, and it must be architected in such a way that its structure holds under conditions of execution, adaptation, and alignment.

1.7.1. Craft

The *craft* of a unit refers to its structural and technical form. It includes the roles, workflows, interfaces, storage, control paths, and technical assets through which the unit functions. Craft is what enables the unit to operate with repeatability and to be externally legible. If the structure is not sound, the unit becomes fragile: processes break, interfaces degrade, and accountability diffuses. Good craft ensures that the unit has shape, that its operations are traceable, and that its boundaries are real rather than rhetorical.

1.7.2. Crew

The *crew* comprises the human and artificial agents who act within the structure: those who interpret purpose, exercise judgment, and take responsibility for execution. A unit cannot be architected around software or policies alone. Without humans to instantiate roles, respond to ambiguity, and maintain alignment, the unit becomes a mechanistic shell. Crew is not simply a list of job titles. It is the living dimension of the unit—the source of both decision-making and adaptability. Well-composed units explicitly link structure to roles and roles to responsibilities. The crew is embedded in the craft, not layered on top of it.

1.7.3. Cohesion

Cohesion is what binds the craft and crew into a functioning whole. It is the internal logic of alignment—between purpose and value, function and structure, decisions and memory. A unit without cohesion may have tools, people, and processes, but it lacks integration. Signals are missed, priorities shift unpredictably, and responses fracture under pressure. Cohesion ensures that the unit's components work together rather than in parallel, that the structure supports the crew, and that actions converge toward purpose without constant managerial intervention.

Cohesion gives the unit its *integrity*. A well-composed unit is

- *Purposeful*: Oriented around a mission that is externally relevant and internally actionable
- *Structured*: Designed to allocate roles, support processes, and maintain accountability
- *Coherent*: Able to coordinate internally without reliance on external micromanagement
- *Interoperable*: Capable of integrating with other units through clear, bounded interfaces
- *Adaptable*: Able to learn, remember, and revise its operations over time

These are not aspirations. They are architectural consequences, as summarized in Table 1-1. Without them, the unit becomes brittle, opaque, or dependent. With them, the unit becomes structurally legitimate—able to contribute meaningfully to the coherence of the larger system.

Table 1-1. *Craft Provides the Vessel, Crew Provides Agency, and Cohesion Provides Alignment—Together Forming a Unit Capable of Purposeful Action*

Element	Focus	Relationship to the whole
Craft	Structural and technical form—the architecture, tools, and processes that enable operation	Provides the vessel through which Purpose is enacted
Crew	Human and artificial agents who interpret Purpose and operate the Craft	Animates and maintains the unit's functionality
Cohesion	The alignment of Craft and Crew around shared Purpose	Ensures the unit acts as an integrated whole

Composing a unit is not about assembling inputs. It is about constructing a system capable of sustaining purpose. *Purpose Driven Design* provides the grammar required to do so: not a set of templates but a compositional structure. The chapters that follow elaborate this grammar—nine principles through which units gain internal strength, external alignment, and the capacity to evolve without collapsing. Cohesion is not an outcome of scale. It is an outcome of design.

1.8. Triadic Evolution and Fleet Viability

Every viable fleet is the result of aligned evolution—not only of its vessels but of the structural forces shaping their design. These forces are not technological alone. Every unit in a sociotechnical system must navigate three interdependent evolutionary domains: the *social*, the *physical*, and the *digital*. Together, they form the triad through which enterprise design must be interpreted and composed. A fleet that ignores this triad becomes unstable. A unit that misaligns with it becomes irrelevant.

Triadic Evolution frames these domains not as abstract influences, but as concurrent trajectories of development that structure human systems:

- The *social domain* evolves through norms, roles, accountability, trust, and institutional purpose.
- The *physical domain* evolves through material constraints, energy, time, geography, and embodiment.
- The *digital domain* evolves through abstraction, automation, connectivity, and computation.

These domains are not interchangeable. They operate at different paces, obey different logics, and introduce different risks. Yet modern enterprises routinely treat them as equivalent, or worse, subordinate one to another—typically privileging the digital over the social or the physical. The result is a proliferation of systems that are technologically sophisticated but socially incoherent and physically misaligned.

A structurally viable fleet must therefore ensure that each unit is *triadically composed.* No vessel in a real-world fleet is purely digital, purely physical, or purely social. Each integrates all three domains—whether visibly or through embedded dependencies. A logistics unit may appear operationally physical, but its coordination relies on digital infrastructure and social contracts. A compliance unit may appear social, but its enforcement depends on physical workflows and digital traceability.

When one domain dominates unaccountably, misalignment follows. Consider digital-first initiatives that abstract away physical constraints—only to discover that supply chains, human attention, and legal jurisdictions are not virtualizable. Or consider social efforts, such as culture programs or purpose statements, that fail because digital systems cannot express or reinforce the intended norms. Or physical infrastructure designed without anticipating the way humans interact with it or algorithms govern access.

Triadic misalignment is not cosmetic. It is architectural. It results in broken coordination, unpredictable latency, governance opacity, and eventual loss of legitimacy. In a fleet, where interoperability, reliability, and navigability are paramount, this misalignment is not localized—it propagates. When a financial institution launches an AI-driven lending platform without aligning digital models with physical operations or social accountability, the results are immediate: algorithms approve loans faster than compliance teams can audit, front-line staff cannot explain outcomes to customers, and regulators lose trust in the system's integrity. A single misalignment in one domain cascades across all three, undermining both efficiency and legitimacy.

The principle of *triadic coherence* does not imply that each unit must equally balance all three domains. Rather, it asserts that *each domain must be explicitly considered in design.* A structurally viable unit does not have to simulate balance—it must *acknowledge dependency* and avoid blind spots. This means recognizing where decisions are social but implemented digitally; where physical constraints shape digital operations; and where digital abstractions obscure physical consequences.

Triadic Evolution also reinforces the need for *architectural separation of concerns.* If a unit becomes too digitally abstracted, it may require structural anchoring in adjacent units with physical or social emphasis. The fleet becomes viable not because each unit is triadically balanced in itself but because the system as a whole is *triadically integrated*—across its structure.

Purpose Driven Design incorporates this perspective as foundational. It ensures that units are not optimized for performance in a single domain but composed for resilience and accountability across all three. Evolution does not privilege simplicity. It privileges fitness. And fitness in complex systems is always multidimensional.

The Fleet, understood through this triadic lens, is more than a distributed system. It is an evolutionary architecture—one that structures itself not against collapse, but toward sustained viability in a world shaped by digital acceleration, physical constraints, and social complexity. Designing units without this lens is not only shortsighted—it is structurally negligent.

1.9. From Belonging to Legitimacy

Structure alone does not earn trust. In any sociotechnical system, legitimacy is not granted by design—it is earned through alignment. A unit belongs not merely because it exists within a fleet, but because its purpose, structure, and conduct are recognized as meaningful and necessary by its environment. Belonging is architectural; legitimacy is systemic. And in Purpose Driven Design, *legitimacy emerges only when belonging is both structurally encoded and socially acknowledged.*

This structural belonging has two dimensions:

- *Systemic belonging*: The unit fits within the broader architecture of the enterprise. It has a defined role, bounded interfaces, and intelligible outputs. It is not a shadow operation, a legacy holdover, or an exception in search of justification.
- *Environmental belonging*: The unit is responsive to—and recognized by—the external entities it exists to serve. These may be consumers, regulators, citizens, partner systems, or other stakeholders. The unit does not operate in abstraction; it participates in a context.

Together, these dimensions determine whether a unit can be said to *deserve its place.* In coherent systems, legitimacy is not imposed—it is *perceived.* A unit is seen as trustworthy when it is understandable, governable, responsive, and aligned with a purpose that matters.

But modern enterprises often sever the link between belonging and legitimacy. Units are created without clear purpose. They duplicate, overlap, or drift from their original role. They become optimized for internal metrics rather than external contribution. They are governed through policy but not held accountable through structure. This leads to legitimacy erosion—not only for the unit but for the system as a whole.

In Tower-based systems, legitimacy is often borrowed from the top—conferred by fiat, brand, compliance, or precedent. But in Fleet-based systems, *legitimacy is compositional.* It arises from structural clarity, architectural coherence, and the demonstrable ability of each unit to carry out its purpose in alignment with both internal systems and external needs.

A unit composed with Purpose Driven Design is not legitimate by definition—it is *designed to become legitimate through function.* Its structure makes it auditable. Its memory makes it accountable. Its interfaces make it governable. Its culture makes it dependable. Its adaptability makes it resilient. These qualities do not guarantee performance, but they enable trust.

More importantly, legitimacy is what allows a system to evolve. Units that are opaque, misaligned, or externally irrelevant cannot adapt—they entrench. They protect themselves rather than serve. But units that are structurally composed to belong—systemically and environmentally—can be reconfigured, repositioned, or retired without destabilizing the system. Legitimacy allows change without collapse.

The goal of Purpose Driven Design is not simply to create units that function. It is to create units that *justify their continued existence through value, accountability, and alignment*. The architecture of belonging is not ornamental—it is what makes structural legitimacy possible.

In the next and final section of this chapter, we turn to the stakes of this work. If legitimacy is what allows systems to change responsibly, what happens when architecture is neglected? What does it mean to evolve without collapsing—and why must we act now?

1.10. What Purpose Driven Design Offers

Purpose Driven Design begins where the Tower ends.

It assumes the collapse has already happened—not only historically, but structurally, metaphorically, and methodologically. Systems designed for control have failed to adapt. Structures optimized for growth have become illegible. Enterprises built to scale have become disconnected from the purposes they were meant to serve. What remains is not emptiness, but the opportunity to rebuild differently.

Unit Oriented Enterprise Architecture responded to this failure by decomposing the Tower. It replaced centralized ambition with distributed orientation. It introduced the Fleet as a structural alternative—a network of bounded, purpose-aligned, interoperable units. But decomposition alone is insufficient. The success of the Fleet depends on the integrity of its vessels.

Purpose Driven Design begins here: with the composition of each sociotechnical unit. It offers a grammar—a set of architectural principles through which craft, crew, and cohesion are designed into every unit. These principles are not isolated techniques or best practices. They are interdependent structural components that ensure each unit is

- Oriented by a clear and relevant *Purpose*
- Justified through the *Value* it produces and delivers
- Governed by actionable *Functions*

- Executed through legible *Process*
- Held together by a coherent *Structure*
- Constrained by explicit *Order*
- Enabled by a sustainable *Culture*
- Grounded in operational *Memory*
- Supported by durable *Storage*

Together, these principles define what it means for a unit to be viable—not just now but as conditions change. Purpose Driven Design does not predict the future; it makes the future *navigable*. It gives architects, strategists, and builders the means to design systems that evolve without collapsing.

This book is not a manual for controlling complexity. It is a framework for constructing the kind of structure *in which complexity can be organized, governed, and adapted*—without sacrificing accountability, coherence, or legitimacy. Each chapter that follows will examine one component of this grammar. Not as an abstract idea but as a structural requirement.

Architecture is no longer optional. It is what allows the system to hold together—not despite complexity but because of it. And it is what enables each unit, and each enterprise, to become not only operational—but rightful.

With the architectural need established, we now turn to the structural grammar that makes each unit intelligible, composable, and evolution-ready.

1.11. Postscript: On Comprehensibility and Control

Not all systems are meant to be understood. Some are optimized for density, not transparency; for throughput, not accountability. A system can function without being seen—can grow, process, and even learn—without ever revealing its internal structure to those who depend on it. But when the stakes are evolutionary, *opacity is no longer acceptable.*

Consider the example of *Chaotic Storage*. In warehouse logistics, this technique dispenses with intuitive organization. Items are placed wherever space is available—randomly, algorithmically, invisibly. There are no labeled sections, no lighting or heating requirements, and no human-accessible maps. The result is undeniable efficiency: warehouses using chaotic storage can hold *five to ten times* more items than traditionally organized counterparts. But there is a cost. *No human can enter and comprehend the system without mediation.* The structure serves optimization, not understanding.

This may be acceptable for shelving inventory. It is not acceptable for *governing the future of human civilization.*

Sociotechnical systems are not neutral tools. They carry values, shape choices, and influence what futures are possible. If humans are to maintain control over the trajectory of their evolution, they must retain control over their technological extensions. And control requires more than access. It requires *comprehensibility, legibility, and structural traceability.*

The goal of architecture, then, is not merely to make systems function. It is to make them *governable by design.* That means they must be organized in ways that are understandable—not only by their builders, but by those entrusted to steward their purpose. This is the cost of responsibility: *clarity over convenience, structure over obscurity, design over improvisation.*

Chaotic systems may outperform their organized counterparts in raw metrics. But organized systems enable *accountability, continuity, and change without collapse.* They give structure to judgment. They allow errors to be traced, interventions to be made, and futures to be shaped with intention.

Architecture is expensive. But disorganization is not free. It defers the cost—until comprehension is needed and cannot be retrieved.

In this book, architecture is treated not as preference but as ethical necessity. A world composed of opaque systems, however efficient, is a world in which humans become peripheral. A world composed of structurally coherent, purpose-aligned, and legible systems is one in which humans remain *responsible agents of their own direction.*

You get what you pay for. Choose between architecture and chaos. Choose wisely.

CHAPTER 2

How Things Work—The Inner Mechanics of Purpose

Before designing systems that evolve, scale, or coordinate, one must first understand what *makes a unit real.* A unit is not a team, a project, or a platform module—it is a structured sociotechnical entity: a composition of physical crafts, digital crafts, and, above all, crews. Each unit exists as a coherent vessel of purpose—its physical craft providing protection and integrated tools, its digital craft enabling computation, communication, and coordination, and its crew interpreting purpose, exercising judgment, and maintaining alignment under pressure. As introduced in the previous chapter, Craft, Crew, and Cohesion form the architectural triad of this composition. On a cruise ship or an airship, for instance, the vessel, its embedded digital systems, and its crew operate as one: the craft sustains environment and motion, the digital systems sense and regulate, and the crew navigates, decides, and acts. The same principle defines every viable unit in an enterprise—physical and digital crafts supporting a crew that enacts purpose through structure and coordination.

Purpose Driven Design provides the internal mechanics that make such coherence possible. It offers a set of *constructs* that are not merely conceptual—they are structurally actionable. They can be represented in interfaces, implemented in workflows, encoded in policies, and retained in memory systems. These constructs define *how things work* inside a unit—not in metaphorical terms but in operational detail. What follows is a breakdown of this design grammar, starting with the architectural logic of the unit itself.

A. Milchman, *The Purpose Driven Design Framework for Modern Enterprises*,
https://doi.org/10.1007/979-8-8688-2223-0_2

2.1. From Metaphor to Model: Units As Ships in a Fleet

The design of coherent systems begins with the composition of bounded entities—*units*—that can function, adapt, and coordinate without central dependency. In the Purpose Driven Design framework, each unit is modeled not as an abstract subsystem or organizational placeholder, but as a *ship within a fleet*. This metaphor is not ornamental; it encodes architectural logic. A ship is bounded, operable, memory-equipped, purpose-driven, and capable of maintaining internal coherence while participating in a larger formation. These same properties are essential for any sociotechnical unit designed to operate at scale.

The metaphor translates directly into structural expectations. Just as a ship must have propulsion, navigation, roles, and compartments, a unit must have *Purpose, Process, Structure, Order, Memory,* and other design elements that support autonomous operation and external interoperability. The fleet metaphor allows system designers to reason compositionally—seeing systems not as static organizations but as *composable constellations* of purposeful actors, each architected for legibility and response.

The ship metaphor clarifies the structural properties of a viable unit—but turning metaphor into method requires a layered design approach. In practice, this means translating the metaphor's physical, social, and digital dimensions into actionable architectural planes. To make this metaphor operational, Purpose Driven Design employs a *multiplane architecture* that allows the designer to view each unit through distinct but interconnected layers:

- The *Unit Plane* defines the internal architecture of the unit using nine interdependent constructs—Purpose, Value, Function, Process, Structure, Order, Culture, Memory, and Storage. This plane ensures that the unit can act meaningfully and remain internally aligned.
- The *Social Design Plane* addresses human and hybrid agency: who acts, decides, interprets, and maintains accountability. This includes role design, decision-making structures, team norms, and interpersonal interfaces.
- The *Digital Plane* maps constructs to technological implementations: APIs, workflows, UI components, databases, rules engines, cloud storage, and communication layers. It ensures that what is modeled conceptually is instantiated concretely.

- The *Physical Plane,* while outside the scope of this book, governs interactions with the built environment, robotics, logistics infrastructure, and manufacturing systems. In physical-digital hybrids, it relies on established architectural and engineering practices.

Each plane supports the others. The Social Design Plane enables the Unit Plane to function meaningfully; the Digital Plane enables its execution and evolution; and the Physical Plane grounds it in embodied infrastructure when applicable. This multiplane perspective ensures that a unit is not only conceptually sound but *implementable across technical and social dimensions,* as summarized in Table 2-1.

Table 2-1. *Interdependence of Architectural Planes in Purpose Driven Design*

Plane	Primary role	Enables/supports
Unit plane	Defines the internal architecture of the unit (Purpose, Value, Function, etc.)	Requires social interpretation and digital/ physical realization
Social design plane	Configures human and hybrid agency (roles, accountability, governance)	Gives meaning and legitimacy to the Unit Plane; guides the Digital and Physical Planes
Digital plane	Implements and evolves the unit through software, workflows, and data systems	Executes structures defined in the Unit and Social Design Planes
Physical plane	Provides embodied environments and infrastructure when applicable	Grounds operations and interfaces with the tangible world

In this model, a unit is never just a department, a feature set, or a team. It is a designed structure—*a sociotechnical system with agency, accountability, and a reason to persist.* What follows in this chapter is a detailed exploration of the constructs that bring this structure to life, with practical examples and implementable patterns that demonstrate how the internal architecture of the unit can be generated, deployed, and evolved using modern tools.

2.2. Overview of the PDD Constructs

The internal coherence of a unit is not achieved through ad hoc configuration or inherited routines. It must be *composed*—structurally and deliberately—using a set of interlocking architectural elements. Purpose Driven Design defines this internal composition through *nine constructs* that together form the internal logic, operational capacity, and adaptive integrity of the unit.

Each construct addresses a specific dimension of internal functionality, but none operate in isolation. They form a chain of interdependence, beginning with *Purpose* and culminating in the delivery and preservation of *Value*. When aligned, these constructs produce a unit that is not only intelligible and effective but also capable of adapting without disintegration.

The nine constructs are as follows:

- **Purpose**

  The orienting logic of the unit. Purpose defines *why the unit exists, whom it is meant to serve,* and *what it intends to contribute* to its environment. It acts as the architectural anchor for all design decisions.

- **Value**

  The tangible output or benefit the unit offers in exchange for interaction—whether in the form of a product, service, or capability. Value is how Purpose is *expressed* and *validated* in practice.

- **Function**

  The set of interfaces through which the unit engages its environment. Function makes Value *accessible* to consumers and partners. It includes UI elements, APIs, service portals, and physical or procedural access points.

- **Process**

  The internal sequence of coordinated actions—manual, automated, or hybrid—that enable the unit to generate or deliver Value. Process translates Purpose into operational reality.

- **Structure**

  The configuration of human and artificial agents within the unit. Structure defines *who or what does the work*, how responsibilities are distributed, and how agents relate to one another and to their tools.

- **Order**

  The explicit and implicit rules that govern activity. Order includes policies, business rules, protocols, permissions, and escalation mechanisms that *constrain, enable*, or *guide* behavior.

- **Culture**

  The normative and emotional atmosphere of the unit. Culture shapes how agents behave under ambiguity, how decisions are interpreted, and how alignment is sustained over time.

- **Memory**

  The record of past actions, decisions, inputs, and states. Memory allows the unit to *learn, self-correct*, and *demonstrate accountability*. It is essential for adaptation and traceability.

- **Storage**

  The domain in which products, documents, assets, or data are retained for use, transfer, or transformation. Storage ensures that what the unit produces is *preserved* and *delivered* in usable form.

Together, these constructs enable a unit to move from static intention to dynamic participation. They make the unit *functionally complete*: capable of interacting, producing, remembering, evolving, and contributing within a larger system of coordination.

In the sections that follow, each construct will be examined not only in conceptual terms, but also through *implementable prompts* and *system-level design patterns*, demonstrating that these constructs are not aspirational—they are buildable. They can be expressed in code, manifested in workflows, embedded in interfaces, and governed across time. The result is an internal architecture that is as operational as it is architectural.

2.3. Construct Realization: From Theory to Code

The constructs of Purpose Driven Design are not abstract categories or philosophical commitments. They are *implementable structures*—each with recognizable instantiations in digital systems, workflows, and interfaces. In this section, each construct is demonstrated through a combination of architectural description and *practical prompts*, showing how these elements can be generated, configured, or orchestrated using current tools and platform capabilities.

The aim is not to simulate design, but to *produce it*: to show that units can be created and composed through prompts that generate web interfaces, application workflows, data schemas, governance rules, and other system artifacts.

2.3.1. Purpose

Defines the reason the unit exists and what it aims to offer or attract.

- **Prompt:**

  "Generate a landing page and a social media bio for a unit that offers AI-powered résumé review services to job seekers."

- **Output:**

  HTML for landing page with service overview, CTA buttons, lead capture form, and identity copy; formatted text for use on LinkedIn and Twitter.

2.3.2. Value

Specifies the product or service the unit delivers—what it exchanges for interaction.

- **Prompt:**

  "List five service offerings (free and paid tiers) for a résumé review platform, each with expected delivery format and SLA."

- **Output:**

  Product catalog with JSON definitions: pricing, delivery channel (PDF, email, dashboard), response time, and payment structure.

2.3.3. Function

Describes how users or systems interact with the unit to request or receive Value.

- **Prompt:**

  "Generate a front-end interface in React that allows users to upload a résumé, select a review plan, and receive feedback."

- **Output:**

  React component structure with file upload, selection dropdowns, and result viewer; interface code integrated with RESTful endpoints.

2.3.4. Process

Outlines the sequence of steps by which Value is produced or delivered, often spanning multiple agents or tools.

- **Prompt:**

  "Generate a jBPM workflow for résumé review: document upload → preprocessing → AI-based evaluation → human review (optional) → delivery."

- **Output:**

  A deployable BPMN model file compatible with jBPM, and optionally, a working Spring Boot application scaffold with jbpm.jar integrated to run the workflow locally or in a cloud container.

2.3.5. Structure

Specifies who or what performs work within the unit: humans, AI agents, or hybrid arrangements.

- **Prompt:**

  "Define the internal team structure for the résumé review unit, including AI agents for evaluation, editors for refinement, and service coordinators."

- **Output:**

  YAML or JSON structure for roles and responsibilities, with assignment logic for different workflow stages; integration templates for human-in-the-loop processes.

2.3.6. Order

Defines the constraints and rules under which processes and interactions are governed.

- **Prompt:**

  "Create policy rules that prevent résumé evaluation from proceeding until user consent is recorded and payment is verified."

- **Output:**

  Policy logic in Drools (for BRMS) or JSON rules for integration with authorization middleware; rule chaining logic for enforcement.

2.3.7. Culture

Shapes informal behavior, tone, and emotional climate inside the unit; especially relevant in teams with persistent human cores.

- **Prompt:**

  "Generate a culture statement for a high-integrity, feedback-oriented résumé review unit focused on trust, quality, and continuous learning."

- **Output:**

  A values manifesto used for onboarding, team alignment, and interface tone-setting; also usable as meta-content in user-facing materials to signal brand ethos.

2.3.8. Memory

Captures and retains structured historical traces for learning, accountability, and continuity.

- **Prompt:**

  "Create a NoSQL schema for time series records capturing upload time, evaluation steps, reviewer interactions, and delivery timestamps per résumé."

- **Output:**

  MongoDB or Firestore schema design; optionally, pre-generated scripts to deploy a clickstream logging service with timestamped entries, searchable metadata, and performance metrics for each unit of Value delivered.

2.3.9. Storage

Ensures that outputs—products, intermediate files, or documentation—are retained, versioned, and accessible.

- **Prompt:**

  "Design a cloud-based storage layout (GCS or S3) to hold original uploads, evaluation results, annotated versions, and delivery artifacts per customer and version."

- **Output:**

  Storage architecture blueprint (folder hierarchy, metadata conventions, access control list templates), ready for deployment via Terraform or CLI tooling.

These examples demonstrate that each construct is not only conceptually defined, but *technically generable*. Prompts can produce not only visual interfaces or written artifacts, but also code, policies, storage plans, and workflows. The unit can be instantiated—digitally and socially—as a *working sociotechnical system* that persists, adapts, and operates under architectural constraints.

2.4. Coherence As Design Property

In complex systems, functionality alone is insufficient. A unit may deliver outputs, respond to inputs, and appear operational while remaining fundamentally unstable or directionless. *Coherence*—the alignment and mutual reinforcement of internal constructs—is what differentiates a unit that merely functions from one that can *endure, evolve,* and *act with integrity*. Coherence is not an aesthetic ideal; it is a structural condition that determines whether a unit remains legible, trustworthy, and governable over time.

In Purpose Driven Design, coherence emerges when each construct—*Purpose, Value, Function, Process, Structure, Order, Culture, Memory,* and *Storage*—is present, properly configured, and *aligned with all others*. No construct is autonomous. Each depends on others to fulfill its role effectively. Purpose guides Function; Function initiates Process; Process depends on Structure; Structure operates under Order; Order and Culture shape behavior; Memory records interaction; Storage retains deliverables. When any construct is misaligned, missing, or in conflict, coherence deteriorates—and system fragility increases.

For example:

- A unit with clear Purpose but inconsistent Function will attract attention but fail to deliver.
- A unit with strong Process but ambiguous Structure will execute tasks without accountability.
- A unit with structured Memory but no Culture may retain data but forget meaning.
- A unit with well-designed Function and Storage but no clear Purpose may deliver outcomes that are efficient yet irrelevant.

Coherence is also *recursive*. It is not merely the result of well-configured parts, but of their dynamic interaction. As units evolve, coherence must be *actively maintained*. Adding new services, extending Function, or reconfiguring Structure may require adjustments to Process, Order, or Memory. For example, when a hospital introduces an AI-based triage system, its Function expands digitally, but to preserve coherence, processes for patient intake must change, new accountability roles must be defined, and decision logs added to institutional Memory. Without an architectural logic like Purpose Driven Design, such changes risk creating local optimizations that destabilize the whole.

Moreover, coherence enables *observability*. A coherent unit exposes the logic of its operations: what it is doing, why it is doing it, and whether it is succeeding. It can be understood, audited, and adjusted. Incoherent units, by contrast, become opaque. Failure modes accumulate silently, and accountability becomes diffuse.

Purpose Driven Design treats coherence as a first-class design property—not a side effect of efficiency but a precondition for sustainable participation in larger systems. Units that are coherent can be composed into systems of systems—*fleets, flotillas, constellations*—without producing cascading fragility. They serve as trustworthy, interoperable components in architectures that must operate under pressure and evolve without collapse.

2.5. Architectural Implications

Treating Purpose Driven Design constructs as real and implementable entities carries specific architectural consequences. It repositions the unit not as a soft organizational abstraction but as a *composable, observable, and evolvable sociotechnical object*—one whose internal logic can be rendered in code, workflows, configurations, policies, and storage patterns. This reframing introduces four key architectural implications.

1. **Units Become Addressable and Composable**

   When a unit is defined by a coherent set of constructs, it becomes a self-contained architectural object. It can be composed into larger systems without extensive adaptation or custom integration. This modularity allows units to participate in *fleet-based architectures*—systems of systems in which units are arranged, governed, and evolved collectively without sacrificing autonomy. Units become interchangeable at the level of capability, not just role, because their internal structure is legible and consistent.

2. **Constructs Map to Implementation Surfaces**

   Each construct has clear instantiations across platforms and services:

   - *Function* aligns with interfaces (e.g., APIs, web portals)
   - *Process* aligns with orchestration layers (e.g., jBPM, workflow engines)

- *Structure* maps to organizational schemas or role definitions
- *Order* maps to rules engines or policy enforcement layers
- *Memory* connects to logging services, time series databases, or NoSQL event stores
- *Storage* links to cloud object storage or versioned repositories

These mappings mean that architectural and software tooling can be organized around the PDD construct model, supporting design consistency and automation across systems.

3. **Architectural Planes Must Remain Synchronized**

   The multiplane architecture—*Unit Plane, Social Design Plane, Digital Plane,* and *Physical Plane*—introduced in the section "From Metaphor to Model: Units As Ships in a Fleet" requires ongoing alignment. For instance:

   - If a new Function is added at the Digital Plane, the Social Design Plane must assign responsibility and ownership for its execution.
   - If Memory is enhanced with new analytics capabilities, Order may need to be updated to define how those insights influence decisions.
   - If a role in Structure is automated by an AI agent, Culture and Order must evolve to accommodate new patterns of coordination and escalation.

   Without cross-plane synchronization, the coherence of the unit degrades over time—even if individual features are well-executed.

4. **Design Tools and Prompt Systems Become Architectural Instruments**

   As demonstrated in the section "Construct Realization: From Theory to Code," prompts can be used to generate and refine each construct. This implies that:

   - *Prompt-based design tools* can serve as first-stage compilers for sociotechnical architecture.

- *Configuration grammars* can emerge to capture unit definitions across environments (e.g., YAML templates defining Purpose, Process, and Structure in a single artifact).
- *Platform automation* can be built around unit-aware scaffolds, provisioning Function, Memory, and Storage automatically based on declared Purpose and Value.

These capabilities transform architecture from static documentation to *live system definition*—executable, inspectable, and incrementally evolvable.

Taken together, these implications position Purpose Driven Design not as a design metaphor, but as a *deployable architectural grammar* for constructing structurally aligned, scalable, and ethically governable systems. In the chapters that follow, each construct will be examined in depth—clarifying its architectural role, operational footprint, and integration logic—ensuring that units are not only designed well but designed to endure.

2.6. Summary: From Abstract to Executable

Purpose Driven Design transforms the internal architecture of a unit from a conceptual model into a *technical blueprint*. By defining nine constructs—*Purpose, Value, Function, Process, Structure, Order, Culture, Memory,* and *Storage*—it provides a design language that is both intellectually rigorous and practically executable. These constructs are not philosophical themes or management metaphors; they are *architectural surfaces* that can be specified, built, versioned, and evolved.

The unit, as defined within this framework, is not a static role group or service cluster. It is a *bounded sociotechnical system*, capable of acting, adapting, and being held accountable. When constructed through Purpose Driven Design, the unit becomes an object with recognizable structure, legible intent, and composable functionality—suitable for participation in larger system formations such as fleets or constellations.

Each construct maps directly to implementation surfaces in modern architectures:

- *Purpose* to landing pages, mission prompts, and audience-facing identity
- *Function* to APIs, portals, and interaction interfaces

- *Process* to orchestrated workflows, such as BPMN models deployed in Spring Boot environments
- *Structure* to agent configurations, role matrices, and automation layers
- *Order* to executable policies and rules frameworks
- *Culture* to social protocols, onboarding materials, and default behavior expectations
- *Memory* to NoSQL time series logs, annotated records, and decision trails
- *Storage* to versioned digital repositories or object stores linked to Function

What this chapter demonstrates is that *prompt-based generation, low-code orchestration,* and *modular infrastructure* can all be used to instantiate and evolve these constructs. Readers are not left to imagine how Purpose or Memory might be represented—they can generate working interfaces, workflow engines, policy files, and schema definitions that give these constructs operational form.

At its core, Purpose Driven Design offers a response to a foundational architectural problem: systems today are built faster than they can be understood, governed, or adapted. This framework allows designers and technologists to regain structural control—not by centralizing complexity but by embedding coherence into every unit from the inside out.

In the chapters that follow, each construct will be unpacked in full detail. The focus will shift from the *framework as a whole* to the *design surface of each individual construct*—its rationale, its failure modes, its implementation pathways, and its integration with the rest of the unit. This begins with the construct that defines the direction of all others: *Purpose.*

2.7. Postscript 1: The Unit As a Mansion

In Chapter 1, the *Fleet* metaphor provided a structural model for system-level design—illustrating how bounded, purpose-aligned units compose into adaptive sociotechnical formations. That metaphor captured the architecture of coordination. In this Postscript, a complementary metaphor is introduced to aid comprehension of *internal composition*: the *Mansion.*

This metaphor does not replace the construct-based grammar of Purpose Driven Design. Rather, it offers an interpretive scaffold for understanding how applications and microservices may be arranged within a unit in a way that reflects functional cohesion, structural clarity, and operational readiness. Where the fleet enables navigation, the mansion enables habitation.

2.7.1. The Mansion As a Representation of a Software-Intensive Unit

A mansion is not merely a set of rooms—it is a composition of inhabitants and spaces, structured to support purposeful living. Each area serves a defined role, and each occupant engages with that structure in specific ways. Similarly, a well-designed unit is not just an arrangement of software components. It is a sociotechnical system composed of human and artificial agents interacting with distinct, interdependent functions.

In this metaphor, the Family represents the social core of the unit—those who inhabit and activate its structure. The rooms that follow correspond to classes of applications, services, or interface layers that these agents operate, maintain, or experience.

Family—Human and Artificial Agents (Social Core of the Unit)

The Family represents the *inhabiting social agents* of the mansion—both human and artificial—who animate its structure, interpret its purpose, and make meaningful use of its capabilities. They reflect the Structure construct of Purpose Driven Design, ensuring that the mansion is not merely functional, but lived in, adaptive, and purpose-aligned.

- *Head(s) of Household*: Interpret Purpose and maintain coherence across rooms and functions; responsible for direction, delegation, and stewardship.
- *Caretakers/Operators*: Manage workflows, maintain systems, and uphold norms across service areas and private quarters.
- *Residents*: Individuals whose roles, preferences, and activities shape the functioning and configuration of the mansion.
- *Guests*: External participants with bounded access and role-specific interactions—such as partners, clients, or collaborators.

- *Pets***:** Dependent agents (natural or artificial) that influence behavior, require support, and occupy meaningful space within the system. Examples include robotic assistants, automated companions, or auxiliary subsystems whose presence shapes processes, configurations, or memory—even if they are not responsible for Purpose or execution.
- *Household Agents***:** Digital or robotic assistants performing operational or support functions (e.g., scheduling, security, climate control).

Primary Living Areas—Core Business Operations Application Suite

- *Foyer: Authentication and Access Control App*

  The threshold through which identity is verified before participation is allowed.

- *Formal Dining Room: CRM and Customer Engagement App*

  A place designed for structured interaction with consumers and clients.

- *Lounge: Dashboard and Reporting App*

  Where information is gathered, contextualized, and made available for inspection.

Private Quarters—User-Centric Application Suite

- *Master Suite: Personalization and Preferences App*

  Where user-specific configurations, roles, and tailored experiences reside.

- *Second Bedroom: Employee Management App*

  Where internal roles and responsibilities are maintained and adjusted.

- *Guest Bedroom: External Partner Portal App*

  A controlled interface for third parties and external collaborators.

Service Areas—Data Handling and Operations Application Suite

- *Kitchen: Data Processing and Workflow Automation App*

  Where raw inputs are processed, transformed, and operationalized.

- *Pantry: Data Storage and Archiving Microservice*

  An organized environment for storing what is regularly used.

- *Laundry Room: System Maintenance and Error Handling Microservice*

  Handling cleanup, exception management, and background routines.

Entertainment and Leisure Spaces—Collaboration and Engagement Application Suite

- *Library: Knowledge Management and Documentation App*

  A structured repository of memory and reference.

- *Game Room: Gamification and Motivation App*

  Tools to incentivize participation, learning, or behavior.

- *Home Theater: Multimedia and Streaming App*

  Interfaces for delivering rich content and visual materials.

Wellness and Fitness Areas—Performance and Security Application Suite

- *Gym: Load Testing and Scalability App*

  Where resilience and performance under stress are evaluated and reinforced.

- *Spa: Security and Compliance Microservice*

  A service designed for safety, integrity, and regulatory alignment.

Work and Study Areas—Productivity and Creation Application Suite

- *Home Office: Project Management and Collaboration App*

  Where structured planning and execution occur.

- *Studio: Design and Prototyping App*

  The creative environment for shaping future functionality.

Storage and Utility Spaces—Infrastructure and Backup Application Suite

- *Walk-in Closet: File Storage and Backup Microservice*

  High-accessibility storage, organized for regular retrieval.

- *Basement: Legacy Systems and Archived Data App Suite*

  Outdated but necessary systems retained for compliance or reference.

- *Garage: Deployment and CI/CD Pipeline App*

  The launchpad for releasing, staging, or integrating new capabilities.

Outdoor Spaces—External Connectivity Application Suite

- *Courtyard: API Gateway and Integration Services*

  A controlled perimeter for external interface and data exchange.

- *Terrace: Partner Ecosystems and Marketplaces*

  Where the unit expands its functionality through external platforms and ecosystems.

2.7.2. Compositional Insights

This analogy makes one principle clear: the unit is not monolithic. It is an integrated residence for interaction, processing, and coordination. Some components form full application suites (e.g., primary living areas); others are microservices (e.g., pantry, spa). Each must serve its purpose while contributing to the overall legibility and operational readiness of the unit.

Moreover, the mansion makes visible the *necessity of differentiation*. A system composed entirely of kitchens is not operable. One with no entry point cannot engage. Architecture, like habitation, requires role-specific zones that are internally cohesive and externally coherent.

2.7.3. Structural Clarification, Not Decoration

This is not a metaphorical flourish. As stated earlier, metaphoric modeling serves a functional role: compressing structural logic into cognitively accessible patterns. For many readers—especially those engaged in platform development or application design—the mansion provides a mental model for how PDD constructs may manifest as application layers and services within a single unit.

It underscores the central argument: that units are not abstract containers. They are habitable systems—coherent, purposeful, composable—designed to deliver value with structural integrity. Whether they are deployed in serverless environments or operated by hybrid teams, their viability depends on this internal architecture.

Just as no mansion is functional without a foyer, kitchen, or place to sleep, no unit can operate without Function, Process, Structure, Memory, and Storage. What differs is only the aesthetic. The structure is universal.

2.8. Postscript 2: The Unit As a Naval Ship

In the architectural grammar of Purpose Driven Design, the *unit* is a structurally bounded, purpose-aligned, and operable sociotechnical system. Chapter 2 defined its internal composition using nine constructs; Postscript 1 used the *Mansion* to illustrate modular internal functionality. This Postscript shifts the metaphorical lens from architecture to operation—conceiving the unit as a *Naval Ship*. Where the Mansion illustrated modular *habitation* and internal composition, the Naval Ship emphasizes *operation under pressure*, mission integrity, and coordinated execution—offering a functional view of how the unit performs as a living sociotechnical system.

The metaphor is not arbitrary. It emerges naturally from the *Fleet* metaphor in Chapter 1. If the enterprise is a fleet of interoperable units, then each unit is a ship: self-contained, crewed, mission-capable, and structurally equipped for sustained function under pressure. The internal configuration of this ship mirrors the internal architecture of the unit.

2.8.1. Structural Zones and Application Suites

The naval ship metaphor captures the composition of the unit as both a social and technical system. The Crew represents the agents who operate, interpret, and maintain the system. The decks, bays, and compartments that follow map to functional or infrastructural domains within the unit—each corresponding to application suites or services that these agents activate, monitor, or depend upon. Together, they form an interpretable model of structure-in-operation.

Crew—Human and Artificial Agents (Social Core of the Unit)

The Crew represents the *social structure* of the unit. These agents—human and artificial—are responsible for interpreting purpose, executing processes, enforcing order, and maintaining coherence across operations. They embody the Structure construct of Purpose Driven Design and enable the unit to function as a living sociotechnical system.

- *Captain*: Interprets Purpose and provides directional leadership; accountable for the unit's identity, external orientation, and internal alignment
- *First Officer (XO)*: Oversees operations; aligns Process with Structure; ensures execution adheres to Purpose and Order
- *Specialists*: Role-specific experts (e.g., engineers, analysts, designers) who operate or maintain specific systems or subsystems
- *Watch Officers*: Monitor and manage Memory, Order, and external interfaces; maintain situational awareness and operational discipline
- *General Crew*: Perform operational, maintenance, and coordination tasks; sustain functionality through routine and adaptive actions

Bridge—Core Business Operations Application Suite

- *Captain's Quarters: Authentication and Access Control App*

  Where high-level identity, access, and command rights are managed.

- *Officer's Mess: CRM and Customer Engagement App*

  The setting for structured, high-stakes communication and interaction.

- *Navigation Deck: Dashboard and Reporting App*

  Instruments for direction, oversight, and decision support.

Crew Quarters—User-Centric Application Suite

- *Captain's Cabin: Personalization and Preferences App*

  Tailored configurations and views for top-level roles.

- *Crew Bunks: Employee Management App*

  Assignments, shifts, schedules, and access levels for operational personnel.

- *Guest Quarters: External Partner Portal App*

  Interface for regulated collaboration with visiting teams or external agents.

Operational Decks—Data Handling and Operations Application Suite

- *Galley: Data Processing and Workflow Automation App*

  Where raw inputs are transformed into actionable outcomes.

- *Supply Storeroom: Data Storage and Archiving Microservice*

  An organized, accessible data repository for recurrent use.

- *Maintenance Bay: System Maintenance and Error Handling Microservice*

  Diagnosing faults, handling exceptions, and ensuring system continuity.

Recreation Deck—Collaboration and Engagement Application Suite

- *Ship's Library: Knowledge Management and Documentation App*

  Indexed, searchable references and guidance materials.

- *Training Simulator: Gamification and Motivation App*

  Tools for readiness, learning, and engagement.

- *Media Room: Multimedia and Streaming App*

  Interface for delivering briefings, media assets, and asynchronous content.

Engineering Bay—Performance and Security Application Suite

- *Engine Room: Load Testing and Scalability App*

  Infrastructure for stress testing, tuning, and system health verification.

- *Security Control Room: Security and Compliance Microservice*

  The embedded enforcement layer for rules, threat detection, and alignment.

Mission Control—Productivity and Creation Application Suite

- *Strategy Room: Project Management and Collaboration App*

  Where planning, execution, and task orchestration are conducted.

- *Design Lab: Design and Prototyping App*

  Space for experimental development and interface prototyping.

Storage Deck—Infrastructure and Backup Application Suite

- *Equipment Locker: File Storage and Backup Microservice*

  Ready access to necessary tools, records, and data payloads.

- *Archive Hold: Legacy Systems and Archived Data App Suite*

  Decommissioned systems or retained data for compliance and review.

- *Hangar Bay: Deployment and CI/CD Pipeline App*

  The launch zone for software releases, updates, and integrations.

Communication and External Operations—External Connectivity Application Suite

- *Radio Room: API Gateway and Integration Services*

  Regulated channel for controlled external communication and data exchange.

- *Port Dock: Partner Ecosystems and Marketplaces*

  Entry point for provisioning, exchange, and cross-organization collaboration.

2.8.2. Why the Naval Ship Works

Unlike the mansion, which emphasizes modular livability, the naval ship is a functional metaphor for operational integrity. Each component exists not for comfort but for performance. Every room supports readiness, resilience, and directed activity—mirroring the properties demanded of a unit in a sociotechnical system.

The ship also introduces additional dimensions of coordination under pressure, escalation protocols, mission readiness, and chain-of-command clarity—useful associations for understanding not only how software components interoperate but how they *govern* and *evolve*.

A unit, like a ship, must operate as a *living vessel*—not a static platform. It must deliver value, process demands, respond to exceptions, and navigate complex environments. To design one is not merely to assemble software—it is to engineer a structure capable of direction, adaptation, and endurance. This metaphor, when paired with the mansion, gives multiple conceptual footholds for grasping how structural design becomes operational capacity.

2.9. Postscript 3: Enterprise Software Using Purpose Driven Design

Purpose Driven Design (PDD) defines the internal architecture of a unit through a structural grammar composed of eight interdependent constructs: Purpose, Function, Process, Structure, Order, Culture, Memory, and Storage. Each construct can be implemented using four classes of deployable building blocks: *Agents, Suites of Applications, Applications, and Microservices.* This postscript outlines how enterprise software can be architected through PDD by aligning each construct with a corresponding Application Suite composed of these structural elements.

2.9.1. Purpose—Facade and Engagement Application Suite

Represents the surface through which the unit communicates its identity, intent, and relevance.

- *Purpose Declaration App*: Presents mission, commitments, and orientation via microsites or static content engines
- *Engagement Monitoring App*: Tracks consumer sentiment, feedback, and inbound signals to assess purpose resonance
- *Reputation Microservice*: Evaluates trust metrics, reputational scores, and purpose adherence
- *Public Identity Agent*: Oversees narrative alignment, branding integrity, and stakeholder perception

2.9.2. Function—Transaction and Interaction Application Suite

Defines how users, systems, and devices interact with the unit to request or consume Value.

- *Digital Access Gateway App*: Manages registration, authentication, and authorization flows
- *Self-Service App*: Enables browsing, filtering, purchasing, and transaction execution across devices
- *Support Interface App*: Provides chatbot-based guidance and escalation pathways
- *Interaction Routing Microservice*: Mediates between front-end actions and backend service invocation
- *Interaction Agent*: Human or AI entity assisting in complex or exceptional interaction cases

2.9.3. Process—Business Process Management Application Suite

Coordinates internal activities and logic necessary for Value production and delivery.

- *Orchestration App*: Executes multistep workflows and business logic
- *Decision Support App*: Applies business rules and conditional logic in operational contexts
- *Process Monitoring Microservice*: Captures run-time metrics, exceptions, and performance indicators
- *Process Execution Agent*: Enacts tasks, manages escalations, and interprets workflows dynamically

2.9.4. Structure—Staff and Workforce Application Suite

Represents the configuration of human and artificial agents responsible for executing and interpreting functions.

- *Structural Agent*: Human or AI resource assigned to specific operational responsibilities
- *Workforce Management App*: Allocates responsibilities, shifts, and access across internal roles
- *Agent Directory App*: Registers and tracks artificial agents, roles, and workload balancing
- *Role Enforcement Microservice*: Ensures proper execution of tasks within defined scopes

2.9.5. Order—Governance and Compliance Application Suite

Encodes the policies, constraints, and escalation mechanisms that ensure procedural integrity.

- *Policy Management App*: Authoring, updating, and publishing executable rules
- *Compliance Oversight App*: Monitors adherence to external and internal obligations
- *Enforcement Microservice*: Evaluates runtime conditions and applies regulatory logic.
- *Governance Agent*: Reviews exceptions, handles escalations, and interprets conflicting directives

2.9.6. Culture—Norms and Enablement Application Suite

Shapes informal behavior and shared expectations among agents within the unit.

- *Onboarding and Enablement App*: Guides new agents through orientation, values, and role expectations
- *Cultural Reinforcement App*: Promotes preferred behaviors through feedback loops and gamified incentives
- *Norm Evaluation Microservice*: Detects patterns, inconsistencies, or anomalies in agent behavior
- *Cultural Agent*: Acts as a tone-setting, mentoring, or values-enforcing participant

2.9.7. Memory—Traceability and History Application Suite

Maintains structured records of activity, decisions, and interactions.

- *Decision Logging App*: Captures rationale, conditions, and actors involved in key decisions
- *Activity Timeline App*: Visualizes workflows and transactional histories
- *Retention Microservice*: Stores immutable logs and event traces
- *Memory Agent*: Queries and interprets past activity for auditing, analysis, or learning

2.9.8. Storage—Asset and Output Retention Application Suite

Holds the tangible outputs, artifacts, and data generated or consumed by the unit.

- *Output Repository App*: Hosts final deliverables, artifacts, reports, or contracts
- *Asset Vault App*: Manages reusable templates, documentation, and shared materials

- *Archival Microservice*: Moves outdated content to long-term or compliance-grade storage
- *Storage Agent*: Monitors availability, life cycle states, and access control for stored assets

This mapping translates abstract constructs into software composition strategies. Each unit is not simply a software deployment—it is a system of structurally interrelated Agents, Application Suites, Applications, and Microservices. Purpose Driven Design provides the architectural discipline necessary to ensure that these components do not accumulate arbitrarily, but instead form a coherent, adaptive, and governable whole.

2.10. Postscript 4: Why Application Is the Fundamental Building Block

In the context of Purpose Driven Design (PDD), each unit must be internally coherent, operationally autonomous, and architecturally composable. Among the available software constructs—agents, microservices, services, platforms, workloads, and gateways—the *Application* emerges as the most appropriate, structurally sound business building block. This postscript clarifies why the Application should be treated as the digital equivalent of a Room in the Mansion or a Compartment in the Ship: independently governed, distinctly purposed, and supported by underlying services.

The Hierarchical Composition

PDD recognizes the following compositional hierarchy:

- *Application Suite*: A cohesive collection of Applications aligned to a common construct (e.g., Function, Process).
- *Application*: A deployable, purpose-specific component with UI, logic, or interface layers that can stand alone or integrate with others.
- *Microservice*: A deployable application that encapsulates a specific subdomain or bounded responsibility, typically exposed via APIs. It is supported by its own backing services and independently versioned and operated. While microservices effectively decompose monolithic applications into manageable components, they are

often deployed without an architectural grammar for recomposing them into a meaningful, user-facing whole—resulting in operational fragmentation rather than purposeful structure.

- *Agent*: A human or artificial entity operating on or through the Application layer.

While all four building blocks are essential and deployable, the **Application** is distinguished by its balance of operational autonomy, business-level expressiveness, and architectural coherence.

2.10.1. Application As the Digital Room

Much like a *Room* in a Mansion, the Application is

- *Functionally bounded*: Each application serves a distinct operational purpose (e.g., booking, payroll, document signing).
- *Independently accessible*: Users or systems enter and interact with the application through defined interfaces.
- *Backed by services*: Heating, plumbing, and electricity in a room correspond to storage, security, and compute backing services in an application.
- *Governed by design and role*: Rooms and applications are both designed to be operated by agents performing specific roles.

This structural analogy is not stylistic. It is a modeling choice that enforces architectural clarity. If the Mansion is the Unit, Applications are its Rooms—primary, governable surfaces of action and value realization.

2.10.2. Clarifying Microservices: Applications Without Context

Microservices are frequently introduced as a remedy for monolithic complexity. Each microservice encapsulates a narrowly defined subdomain, adheres to the single-responsibility principle, and includes its own infrastructure dependencies. They are autonomous at the runtime level but not meaningful in isolation.

The prevailing problem is *decomposition without recomposition*: microservices are deployed independently but lack structural integration into coherent applications. Without an organizing layer, they fail to reflect business intent or enable navigable experiences. While many architectural styles can address these challenges effectively, Purpose Driven Design represents one such approach—grounded in systems thinking that restores coherence by enforcing the concept of the whole. It ensures that decomposition serves composition, preserving architectural integrity as systems evolve.

2.10.3. Clarifying Industry Terminology

The software industry has long struggled with concept overload and inconsistent terminology:

- *Services* often refer ambiguously to backend logic, support functions, or infrastructure endpoints.
- *Platforms* refer to foundational environments (e.g., Kubernetes, Salesforce, Databricks) but are often conflated with product layers.
- *Workloads* describe resource consumption patterns but lack architectural clarity.
- *Gateways* (e.g., API gateways, service meshes) act as traffic routers, not architectural units.

This inconsistency leads to poor composability, unclear ownership, and mismatched governance scopes. Applications, in contrast, are universally understood as functional, deployable, and user- or system-facing entities.

2.10.4. Toward a Universal Classification

Purpose Driven Design introduces a structured classification framework to bring coherence to the software architecture domain:

- *Applications* are the primary architectural building blocks of a unit.
- *Application Suites* organize Applications by construct or capability cluster.

- *Microservices* are deployable subdomain-specific apps that support Applications but do not define business structure alone.
- *Agents* interact with Applications to execute purpose and deliver value.
- *Platforms, Services, Gateways, and Workloads* are infrastructural or systemic enablers—they support but do not define the structure of the unit.

This classification offers a unifying language and a blueprint for system design, implementation, and governance. It bridges the current industry gap where architecture is often either over-abstracted or arbitrarily granular.

The Application stands as the unit's fundamental digital building block because it balances *functional purpose, operational independence*, and *architectural coherence.* Its role is neither incidental nor interchangeable. It is the room where purpose is lived, value is produced, and structure becomes actionable.

CHAPTER 3

Purpose

In 1998, two graduate students at Stanford launched a deceptively simple search engine with an unusually ambitious mission: *to organize the world's information and make it universally accessible and useful.* There was no product catalog, no monetization engine, and no clear roadmap to profitability. There was only a single text box—an interface so austere it appeared unfinished.

But behind this sparse facade was an architectural insight. Google was not designed to *sell* something. It was designed to *attract.* It did not start with a customer but with a consumer: a person in need of direction, understanding, or relevance. Its initial goal was not revenue, but resonance. And that goal gave the system its architecture.

Google's early design exemplified the Principle of Purpose in sociotechnical terms. It created a high-utility Function (search), aligned with a clear Purpose (universal access to information). That Function was *free*, and more importantly, *good.* Its effectiveness built

A. Milchman, *The Purpose Driven Design Framework for Modern Enterprises*,
https://doi.org/10.1007/979-8-8688-2223-0_3

trust, habitual use, and scale. And with scale came a new architectural opportunity: the emergence of a second group—*customers*—who were not the primary users, but were willing to pay for access to those users' attention, intent, or queries.

Thus emerged a two-sided structure:

- *Consumers* interacted with Google through search, expecting quality, neutrality, and speed.
- *Customers*—advertisers, partners, and data buyers—engaged through a monetization layer that was spatially and conceptually distinct from the core Function.

Critically, both sides were held together by the original Purpose. The advertising model could not degrade the experience without undermining the system itself. Ads had to be relevant, minimally intrusive, and algorithmically justified. The system was constrained by its originating intent.

What made Google architecturally significant was not the use of algorithms or cloud infrastructure. It was the alignment of structure, experience, and sustainability through Purpose. Every addition—maps, email, documents—extended the same structural logic: provide high-utility consumer services that attract engagement, and monetize *that attention* through non-disruptive, purpose-aligned channels.

This model shaped everything: interface minimalism, backend indexing, ad relevance scoring, and even the content of internal culture (famously captured by the phrase "don't be evil"). Purpose was not abstract—it was operational. It gave the system its form, its limits, and its growth path.

Over time, this structure allowed Google to evolve into Alphabet: a constellation of purposeful units, each with its own mission—Waymo, DeepMind, YouTube—held together not by a single operating model but by a compositional grammar rooted in architectural clarity. What began as a simple search bar became a reference model for structural purpose: attraction before monetization, Function before revenue, trust before targeting.

In the context of Purpose Driven Design, Google's origin story is not an example of innovation. It is an example of *alignment*. It shows how Purpose, when clearly defined and structurally encoded, can generate not only consumer engagement but also architectural resilience, ecosystem viability, and economic sustainability.

This chapter will explore Purpose as the orienting construct of every viable unit—how it is defined, how it functions, and how it constrains and enables everything that follows.

Google's early architecture illustrates a central insight of Purpose Driven Design: a unit's Purpose is not a tagline—it is an architectural anchor, a structural element that orients design decisions, governs composition, and holds the system together as it scales. For example, Google's mission to "organize the world's information" shaped not only its user interface but the architecture of its indexing, ranking, and data infrastructure—demonstrating Purpose as an organizing constraint rather than a marketing statement. Purpose defines the reason the unit exists, whom it serves, and what conditions its legitimacy. Without a clearly articulated and structurally embedded Purpose, a unit may perform activities or generate outputs, but it cannot sustain alignment, adapt meaningfully, or justify its participation in a larger system. In Purpose Driven Design, Purpose is the first and governing construct in the design of a unit. It shapes not only what the unit offers, but how it organizes itself internally and interacts externally. What follows is a detailed examination of Purpose as a design surface: its definition, properties, architectural implications, and implementation patterns.

3.1. Conceptual Foundations of Purpose

Before any construct can be designed, deployed, or coordinated, there must be a reason for its existence. In Purpose Driven Design, that reason is not symbolic or motivational—it is structural. This section defines Purpose as the architectural origin of all unit composition, establishing its conceptual necessity, systemic role, and historical significance as the first principle in viable sociotechnical design.

3.1.1. Defining Purpose Architecturally

In Purpose Driven Design, *Purpose* is not a statement of aspiration—it is an architectonic construct. It serves as the structural anchor of the unit, defining not only why the unit exists but also for whom and under what conditions its existence is justified. Purpose determines the unit's reason for inclusion in the system and its relevance to the environment it interfaces with. Without Purpose, a unit has no direction, no evaluative frame, and no structural legitimacy.

Architecturally, Purpose is the point of origin from which all other constructs derive their orientation. Function exists to enact Purpose through interaction. Process is designed to fulfill it. Structure configures agents to carry it out. Order and Culture stabilize its execution. Memory and Storage preserve its trajectory. When Purpose is absent or misaligned, coherence collapses across these interdependent domains.

A well-formed Purpose possesses the following architectural characteristics:

- *Externally oriented*: Purpose exists in relation to an environment. It specifies who or what the unit is designed to serve—customers, regulators, partners, systems, or broader societal roles. It cannot be validated internally; it must be legible and relevant from the outside.
- *Structurally constraining*: Purpose is not an open-ended aspiration. It narrows design choices by defining what kinds of functions are permissible, which processes are necessary, and what kinds of outputs constitute legitimate value. For instance, in a public health unit whose Purpose is to prevent disease rather than treat it, the architecture must favor early detection, education, and data monitoring functions over clinical treatment workflows—ensuring that every structural element reflects preventive intent.
- *Durable yet revisable*: While Purpose must provide continuity and coherence, it is not immutable. It can evolve, but only through managed revision. Structural adaptation must follow changes in Purpose—not precede them arbitrarily.
- *Architecturally expressive*: Purpose manifests in observable surfaces—landing pages, system identifiers, engagement interfaces, and narrative framings. These are not cosmetic elements; they are structural declarations that initiate interaction and signal accountability.

In this framework, Purpose is a design primitive, not an abstraction—it is a foundational element from which other constructs derive, not a generalized idea imposed afterward. It is the first structural decision in any viable unit composition. If Purpose is unclear, misdirected, or undefined, subsequent constructs cannot align. Clarity at the origin enables coherence in execution.

From an implementation perspective, Purpose should be concretized in the unit's identity artifacts and observable interfaces. It should guide decision-making in architectural planning, interface design, service configuration, and external coordination. Every downstream construct—Function, Process, Structure, and beyond—must be interrogated for its alignment with Purpose.

To design a unit, one must first define what it is *for, whom it serves,* and *why it matters*. These are not philosophical questions. They are architectural requirements.

3.1.2. What Problem Purpose Solves

Purpose addresses a foundational architectural problem: the absence of directional coherence in sociotechnical systems. Without a clearly articulated and structurally embedded Purpose, units drift into misalignment—internally fragmented, externally illegible, and functionally unstable. Purpose provides the structural mechanism that prevents this fragmentation by answering the essential question: *Why does this unit exist, and for whom?*

The primary problem Purpose solves is *design without orientation*. In large systems, units are often created to fulfill tactical needs—process automation, data capture, service provision—without an explicit understanding of their systemic role or external justification. This leads to hollow structures: units that execute tasks but cannot be evaluated for relevance, legitimacy, or contribution. Purpose introduces a directional anchor that constrains arbitrary expansion and aligns all internal constructs toward a meaningful outcome.

Specifically, Purpose solves the following architectural deficiencies:

- *Lack of external relevance*: Units often become self-referential, optimized for internal metrics or efficiency without producing recognizable Value for their environment. Purpose re-establishes the environment-facing orientation necessary for legitimacy and trust.
- *Incoherent system behavior*: In systems where units operate without aligned Purpose, coordination becomes erratic. Interactions may occur, but their meaning and utility degrade over time. Purpose provides the semantic coherence needed for reliable integration.

- *Structural ambiguity*: Without Purpose, design decisions become arbitrary. Functions are added without rationale, processes accumulate without clear outcomes, and structural roles proliferate without accountability. Purpose introduces constraint and justification across the design chain.
- *Impeded adaptation*: When Purpose is unclear, adaptation cannot be intentional—it defaults to reactive change. Feedback becomes meaningless without an anchor to assess directionality. Purpose enables structured learning, recalibration, and purposeful evolution.

In summary, Purpose solves the problem of *unanchored complexity*. It converts a loosely coupled collection of features or roles into a directional system capable of delivering consistent Value under environmental pressure. It transforms the unit from a technical configuration into a legible architectural object—one that exists *for a reason* and can evolve *with a reason*. Without Purpose, design is possible, but coherence is not.

3.1.3. Historical Patterns and Structural Evolution

The architectural role of Purpose has evolved in parallel with the development of sociotechnical systems. In earlier eras, Purpose was often implicit—embedded in religious mandates, social hierarchies, or institutional missions. These structures provided continuity but lacked adaptability. As systems grew more complex, especially in industrial and digital domains, Purpose became increasingly abstracted, eventually displaced by operational targets, compliance metrics, or scale-driven ambitions. This evolution has produced recurrent patterns of structural failure—systems that grow in size and sophistication but collapse in meaning.

Historically, three dominant patterns illustrate the mismanagement or erosion of Purpose:

- *Canonical unification (pre-modern structures)*: In early civilizations, Purpose was inseparable from identity—social, spiritual, or political. Temples, guilds, and courts operated with Purpose as a given. Systems were not architected around Purpose because it was embedded culturally and ritually. This ensured cohesion but limited pluralism and structural differentiation.

- *Functional fragmentation (industrial-modern systems)*: The industrial age introduced specialization and hierarchical control. Purpose was translated into productivity, efficiency, or compliance, often measured in output volume or labor optimization. This made systems scalable but disconnected from external interpretation. Purpose, in this model, became subordinate to operational goals, and misalignment emerged between roles, outcomes, and societal value.
- *Procedural abstraction (digital-contemporary systems)*: In the digital era, systems became increasingly virtualized, modularized, and orchestrated through software, platforms, and interfaces. While these systems excel in connectivity and automation, they often suffer from directional opacity. Purpose is either reduced to branding rhetoric or fragmented across siloed subsystems. Coordination persists, but coherence degrades. Decisions are executed faster than they are understood.

Across these phases, the pattern is clear: as systems scale and abstract, structural Purpose becomes harder to maintain unless it is explicitly designed. The failure to do so produces outcomes that are technically integrated but semantically incoherent. This culminates in phenomena such as

- *Technocratic monologs*: Where a single logic (financial, procedural, algorithmic) dominates, suppressing alternative interpretations of Purpose.
- *Organizational drift*: Where units persist beyond their relevance due to inertia, legacy entitlements, or lack of evaluative anchors.
- *Architectural debt*: Where the absence of Purpose leads to ad hoc layering of processes, roles, and interfaces—none of which can be rationalized systemically.

Purpose Driven Design responds to these failures by reintroducing Purpose as an explicit, composable architectural primitive. Unlike historical models that treated Purpose as an assumption or symbolic gesture, this framework operationalizes Purpose as the first structural element in unit design. It marks a return not to tradition, but to structural intelligibility—where Purpose becomes both a directional force and a design constraint.

This shift aligns with broader evolutionary transitions—from hierarchical monuments to adaptive networks; from universal logics to contextual legitimacy; from central orchestration to distributed purpose alignment. In this trajectory, Purpose is no longer a background principle. It is a foregrounded architectural necessity—required not only to justify existence but to enable responsible evolution within increasingly interdependent systems.

3.2. The Functional Role of Purpose in Sociotechnical Units

Purpose is not merely a static declaration—it is a functional element that stabilizes internal design and guides external interaction. It enables units to operate coherently, adapt responsively, and maintain structural alignment under real-world conditions. The following sections clarify how Purpose functions both as an internal architectonic force and as an external orienting interface.

3.2.1. Internal Role: Structural Stabilization

Within a sociotechnical unit, Purpose functions as the primary stabilizing element. It aligns internal constructs—Function, Process, Structure, Order, Culture, Memory, and Storage—by establishing a directional logic that informs their configuration and coherence. Without Purpose, these constructs risk operating in isolation, accumulating local optimizations that undermine system-wide intelligibility and responsiveness. Purpose prevents this fragmentation by serving as the common referent against which architectural decisions are evaluated and interdependencies are structured.

Structurally, Purpose achieves stabilization through three mechanisms:

- *Orientation of Design Constructs*: Purpose provides the basis for determining whether a unit's configuration is coherent. It clarifies what the unit is for, which enables precise design of how it should operate. For example, Function must expose the pathways through which Purpose is realized; Process must be constructed to enact that Purpose consistently; and Structure must assign roles in a way that supports both Purpose fulfillment and internal accountability.

- *Constraint of Arbitrary Expansion*: In the absence of a clear Purpose, units tend to expand reactively—accreting features, responsibilities, or workflows in response to short-term demands. Purpose introduces constraint by filtering these additions through a structural question: *Does this enable or dilute the unit's reason for existence?* This constraint mechanism ensures bounded growth and preserves functional identity over time.
- *Alignment of Feedback and Adaptation*: Internal learning systems—such as Memory, feedback loops, and retrospective analysis—require a reference frame to assess relevance and direction. Purpose supplies that reference. It allows the unit to evaluate whether outcomes are aligned with intent, whether adaptation is warranted, and how structural adjustments should be made. Without Purpose, feedback lacks interpretability and change becomes directionless.

The result is a unit whose architecture is not just operationally viable but structurally intelligible. Each component exists in deliberate relation to the Purpose of the unit, reducing internal contradiction and increasing the likelihood of sustainable operation. Stabilization, in this context, does not imply rigidity. Rather, it denotes compositional clarity—ensuring that as the unit evolves, its internal architecture continues to express a coherent reason for existing.

3.2.2. Purpose As Systemic Fit

In complex sociotechnical systems, individual units do not operate in isolation. Their viability depends not only on internal coherence or external alignment but also on their ability to function as meaningful, non-fragmenting components within a larger system. *Systemic fit* refers to this architectural requirement. It ensures that each unit's Purpose is not only structurally sound on its own terms but also intelligible, non-redundant, and composable within the broader organizational architecture.

From Self-Containment to Composability

While internal stabilization addresses the self-sufficiency of a unit, and environmental fit concerns external relevance, systemic fit ensures *relational integrity*—that the unit contributes productively to system-level goals, structures, and flows. Purpose achieves systemic fit by

- *Establishing boundaries of responsibility*: Clearly articulated Purpose prevents overlap and conflict between units. It delineates scope and avoids duplication by defining what a unit does *that no other unit should.*
- *Enabling architectural composability*: A unit's Purpose must allow it to interface meaningfully with other units—through aligned functions, coordinated processes, and compatible expectations. Without composability, systems cannot scale without entanglement.
- *Supporting federated coherence*: In distributed systems (e.g., platform ecosystems, data mesh architectures), systemic fit ensures that diverse units operate under local autonomy while still contributing to collective directionality. Purpose serves as the logical anchor for that dual autonomy and coordination.

Failure Modes of Poor Systemic Fit

When Purpose is defined without consideration of systemic fit, the following structural breakdowns occur:

- *Redundant units*: Multiple units address similar needs with conflicting or overlapping Purposes, creating fragmentation or internal competition.
- *Uncoordinated behaviors*: Without systemic anchoring, units interpret policies, roles, or objectives inconsistently—leading to misaligned implementations or incoherent user experiences.
- *Architectural drift*: Units evolve in isolation, gradually losing compatibility with system-wide priorities or dependencies.

These failures do not arise from lack of local design effort but from the absence of a compositional logic—one that Purpose is uniquely positioned to provide.

Purpose As a Structural Contract

In systems designed for adaptability and modularity, Purpose must act as a *structural contract*—declaring what the unit is accountable for within the system and how it interacts with others. This contract supports

- *Governability*: Enabling oversight, coordination, and responsibility assignment across units.
- *Substitutability*: Allowing units to be upgraded, replaced, or retired without systemic collapse.
- *Traceability*: Making it possible to understand how system-wide behavior emerges from the interaction of distinct, purpose-driven components.

In this framing, a well-defined Purpose is not simply a unit-level necessity—it is a system-level safeguard. It ensures that the system remains decomposable, navigable, and resilient under pressure.

Integrating Systemic Fit into Purpose Definition

To achieve systemic fit, Purpose must be defined not only in relation to *what the unit does* but in relation to

- What other units do (to avoid duplication)
- How systemic value is composed (to ensure contribution)
- What structural gaps exist (to identify necessary roles)
- How adaptation propagates (to avoid unintended consequences)

In this sense, Purpose is not only the anchor of the unit—it is the logic by which the unit justifies its inclusion in the system. When articulated with systemic fit in mind, Purpose makes large, complex architectures possible without reverting to centralization or brittle integration.

Systemic fit is the architectural bridge between unit autonomy and system coherence. Purpose is what makes that bridge structurally sound.

3.2.3. External Role: Attraction and Alignment

While Purpose stabilizes the internal configuration of a unit, its equally critical role lies in establishing the unit's relationship to its external environment. A unit is not structurally complete unless it is legible and relevant to the entities it serves—customers, partners, regulators, or broader societal systems. Purpose enables this external coherence by functioning as both an attractor and an alignment mechanism.

Attraction Through Orientation

Purpose attracts external engagement by signaling what the unit offers, whom it serves, and why its existence matters. This attraction is not rooted in promotional language or aspirational branding but in architectural clarity. A unit that articulates its Purpose in structurally visible forms—such as service descriptions, interface prompts, or engagement flows—allows consumers and collaborators to recognize its relevance and initiate interaction.

This attractor function is especially vital in modular, distributed systems where coordination is decentralized. In such systems, interaction is voluntary and value-based. Purpose ensures that external actors can discover the unit, interpret its role, and determine whether it satisfies a real need. Without this signal, units become indistinguishable or irrelevant—structurally present but functionally unused.

Alignment With Environmental Demands

Beyond attraction, Purpose facilitates alignment between the unit and the environment in which it operates. This includes not only consumer expectations but also regulatory frameworks, ethical norms, economic conditions, and ecosystem dependencies. Purpose acts as a structural contract, ensuring that the unit's internal design corresponds to the external commitments it must uphold.

Alignment occurs across multiple dimensions:

- *Semantic alignment*: Purpose provides a shared vocabulary that allows external actors to interpret the unit's function and outcomes.
- *Behavioral alignment*: It ensures that the unit behaves in ways that are intelligible, predictable, and trustworthy to its stakeholders.
- *Temporal alignment*: Purpose guides long-term orientation, ensuring that the unit remains relevant as external conditions evolve.

Failure to align structurally with the environment results in systemic fragility. Units may appear functional internally but generate friction, mistrust, or noncompliance externally. In contrast, when Purpose is well-defined and consistently expressed, the unit becomes a reliable participant in its ecosystem—able to integrate, adapt, and contribute without requiring excessive coordination overhead.

In sum, Purpose is the primary interface between the internal logic of a unit and the external systems that confer legitimacy, trust, and utility. It enables engagement not by imposing control, but by making the unit discoverable, interpretable, and compatible with its environment. This attraction and alignment function transforms the unit from a technical object into a meaningful actor within a broader sociotechnical system.

3.2.4. Purpose As Environmental Fit

For a sociotechnical unit to be viable, it must not only operate coherently within itself and interact intelligibly with others—it must also fit its environment. Purpose serves as the primary mechanism by which this fit is achieved. It positions the unit within a broader ecological, institutional, or societal context, aligning its internal configuration with external demands, constraints, and opportunities. In this role, Purpose functions as the architectural logic that links systemic identity to environmental relevance.

Environment As Structural Context

Every unit exists within an environment composed of expectations, norms, feedback mechanisms, and interdependencies. This environment may include

- Consumers and end users with evolving needs
- Legal and regulatory bodies imposing structural constraints
- Partner systems and data sources requiring coordination
- Social and cultural narratives shaping perceived legitimacy
- Market forces and operational dependencies determining viability

Purpose interprets these environmental signals and encodes them into the structural composition of the unit. It ensures that the unit is not self-referential but outwardly oriented—designed to contribute something meaningful, expected, and valuable to its surroundings.

Structural Calibration Through Purpose

Environmental fit requires continuous calibration. A unit designed without consideration for its environment may function efficiently but fail to produce meaningful outcomes, attract usage, or earn trust. Conversely, a unit that over-conforms to external expectations may sacrifice internal coherence and operational resilience. Purpose enables calibration by

- *Selecting the right boundaries*: Ensuring the unit is neither over-scoped nor under-scoped relative to what the environment demands.
- *Prioritizing features and capabilities*: Filtering potential functions and processes to those most aligned with external relevance.
- *Structuring accountability*: Defining clear interfaces, metrics, and roles that reflect environmental expectations (e.g., compliance, service-level agreements, or ethical standards).

Purpose thus mediates between internal autonomy and external legitimacy, ensuring that the unit can act independently while remaining contextually appropriate.

Adaptive Viability

Environmental fit is not a static condition. Ecosystems change, regulations evolve, and stakeholder needs shift. Purpose allows the unit to adapt structurally without collapsing—by serving as the anchor against which such adaptation is interpreted and managed. When environmental feedback is received, Purpose determines whether the unit should extend its capabilities, contract its scope, or reconfigure its structure. Without this anchor, adaptation becomes reactive and incoherent.

A unit that lacks environmental fit becomes a structural orphan—present but disconnected, maintained internally but unused or distrusted externally. In contrast, a unit grounded in Purpose maintains continuous relevance, adapting without losing identity, and evolving without violating the conditions that justify its place in the system.

In this regard, Purpose is not an abstraction. It is the means by which a unit earns the right to persist—structurally calibrated to a world that expects it to contribute, comply, and coordinate. Fit is not granted—it is designed. And Purpose is the structure that makes that fit possible.

3.2.5. Serving Internal and External Needs

A structurally viable unit must serve both *internal* and *external* needs. Purpose, as its orienting construct, must be defined in a way that acknowledges and integrates these two dimensions. Internally, a unit must contribute to the coherence, composability, and evolution of the larger system it belongs to. Externally, it must generate recognizable Value for entities in its environment—customers, regulators, partners, or the broader public. These two roles are not separate—they are concurrent, and their alignment is essential for architectural legitimacy.

Internal Accountability: Serving the System

Within a complex sociotechnical system, every unit functions as a component in a larger architecture. It may consume services, supply data, expose interfaces, or participate in coordinated workflows. Its Purpose must reflect a structural contribution—filling a necessary role, avoiding duplication, and enabling reliable interaction with adjacent units. In this context, the "customers" are other units, platforms, or orchestration layers that depend on the unit's function to perform their own.

Failing to meet internal needs leads to systemic friction: broken dependencies, integration bottlenecks, and loss of compositional integrity. Even if a unit produces externally visible outcomes, it cannot be considered structurally legitimate if it destabilizes the system it belongs to.

External Justification: Serving the Environment

At the same time, every unit must face outward. It must offer something that external entities recognize as meaningful—a product, service, interface, or capability that fulfills a real need. This environmental service is what grants the unit its external legitimacy. It is how trust is earned, relevance is maintained, and interaction is initiated.

Failing to meet external needs results in obsolescence. The unit may be well-integrated internally but will gradually lose justification for its existence if it no longer produces observable Value beyond the system.

Dual-Facing Design

Purpose must therefore be defined as a dual-facing construct. It must be

- *Systemically coherent*: Ensuring that the unit fits, composes, and contributes meaningfully to the system's architecture.
- *Environmentally relevant*: Ensuring that the unit is discoverable, interpretable, and valuable to external actors.

These two facets of Purpose are not competing—they are structurally interdependent. A unit that serves only the system risks becoming insular and brittle. A unit that serves only the environment risks producing side effects that undermine systemic coherence. Architectural viability demands both.

In Purpose Driven Design, the design of a unit begins with a clear articulation of whom it exists to serve *and* how it integrates structurally. These are not sequential considerations—they are simultaneous requirements. Purpose is the only construct that can hold both perspectives together, enabling units that are not only operable and composable but also justified and trusted.

This dual alignment—internal and external—is what transforms a unit from a functional enclosure into a structurally meaningful actor within a dynamic, evolving system.

3.3. Design Principles and Heuristics for Purpose

Purpose must not only be conceptually sound; it must be rendered in forms that constrain design decisions, guide structural alignment, and enable evaluative traceability. The following principles translate Purpose from directional intent into operational structure, ensuring that it is actionable, durable, and governable across time and context.

3.3.1. Principles of Purpose-Oriented Design

Designing Purpose is not a matter of defining vision statements or branding artifacts—it is an architectural activity. In Purpose Driven Design, Purpose is treated as a structural construct that must direct, constrain, and coordinate all other components of the unit. The following principles define what constitutes a well-formed Purpose in a sociotechnical system. These principles are not stylistic preferences; they are architectural conditions for coherence, composability, and legitimacy.

1. **Externally Justified, Not Internally Derived**

   A unit's Purpose must originate from the environment it serves—not from internal capabilities, preferences, or legacy roles. This requires identifying the specific entities—customers, regulators, platforms, communities—for whom the unit produces recognizable Value. Designing from internal abstraction leads to self-referential structures; designing from external justification produces units that are necessary, not just operational.

2. **Structurally Constraining, Not Symbolically Framing**

   Purpose must act as a design constraint. It should eliminate incompatible features, discard unnecessary integrations, and reject non-aligned activities. In practice, this means evaluating every proposed Function interface, process or workflow, rule or constraint, structural configuration, and cultural artifact against the unit's Purpose—and excluding what does not directly advance it. Purpose is not a narrative overlay—it is a filter for architectural consistency. Consequently, this principle guides evaluation at every stage of design, ensuring that each construct, from Function to Storage, can be traced back to its originating Purpose.

3. **Scope-Defining and Boundary-Setting**

   A clearly defined Purpose determines what is *in scope* and what is *outside* the unit's responsibility. It creates a boundary of accountability, ensuring that adjacent units do not overlap, duplicate, or interfere. In this sense, Purpose enables the architectural principle of separation of concerns, without requiring top-down orchestration.

4. **Durable, Yet Adaptable**

   Purpose must remain stable enough to support consistent design, yet flexible enough to evolve with its environment. This balance requires treating Purpose as a versioned architectural object—revisable through structured governance, traceable through Memory, and reflected in evolving Function and Process. Change in Purpose should propagate through the unit deliberately, not retroactively rationalize emergent behavior.

5. **Internally Actionable**

   Purpose must be translated into internal behavior. It should inform how Structure is configured, how Process is executed, how Order is governed, and how Culture is reinforced. If agents within the unit cannot interpret and act on the Purpose through their assigned roles or systems, the Purpose has failed as a design construct.

6. **Systemically Non-Redundant**

   In federated architectures, Purpose must also clarify what differentiates the unit from others. A unit that claims Purpose already fulfilled elsewhere introduces architectural noise, not value. To be valid, Purpose must uniquely position the unit within the broader system—filling a structural gap, enabling a distinct interaction, or absorbing a specific class of constraints.

7. **Externally Legible**

   A well-designed Purpose is observable through the unit's interfaces—its Function, communication artifacts, and service boundaries. It must be interpretable not only by internal actors but by those who interact with the unit externally. This legibility reinforces trust, discoverability, and appropriate use. Purpose hidden in internal documentation or implied in behavior is architecturally weak.

8. **Composable Across Time and Scale**

   Purpose must be resilient to system growth and shifting coordination patterns. It should remain meaningful when composed into larger systems (e.g., fleets, constellations) and when decomposed into subunits or processes. Composability requires that Purpose be neither overfitted to one context nor generalized to the point of abstraction.

Together, these principles ensure that Purpose is not treated as a rhetorical artifact but as a primary design surface—structuring all other elements of the unit and maintaining coherence across contexts. In systems characterized by distributed control, federated structure, and ongoing adaptation, these principles are not optional—they are preconditions for architectural viability.

3.3.2. Architectural Heuristics

While principles articulate what makes Purpose structurally sound, heuristics offer practical guidance for applying those principles under real design conditions. These heuristics help architects and designers translate Purpose into implementable decisions, particularly in complex or evolving environments. Each heuristic serves to reduce ambiguity, prevent misalignment, and reinforce architectural coherence across scale and time.

1. **Start with the External Stakeholder, Not the Internal Capability**

   Before defining any functionality or allocating roles, identify the specific external entity the unit is meant to serve. Ask: *Who needs this unit to exist, and what would break if it didn't?* This framing prevents internal abstraction from drifting into irrelevance. A unit that begins with its own skills or assets often produces features without fit.

2. **Design the Boundary Through Purpose, Not Through Function**

   The perimeter of a unit is not defined by what it *does* but by what it *exists to achieve*. Functions may shift, expand, or contract—Purpose defines the enduring boundary of structural accountability. This prevents architectural sprawl and protects against coordination entanglement.

3. **Use Negative Design to Clarify Scope**

   A clear Purpose should allow designers to specify what the unit *does not do*. Explicitly declaring exclusions (e.g., "This unit does not issue credentials," or "This unit does not process transactions") reinforces separation of concerns and reduces future coupling risks. Negative space clarifies structural intent.

4. **Write Purpose As a Structural Statement, Not a Slogan**

   Purpose statements should be formulated to express structural intent, not motivational tone. Prefer statements like

   - "This unit enables secure onboarding of accredited users in regulated jurisdictions,"

     over

   - "We exist to empower trust and inclusion in digital ecosystems."

     Clarity of structure supports traceability and composability.

5. **Treat Purpose As a Versioned Artifact**

   In adaptive systems, Purpose must evolve deliberately. Versioning the Purpose statement—along with a recorded rationale and observed impact—enables longitudinal coherence, simplifies audits, and improves resilience during organizational or technological transitions.

6. **Make Purpose Observable in Interfaces**

   Ensure that external interfaces (e.g., landing pages, service descriptions, API documentation) express the unit's Purpose explicitly. Interfaces should not only support functional use but communicate what the unit *is for*. Purpose hidden in internal documentation creates architectural opacity.

7. **Check for Purpose–Value–Function Alignment**

   Periodically validate that the unit's Purpose continues to justify its Value and that the Function exposes that Value effectively. Misalignment among these three typically indicates architectural drift, overextension, or underdelivery. This triadic check serves as a diagnostic lens.

8. **Avoid Surrogate Purpose via Performance Metrics**

   Do not allow internal KPIs or throughput measures to become proxies for Purpose. Metrics are performance indicators, not structural justifications. A unit may achieve internal efficiency while failing to fulfill its external reason for existence. Purpose must remain the primary criterion for validity.

9. **Ensure Systemic Non-Redundancy Before Deployment**

   Before finalizing a unit's Purpose, verify that no existing unit already fulfills it. Redundant Purpose leads to coordination friction, governance gaps, and data or capability duplication. In compositional systems, distinct Purpose is a prerequisite for structural composability.

10. **Design Purpose to Survive Process and Platform Change**

    Purpose should remain intelligible even if the unit's technical implementation, workflow, or toolchain changes. If a platform upgrade or process refactoring invalidates the Purpose, it was too narrowly defined. Durability of Purpose across operational change is a test of architectural fitness.

These heuristics do not replace architectural reasoning—they support it. They serve as cognitive instruments for identifying weak Purpose definitions, validating structural design, and sustaining long-term composability in dynamic sociotechnical environments. Used consistently, they help ensure that Purpose remains not only conceptually clear but architecturally real.

3.4. Implementation Patterns for Purpose

With Purpose now defined as a structurally constraining and direction-setting construct, we turn to its concrete realization. Purpose must not remain an abstract declaration—it must be instantiated through observable artifacts, functional interfaces, and system behaviors. The implementation patterns that follow demonstrate how Purpose can be embedded across digital, organizational, and operational layers to ensure alignment, legibility, and traceability. These patterns are not stylistic choices; they are architectural surfaces through which Purpose becomes actionable and governable.

3.4.1. Purpose in the Digital Plane

In a digitally enabled enterprise, sociotechnical units must project their Purpose in ways that are observable, interpretable, and structurally legible. The digital plane provides the surfaces through which a unit's Purpose becomes discoverable and externally meaningful. These surfaces are not implementation details—they are architectural declarations that position the unit within its system and environment.

Purpose in the digital plane is instantiated not through internal code artifacts but through high-level structural representations that expose the unit's identity, intent, and accountability. These surfaces serve both as points of engagement and as signals of alignment with broader system architecture.

High-Level Digital Expressions of Purpose

At the unit level, Purpose is expressed across the following digital artifacts:

- *Dedicated microsites or service portals*: These serve as the unit's outward-facing presence. They communicate whom the unit exists to serve, what Value it offers, and under what conditions interaction is appropriate. A microsite is not a marketing layer—it is an architectural interface.
- *Digital charters or structural manifests*: Often encoded as static content or structured metadata, these documents state the unit's Purpose, boundaries, interfaces, and obligations. They establish the unit's reason for inclusion in the system and clarify its systemic and environmental alignment.
- *Onboarding documentation and integration briefs*: These materials explain to external stakeholders—partners, consuming systems, or federated teams—how the unit operates, how it fulfills its Purpose, and what expectations are attached to engagement. They ensure that Purpose is not only declared but understood.

Required Attributes of Purpose Artifacts

To serve as valid architectural expressions, these digital representations of Purpose must be

- *Externally oriented*: They must clarify who the unit is for, not just what it contains.
- *Purpose-aligned*: All language and structure must be consistent with the unit's declared reason for existence.

- *Non-recursive*: They must describe the unit as a whole, not its subcomponents or technical implementations.
- *Composable*: They must enable the unit to participate in larger digital ecosystems through clearly expressed identity and structural intent.

Architectural Function of Digital Purpose Surfaces

These surfaces are not cosmetic. They provide

- *Structural discoverability*: Making the unit visible within a federation or platform
- *Semantic clarity*: Communicating the unit's boundaries and value proposition
- *Coordination readiness*: Enabling consumers, partners, and integrators to engage with the unit appropriately and with minimal friction

When these surfaces are absent, ambiguous, or overloaded with internal details, the unit loses architectural legibility. Coordination degrades, trust erodes, and the system accumulates structural noise. A unit's Purpose, once made visible at the appropriate level of abstraction, becomes not just an internal constraint but a shared architectural reference.

In summary, Purpose in the digital plane is not implemented through deep instrumentation or component-level prompts—it is projected through high-level structural representations that clarify the unit's role, relevance, and legitimacy. These surfaces enable a unit to participate in a digital ecosystem as a purposeful, bounded, and intelligible entity.

3.4.2. Implementation Surfaces

To function as a coherent architectural construct, Purpose must manifest across clearly defined implementation surfaces. These surfaces are not internal subsystems or granular artifacts but externally visible and structurally bounded points at which the unit declares its identity, communicates its intent, and signals how it contributes to its system and environment. Implementation surfaces translate Purpose from abstract orientation into inspectable, composable form—without decomposing into subunit detail.

A sociotechnical unit expresses its Purpose structurally through three primary implementation surfaces: *identity surfaces, interface surfaces,* and *coordination surfaces.*

1. **Identity Surfaces**

   These surfaces make the unit's Purpose recognizable and intelligible to external stakeholders. They are designed to communicate *what the unit is, why it exists,* and *whom it serves.* Identity surfaces establish initial orientation and initiate trust.

   - *Examples*: Dedicated microsite, service homepage, purpose charter, public profile page in an internal ecosystem registry
   - *Architectural function*: Clarify unit-level relevance; initiate engagement based on structural intent, not internal function

   A well-formed identity surface should convey

   - The Purpose statement
   - The Value the unit delivers
   - The boundaries of its responsibility
   - The external entities it is accountable to

2. **Interface Surfaces**

   These surfaces expose the mechanisms by which the unit makes its Purpose accessible. They are not limited to technical APIs or UI elements but include any structured point of interaction that allows external actors to engage with the unit on the terms its Purpose defines.

   - *Examples*: Interaction portals, service request entry points, business-level integration guides, partnership onboarding materials
   - *Architectural function*: Ensure that the way others interact with the unit reinforces—not contradicts—its stated Purpose

   Interface surfaces must preserve the coherence between the unit's declared Purpose and its functional behavior. Any misalignment (e.g., interfaces that expose functionality irrelevant to Purpose) signals architectural drift.

3. **Coordination Surfaces**

   These surfaces communicate how the unit aligns with its broader system—how it fits, composes, or integrates structurally. They are essential in federated or modular architectures, where units must coordinate without central orchestration.

   - *Examples*: Integration agreements, capability registries, role alignment documentation, structural manifests
   - *Architectural function*: Make explicit how the unit interoperates, under what assumptions, and with which adjacent roles or services

   Coordination surfaces ensure that the unit's Purpose is not only declared but *situated*—that is, intelligibly positioned relative to system-wide Purpose distributions and interface expectations.

Avoiding Surface-Level Misuse

Architectural surfaces must remain at the level of the unit. Misuse occurs when

- Implementation surfaces describe *subcomponents* instead of the unit as a whole
- Identity surfaces adopt *generic language* that obscures specificity
- Interface surfaces *diverge* from Purpose-aligned interaction logic
- Coordination surfaces are treated as *policy appendices*, rather than architectural declarations

Purpose is not fulfilled by the presence of surfaces alone—it is fulfilled when those surfaces are compositionally aligned, intelligible, and anchored in structural intent.

Properly implemented, these surfaces allow a unit to participate in a broader sociotechnical system with clarity, accountability, and minimal coordination overhead. They ensure that Purpose is not only designed, but *realized*—structurally, digitally, and systemically.

Delineating Purpose from Function

While implementation surfaces help manifest Purpose at the unit level, it is essential to maintain a strict boundary between the *Purpose construct* and downstream constructs such as *Function*. Confusing the two results in conceptual overlap, diluted accountability, and implementation ambiguity.

Interface Surfaces: Framing, Not Transacting

Interface surfaces in the context of Purpose are not value-exchange mechanisms. They frame the unit's orientation—what it is for, whom it serves, and what type of engagement is expected. They may include entry points such as service overviews or onboarding introductions, but they do not perform the value delivery itself. That responsibility belongs to Function.

If the surface explains *why* the unit exists and *what it is accountable for*, it reflects Purpose.

If it enables *how* the unit exchanges Value with others, it reflects Function.

Coordination Surfaces: Declaring Position, Not Executing Flow

Coordination surfaces, likewise, declare the unit's role within a larger system. They describe *why the unit is here, what structural gap it fills,* and *how its Purpose aligns with adjacent units.* They do not perform dependency resolution, data routing, or service invocation—those are concerns of Function and Process.

Purpose defines the structural position; Function manages the operational exchange.

By clearly delineating these boundaries, designers preserve architectural coherence and avoid recursive descent into subunit behavior. Purpose must always operate at the level of the *unit-as-whole,* never at the level of its mechanisms. Implementation surfaces serve to reflect that Purpose structurally, not to operationalize what belongs to other constructs.

3.4.3. Realization in Sociotechnical Contexts

While digital surfaces provide accessible representations of Purpose, a sociotechnical unit is not reducible to digital form. Its Purpose must also be realized across social and operational dimensions that give the unit meaning, accountability, and legitimacy in real-world contexts. This realization is not symbolic—it is structural. It ensures that Purpose is not merely visible but *lived* through the unit's practices, expectations, and interactions.

Purpose becomes fully realized only when it shapes how the unit is *interpreted by people, positioned within systems,* and *engaged in context.* This section describes how Purpose is enacted in sociotechnical terms—across organizational alignment, stakeholder relationships, and institutional environments.

Organizational Realization

Within the unit's internal structure, Purpose must be consistently recognized across roles, decisions, and governance routines. It should orient how team members understand their responsibilities, how trade-offs are resolved, and how structural coherence is maintained under pressure.

- *Hiring and onboarding*: New agents should be introduced not only to processes and tools but to the Purpose that anchors the unit's existence.
- *Decision-making frameworks*: Conflicting priorities and design options should be evaluated in terms of alignment with the unit's declared Purpose.
- *Accountability structures*: Roles and escalation paths should reflect the unit's Purpose, ensuring that responsibility is neither abstracted away nor ambiguously distributed.

If internal agents cannot interpret or apply the unit's Purpose meaningfully, the construct has not been realized—it has only been declared.

Systemic Realization

In a broader system of units—such as a fleet, mesh, or federation—the realization of Purpose requires *structural differentiation and role clarity.* A unit's Purpose must justify its presence and define its contribution in relation to others.

- *Non-overlapping mandates*: No two units should share indistinct or interchangeable Purposes. Redundancy erodes system trust and composability.

- *Role-based interaction protocols*: Coordination must be framed not only in terms of interface compatibility but of Purpose compatibility—ensuring that inter-unit dependencies are mission-aligned, not merely technical.
- *Architectural registration*: Within registries, directories, or governance records, each unit's Purpose should be traceable and structurally distinct, not inferred from activity.

Purpose is realized systemically when the unit's function, boundaries, and role are accepted and upheld across the architecture—not when they are simply documented.

Environmental Realization

Sociotechnical units operate in real environments: legal, cultural, market, and institutional. Realizing Purpose in these contexts means embedding it into the unit's posture toward its stakeholders—not only in what it produces, but in how it conducts itself.

- *Stakeholder recognition*: The unit's Purpose should be intelligible and relevant to the external entities it interacts with. These entities must be able to trust, interpret, and evaluate the unit on the basis of its declared intent.
- *Compliance and legitimacy*: The unit must not only meet regulatory obligations—it must express its Purpose in a way that aligns with public, civic, or sectoral expectations. Purpose is part of its social contract.
- *Adaptation pathways*: If environmental signals shift—policy, public trust, partner needs—the unit's Purpose must provide the framework for structured adaptation, not reactive improvisation.

Purpose is realized environmentally when it governs how the unit relates to its world—not just how it explains itself.

Realization vs. Representation

Digital representation alone is insufficient. A unit with a microsite, charter, or registry entry may appear structurally defined, but unless its Purpose is reflected in how it behaves socially, systemically, and operationally, it remains unanchored. Representation specifies how Purpose is documented and displayed; realization determines how that Purpose is enacted through structure, behavior, and interaction. For instance, an organization may publish a sustainability charter and a dedicated webpage describing its environmental commitments, yet if its procurement processes, supplier relationships, and facility operations remain unchanged, its Purpose is represented but not realized. Representation without realization produces surface coherence but internal fragility. Full realization of Purpose demands consistent expression across all sociotechnical dimensions—visible not only in interfaces but in decisions, boundaries, relationships, and adaptation.

In this way, Purpose becomes more than a design input—it becomes a structural truth. A unit has realized its Purpose not when it can declare it but when that Purpose is evident in how the unit operates, coordinates, and sustains its role in a complex and evolving world.

3.5. Integration with Other Constructs

Purpose is the initiating construct of unit design, but its effectiveness depends on how it shapes—and is reinforced by—the other eight architectural constructs of Purpose Driven Design. A unit is not viable if its Purpose is clear but disconnected from its Function, Process, or Structure, nor if its Memory or Storage contradict its declared intent. Let's now examine how Purpose integrates with the broader grammar of the unit, highlighting key coupling points, reinforcing dependencies, and potential misalignments. Integration is not optional—it is what allows Purpose to sustain coherence across time, interaction, and evolution.

3.5.1. Interlocking Dependencies

In Purpose Driven Design, Purpose is the architectonic driver—the initial constraint that orients all other design decisions. However, its role is not unidirectional. While Purpose gives form to the unit, it also depends on the alignment and reinforcement of the other constructs to remain coherent, interpretable, and actionable. These interlocking dependencies form a compositional system in which each construct must structurally support and validate the others. If any construct drifts from Purpose, the entire unit risks fragmentation or loss of legitimacy.

Purpose—Function: Interface Must Express Intent

Function provides the means by which external actors interact with the unit. Its design must be a direct expression of Purpose. If Purpose defines whom the unit serves and why, Function must operationalize *how* that service is made available.

- *Dependency*: Purpose constrains what types of interfaces are appropriate and what classes of interaction should be exposed.
- *Failure Mode*: If Function introduces interfaces that do not align with Purpose, the unit becomes semantically incoherent—usable, but misdirected.

Purpose—Process: Operations Must Be Purpose-Aligned

Process encodes the internal logic that transforms inputs into outcomes. It defines how the unit delivers Value. All Process design—activities, workflows, escalation paths—must derive from the structural requirements of Purpose.

- *Dependency*: Purpose determines which operational activities are required, and which are extraneous or misaligned.
- *Failure Mode*: Purpose drift occurs when Process accumulates unanchored complexity, expanding operational behavior beyond what the unit's Purpose justifies.

Purpose—Structure: Roles Must Serve Mission

Structure specifies the configuration of agents—human and artificial—within the unit. It defines who is responsible for what. Purpose determines the structural shape of the unit: which roles are necessary, how responsibilities are distributed, and how authority flows.

- *Dependency*: Purpose informs which capacities must be embedded in the unit, and how those capacities should be allocated.
- *Failure Mode*: When Structure is defined independently of Purpose, roles become decoupled from mission. Coordination fails, accountability diffuses, and legitimacy erodes.

Purpose—Order and Culture: Governance Must Stabilize Intent

Order formalizes rules, permissions, and escalation logic; Culture embodies informal norms and default behaviors. Together, they govern how the unit maintains coherence under pressure. Purpose must guide both—establishing the logic that Order stabilizes and the values that Culture reinforces.

- *Dependency*: Purpose defines what conduct is appropriate, what exceptions matter, and what behaviors signal alignment.
- *Failure Mode*: If Order enforces constraints irrelevant to Purpose, or Culture reinforces behaviors that conflict with it, the unit becomes internally contradictory—even if externally consistent.

Purpose—Memory and Storage: Continuity Must Be Purpose-Traced

Memory captures what the unit has done and why; Storage retains the tangible artifacts it has produced. Both must reflect and support the unit's Purpose. Purpose gives meaning to what is remembered and stored—it determines what matters.

- *Dependency*: Purpose constrains what histories are retained, what data is preserved, and what traceability is required.
- *Failure Mode*: If Memory is directionless or Storage accumulates irrelevant outputs, the unit cannot learn, evolve, or demonstrate accountability. It becomes opaque and untraceable.

The Grammar of Reinforcement

These dependencies are not optional. Each construct must be *Purpose-aligned* at design time and *Purpose-consistent* at runtime. A change in Purpose—whether due to system evolution, environmental adaptation, or structural realignment—necessitates a reassessment of all other constructs. Similarly, any change to Function, Process, or Structure must be evaluated for Purpose conformity.

Purpose does not operate above the system; it operates *through* it. Its viability depends on how consistently it is expressed, constrained, and validated across the architectural grammar. Without these interlocking dependencies, Purpose remains declarative. With them, it becomes architectural.

3.5.2. Examples of Alignment and Misalignment

Understanding the integration of Purpose with other constructs is best reinforced through concrete patterns of alignment and misalignment. Alignment ensures that Purpose is consistently realized across the unit's structure and behavior. Misalignment, by contrast, reveals structural incoherence—where a unit's declared intent diverges from its operational, functional, or systemic reality. The following examples illustrate both conditions, highlighting how each construct either reinforces or erodes architectural integrity.

Example 1: Purpose–Function Alignment

Scenario (Aligned):

A unit with the stated Purpose *"to provide authenticated access to regulated financial data for approved institutional partners"* exposes a functionally limited, purpose-specific interface that enables credential verification, permission validation, and controlled data retrieval. Access is tightly scoped, and the onboarding process reinforces regulatory constraints.

Scenario (Misaligned):

The same unit exposes a broad data query interface usable by external developers and internal analysts alike. Access control is generic and permission boundaries are ambiguous. The exposed Function exceeds the scope of the declared Purpose, enabling unintended use cases.

Consequence:

The unit appears flexible but structurally incoherent. Consumers may exploit functionality misaligned with the unit's role, creating downstream risk and regulatory exposure.

Example 2: Purpose–Process Alignment

Scenario (Aligned):

A unit designed to *"support internal audit readiness for enterprise systems"* defines processes centered on evidence collection, control validation, and exception tracking. Processes are repeatable, transparent, and tailored to regulatory timelines.

Scenario (Misaligned):

The same unit introduces efficiency-driven automation routines optimized for general reporting. While performant, these processes fail to capture key compliance artifacts and omit exception reasoning.

Consequence:

The unit cannot demonstrate that it is fulfilling its Purpose, even if technically functional. Regulatory findings or external reviews expose gaps in legitimacy.

Example 3: Purpose–Structure Alignment

Scenario (Aligned):

A unit whose Purpose is *"to support multilingual customer onboarding in underserved markets"* includes roles for regional language experts, customer support coordinators, and local compliance officers. Structural decisions reflect the unit's stated focus.

Scenario (Misaligned):

The same unit is staffed exclusively by centralized engineers and generalists. Language and regional expertise are handled informally or outsourced.

Consequence:

Despite the stated Purpose, the internal Structure does not possess the capabilities needed to execute it. Gaps in accountability, cultural understanding, and execution emerge under load.

Example 4: Purpose–Culture Alignment

Scenario (Aligned):

A unit responsible for *"ensuring data minimization in customer analytics"* fosters a culture of restraint in feature development, peer accountability in data use, and openness in raising ethical concerns.

Scenario (Misaligned):

The same unit adopts growth-driven values that reward data volume and predictive reach. Teams optimize for insight density rather than data discipline.

Consequence:

The unit's behavior gradually undermines its stated Purpose. Ethical risks accumulate, and external trust deteriorates despite internal performance.

Example 5: Purpose–Memory Misalignment

Scenario (Aligned):

A unit whose Purpose is *"to document and preserve institutional knowledge related to product safety incidents"* maintains a version-controlled knowledge base, incident rationale logs, and retrospective summaries with structured access.

Scenario (Misaligned):

The same unit stores incident reports in ad hoc email threads and relies on transient knowledge held by key personnel. No persistent Memory layer exists.

Consequence:

The unit becomes structurally fragile. When individuals depart or compliance events arise, the unit cannot fulfill its Purpose or demonstrate accountability.

Interpretation

Alignment confirms that a unit is structurally composed—not merely assembled. Misalignment exposes that Purpose may be declarative but not operationalized, risking incoherence, erosion of trust, and systemic instability.

Diagnosing misalignment requires not only reviewing documentation but observing behaviors, structures, and patterns of interaction. A unit's integrity is preserved only when Purpose consistently governs all other constructs—not symbolically, but compositionally.

3.6. Failure Modes and Antipatterns of Purpose

Even when Purpose is explicitly defined, its structural role can fail to materialize due to misapplication, distortion, or erosion over time. It's time to identify common failure modes and architectural antipatterns associated with Purpose—cases where the construct appears present but is functionally absent, misaligned, or counterproductive. Understanding these patterns is critical not only for preventing structural incoherence but also for diagnosing drift, misrepresentation, and loss of system integrity. Recognizing failure is not a sign of design collapse—it is a prerequisite for architectural recovery.

3.6.1. Common Antipatterns

Antipatterns emerge when Purpose is superficially defined, misapplied, or structurally bypassed. These failures are not merely rhetorical—they represent architectural conditions in which the unit's design, behavior, or interpretation contradicts its declared orientation. Recognizing these antipatterns is critical for preventing the accumulation of architectural debt and systemic incoherence. The following are the most prevalent forms of Purpose misapplication in sociotechnical units.

1. **Purpose Drift**

   Description:

   The unit's activities, interfaces, or outputs gradually diverge from its declared Purpose due to incremental decisions, feature accretion, or shifting stakeholder pressures.

   Symptoms:

   - Expanding scope without formal revision of Purpose
   - Functions or processes that no longer align with original intent
   - Misalignment between what the unit claims to do and what it operationally supports

   Consequence:

   The unit becomes functionally active but structurally unanchored. Trust erodes and governance becomes reactive.

   Example:

   A data platform originally built to support research begins prioritizing marketing analytics, gradually losing its original orientation toward scientific value.

2. **Surface Purpose**

   Description:

   Purpose is articulated as a branding or positioning statement but is not structurally embedded into the unit's design.

Symptoms:

- High-level Purpose declarations with no architectural traceability
- Public-facing materials that present a mission unreflected in Function, Process, or Structure
- Lack of internal accountability mechanisms tied to Purpose

Consequence:

Purpose becomes performative rather than directive. The unit presents clarity while operating incoherently.

Example:

A company proclaims customer-centricity but measures success only by quarterly sales metrics.

3. **Overgeneralized Purpose**

Description:

The unit's Purpose is defined so broadly that it becomes non-discriminating—applicable to any unit and decisive for none.

Symptoms:

- Statements such as "enable innovation" or "deliver customer value" without specifying domain, method, or boundaries
- Difficulty distinguishing the unit from adjacent units based on Purpose
- Inability to use Purpose as a filtering constraint in design decisions

Consequence:

The unit lacks identity, is prone to overlap with others, and cannot be composed meaningfully into a differentiated system.

Example:

An internal analytics unit defines its Purpose as "enabling better decision-making across the enterprise." Because the statement applies to nearly every department, it fails to guide architectural choices, resulting in diffuse tools and uncoordinated data pipelines.

4. **Surrogate Purpose**

   Description:

   Internal metrics, KPIs, or performance targets are treated as the unit's de facto Purpose, overriding its structural intent.

   Symptoms:

   - Optimization for throughput, uptime, or usage that contradicts declared Purpose
   - Decision-making driven by performance indicators rather than mission alignment
   - Success defined by internal metrics, not external value

   Consequence:

   The unit may perform efficiently but fail to justify its existence within the system or deliver on its stated commitments.

   Example:

   A public-service agency measures its success solely by response-time metrics. The original Purpose—to deliver equitable access to services—is gradually displaced by the numerical KPI, prompting teams to optimize speed even when accuracy or fairness suffers.

5. **Repurposed Unit Without Structural Realignment**

   Description:

   The Purpose of a unit is changed—often due to organizational shifts or strategic reorientation—but the internal architecture is not redesigned accordingly.

Symptoms:

- New Purpose statements layered onto legacy structures and behaviors
- Mismatched roles, processes, or interfaces inherited from prior configurations
- Incoherence between declared direction and actual capability

Consequence:

The unit enters a transitional state of contradiction, leading to execution failure, credibility loss, or systemic blockage.

Example:

A manufacturing quality-assurance unit is reassigned to lead sustainability initiatives, but its systems, roles, and reporting lines remain built around defect tracking. Without redesign, the unit continues collecting product-failure data instead of environmental-impact metrics, leaving its new Purpose unrealized.

6. **Misplaced Purpose**

Description:

Purpose is assigned to the wrong level of abstraction—either at too granular a level (e.g., component or task) or too high a level (e.g., enterprise mission instead of unit function).

Symptoms:

- Purpose statements that describe implementation details rather than system orientation
- Confusion between unit Purpose and project objectives or organizational values
- Inconsistent application of Purpose across teams or interfaces

Consequence:

Design decisions become misaligned with the appropriate scope, producing incoherent boundaries, ambiguous accountability, and reduced composability.

Example:

A software engineering unit defines its Purpose as "delivering seamless customer experience," a goal that properly belongs to customer-facing units such as product or service delivery. The engineering unit's structural Purpose is to provide stable, extensible digital capabilities that enable those units to serve customers effectively. By adopting a Purpose that sits outside its domain of service, it blurs accountability, diffuses priorities, and weakens overall architectural coherence.

These antipatterns do not indicate that Purpose was ignored—they indicate that it was misstructured. Recovering from them requires more than rewording a statement; it requires architectural correction. Purpose must not only be declared—it must be embedded, constrained, and validated through every dimension of unit design. When these antipatterns are identified early, they become opportunities for re-alignment. When left unaddressed, they become systemic liabilities.

3.6.2. Diagnostic Indicators

Identifying misalignment or degradation of Purpose in a sociotechnical unit requires more than reviewing documentation. Architectural failure often emerges through subtle inconsistencies in behavior, interpretation, or coordination. Diagnostic indicators are observable conditions that signal structural problems—specifically, that Purpose may be absent, misapplied, or ineffective. These indicators help architects, analysts, and system stewards recognize the early signs of incoherence before functional or systemic failure becomes visible.

The following diagnostic indicators can be used independently or as part of structured evaluation frameworks to assess the integrity of Purpose within a unit.

1. **Incoherent Interfaces**

   Indicator:

   Interfaces expose functionalities that contradict or extend beyond the declared Purpose.

   Interpretation:

   The unit is either over-serving, enabling misuse, or compensating for unclear scope boundaries. This typically indicates Purpose drift or overgeneralization.

2. **Role Confusion or Accountability Gaps**

   Indicator:

   Internal agents cannot clearly explain what the unit is for or who is responsible for its primary outcomes.

   Interpretation:

   Purpose is either missing, non-actionable, or has not been integrated into Structure. This often reflects Surface Purpose or Repurposing without redesign.

3. **Multiplicity of Internal Narratives**

   Indicator:

   Different stakeholders within the unit describe its Purpose differently, or reference incompatible goals.

   Interpretation:

   Purpose has not been internalized as a common referent. This indicates misalignment across Culture, Order, or Memory and points to lack of structural integration.

4. **Value Delivery Without Justification**

   Indicator:

   The unit performs activities or delivers outputs that are used but cannot clearly justify why they exist.

   Interpretation:

   Function may be operational, but it is decoupled from structural intent. This often accompanies Surrogate Purpose, where usage metrics replace architectural rationale.

5. **Excessive Adaptation Without Purpose Revision**

   Indicator:

   The unit changes roles, outputs, or dependencies repeatedly without updating its Purpose statement.

   Interpretation:

   Purpose is either being bypassed or is too weak to guide adaptation. This signals organizational drift and threatens long-term viability.

6. **Boundary Ambiguity in System Context**

   Indicator:

   External stakeholders cannot differentiate the unit from other adjacent or overlapping units.

   Interpretation:

   The Purpose is either overgeneralized or structurally misaligned. The unit lacks a clear systemic role, undermining its composability.

7. **Feedback Loops Without Learning**

   Indicator:

   The unit receives feedback but makes repetitive errors or exhibits no directional change.

Interpretation:

Memory exists, but Purpose is not providing interpretive context. The unit cannot determine which adaptations matter, indicating a breakdown in Purpose-Memory alignment.

8. **Governance Friction**

   Indicator:

   Escalation paths, exception handling, or decision rights are frequently contested or unclear.

   Interpretation:

   Order is not aligned with Purpose. The rules governing the unit are inconsistent with its structural orientation, causing enforcement breakdowns.

Using Diagnostic Indicators Structurally

These indicators are most effective when used *longitudinally* (e.g., over successive design cycles or retrospectives) and *comparatively* (e.g., across units within a system). They do not require access to source code or technical configurations—they rely on **architectural** *observation,* including

- Interviews with role-holders
- Review of public artifacts
- Analysis of behavior under pressure
- Comparison of stated Purpose with enacted behavior

Where multiple indicators are present, the unit should be treated as *architecturally at risk.* Recovery should begin with a Purpose review, followed by reassessment of Function, Process, and Structure for re-alignment.

Purpose failure is rarely immediate. It accumulates. Diagnostic indicators provide the necessary resolution to detect it while recovery is still structurally feasible.

3.6.3. Recovery Strategies

When a unit's Purpose is degraded, misaligned, or structurally invalid, recovery is not a matter of rephrasing intent—it requires architectural re-alignment. Because Purpose governs the unit's structural boundaries and its legitimacy within the system and environment, recovering from Purpose failure involves more than retrospective correction. It entails reconstructing coherence across the unit's internal design, systemic role, and external posture. The following strategies offer structured pathways for recovery.

1. **Conduct a Purpose Traceability Review**

   Objective:

   Systematically assess whether each construct—Function, Process, Structure, Order, Culture, Memory, and Storage—can be traced back to the declared Purpose.

   Method:

   - For each construct, ask: *How does this exist to serve the unit's Purpose?*
   - Identify elements that lack justification in terms of Purpose.
   - Flag any construct that contradicts or bypasses the Purpose.

   Outcome:

   A diagnostic baseline from which alignment gaps can be addressed in a targeted, construct-by-construct manner.

2. **Re-anchor Design Decisions in Purpose**

   Objective:

   Use the declared or revised Purpose as a constraint to revisit and potentially reconfigure architectural decisions.

   Method:

   - Review recent design choices (e.g., feature additions, workflow changes, staffing decisions).
   - Filter each decision through the Purpose constraint: *Does this enable the unit to fulfill its declared role?*

- Reverse or isolate changes that cannot be structurally justified.

Outcome:

Restores Purpose as an active design constraint and halts uncontrolled architectural drift.

3. **Realign Interfaces with Declared Purpose**

Objective:

Ensure that all exposed interaction surfaces reinforce, rather than obscure or contradict, the unit's Purpose.

Method:

- Review identity and interface surfaces (e.g., portals, onboarding pathways, documentation).
- Remove or revise components that expose capabilities not aligned with Purpose.
- Reinforce Purpose explicitly through framing statements and interface logic.

Outcome:

Improves legibility, trust, and external alignment while reducing risk of misuse or misinterpretation.

4. **Reaffirm Purpose Internally Through Role Reorientation**

Objective:

Embed Purpose into the expectations, behaviors, and decision-making frameworks of agents within the unit.

Method:

- Reintroduce Purpose during retrospectives, team charters, onboarding, and governance discussions.
- Ensure all role-holders can articulate the unit's Purpose and interpret how their responsibilities support it.
- Adjust roles and accountabilities that have drifted away from Purpose alignment.

Outcome:

Restores cultural and structural cohesion around a shared architectural anchor.

5. **Reassess Systemic Fit and Boundary Validity**

   Objective:

   Verify that the unit's Purpose remains differentiated and structurally justified within the broader system.

   Method:

   - Compare the unit's Purpose with those of adjacent or interacting units.
   - Identify duplication, boundary overlaps, or dependency conflicts.
   - Modify the unit's scope, relationships, or systemic role as needed.

   Outcome:

   Reestablishes the unit's composability and ensures it occupies a structurally necessary position.

6. **Version and Rationalize Purpose Adjustments**

   Objective:

   Where Purpose must change, ensure that the change is governed, traceable, and reflected across all constructs.

   Method:

   - Treat Purpose as a versioned artifact.
   - Document the rationale for change, including triggers and expected consequences.
   - Coordinate revisions across Function, Process, and Structure to reflect the new orientation.

   Outcome:

   Prevents superficial repurposing and enables sustainable adaptation without architectural incoherence.

7. **Introduce Periodic Alignment Reviews**

   Objective:

   Institutionalize recovery by making Purpose-construct alignment a recurring evaluative practice.

   Method:

   - Schedule structured reviews at defined intervals or change milestones.
   - Use a standard framework to reassess Purpose alignment across constructs and surfaces.
   - Embed results into governance or architectural decision records.

   Outcome:

   Sustains structural integrity over time and ensures that recovery is not a one-time intervention but a continuous discipline.

Recovery is not merely about restoring alignment—it is about reaffirming architectural intent and reestablishing the unit's fitness to operate within its system and environment. When Purpose is actively reinforced and structurally validated, the unit regains not just operational coherence, but structural legitimacy.

3.7. Designing Purpose for Adaptive Capacity

In complex and evolving systems, Purpose cannot be treated as a fixed declaration. While it must provide stability and constraint, it must also support continuous adaptation in response to environmental signals, systemic shifts, or evolving stakeholder needs. Let's now address how to design Purpose with adaptive capacity—preserving its architectural role while enabling structural evolution. Adaptive Purpose is not loosely defined—it is durably anchored and intentionally revisable, supported by mechanisms that ensure coherence over time rather than fragility under change.

3.7.1. Refinement Over Time

Purpose, while architecturally foundational, must not be static. In dynamic sociotechnical systems, environmental conditions, system roles, and stakeholder expectations evolve. A unit's Purpose must be capable of refinement—changing not arbitrarily, but in a controlled manner that preserves structural integrity and systemic fit. Refinement is not redefinition; it is the process of *clarifying, narrowing, or repositioning* Purpose in response to validated change.

Purpose As a Stable Constraint with Revisable Expression

The core function of Purpose is to constrain design and ensure coherence. However, the *expression of that constraint* may shift over time to reflect:

- *Changes in external demands* (e.g., new regulatory frameworks, market shifts, user behavior)
- *Internal learning and feedback* (e.g., operational friction, recurring exceptions, retrospective insights)
- *Systemic evolution* (e.g., redistribution of responsibilities across units, emergence of new coordination models)

The Purpose itself remains structurally consistent in its role as a directional anchor, but the unit may refine how it interprets or applies that Purpose under new conditions.

Structured Triggers for Purpose Refinement

Purpose refinement should not result from informal pressures or local optimizations. It must be structurally justified and procedurally governed. Common triggers include

- *Sustained misalignment* between Purpose and Function or Process, observed through diagnostic indicators
- *Repeated exceptions* to governance rules that suggest the current Purpose no longer supports viable operation
- *System-wide rearchitecture* that alters the unit's role or boundary within the larger context
- *External feedback* from stakeholders who can no longer interpret or engage with the unit as intended

Each trigger must prompt not only a restatement of Purpose but a re-evaluation of the unit's configuration across all constructs.

Patterns of Purpose Refinement

There are three typical forms of refinement:

1. *Clarification*

   The unit's Purpose is made more explicit without materially changing its structural role. Often used to correct ambiguity or resolve interpretive drift.

2. *Narrowing*

   The Purpose is focused to eliminate peripheral activities or reduce scope, often to improve alignment or simplify interfaces.

3. *Repositioning*

   The Purpose is shifted laterally within the system to fill a new structural gap. This requires coordinated changes in adjacent units and system relationships.

Each form requires corresponding updates to Function, Process, Structure, and interface surfaces to maintain architectural coherence.

Risks of Unmanaged Refinement

When refinement is conducted without governance, the following risks emerge:

- *Hidden repurposing*: The unit begins to operate as if its Purpose has changed, but no formal revision has occurred.
- *Structural incoherence*: Downstream constructs are not updated, resulting in misaligned behavior or interaction failure.
- *Loss of trust*: External actors perceive inconsistency or unpredictability, eroding confidence in the unit's legitimacy.

Refinement must be deliberate, justified, and versioned. Without structure, it leads to erosion rather than adaptation.

Embedding Refinement Capacity

To enable sustainable refinement, the unit should embed:

- *Version control for Purpose declarations*, with rationale and structural impact assessment
- *Feedback loops linked to Purpose evaluation*, not only to performance metrics
- *Governance triggers* for initiating re-assessment and coordinating updates across constructs
- *Traceability mechanisms* that connect Purpose changes to observable system behavior

Refinement over time does not weaken Purpose—it preserves its function. A unit capable of refining its Purpose in response to structured signals is not inconsistent; it is *architecturally resilient*. In a system that must evolve without collapsing, the capacity to refine Purpose is not optional. It is a design requirement.

3.7.2. Mechanisms for Continuous Alignment

Designing Purpose for adaptive capacity requires more than the ability to refine it episodically. To maintain structural coherence over time, the unit must implement mechanisms that support *continuous alignment* between its declared Purpose and its evolving behavior, interfaces, and role. Continuous alignment is not automation—it is architectural instrumentation: the intentional embedding of feedback, monitoring, and interpretive processes that ensure the unit remains anchored in Purpose as it adapts.

These mechanisms allow the unit to detect divergence early, interpret signals meaningfully, and guide change in a way that preserves legitimacy and composability.

1. **Purpose-Centered Feedback Loops**

   Function:

   Feedback should not only inform performance or stability—it must inform Purpose relevance and alignment. Feedback loops must be connected to the question: *Does the current behavior of the unit reflect what it exists to do?*

Implementation Considerations:

- Regular stakeholder input tied to Purpose evaluation, not just satisfaction or throughput
- Internal retrospectives structured around Purpose alignment, not only process efficiency
- Environmental scanning mechanisms to detect shifts that may invalidate the current Purpose expression

2. **Alignment Reviews Anchored in the Architectural Grammar**

Function:

Periodic assessment of each construct's dependency on Purpose helps ensure that drift is identified and addressed before it accumulates into structural incoherence.

Implementation Considerations:

- Purpose–Function–Value triad validation as a standing agenda item in governance cycles
- Process reviews that test whether workflows still enact the declared Purpose
- Structural role assessments that confirm Purpose-justified roles and responsibilities

3. **Integration with Memory as Interpretive Context**

Function:

Memory is not merely recordkeeping—it is the source of interpretive continuity. When past decisions, exceptions, and refinements are stored in relation to Purpose, the unit can evaluate whether it is evolving with coherence or accumulating contradiction.

Implementation Considerations:

- Log structural decisions with Purpose rationale and impact traceability
- Maintain Purpose-aligned incident or exception histories to identify recurring mismatches
- Use historical Purpose versions as context during design transitions or audits

4. **Purpose-Visible Governance Triggers**

   Function:

   Certain thresholds or structural events must automatically trigger a reassessment of Purpose alignment.

   Trigger Examples:

   - Introduction of a new stakeholder group or service dependency
   - Recurring escalation of issues not anticipated by the current Purpose
   - Change in external regulation that affects the unit's role or obligation

   Implementation Considerations:

   - Encode Purpose review as a governance condition for change approval
   - Require updated Purpose declarations as part of integration proposals or decommissioning plans

5. **Semantic Monitoring and Structural Observability**

   Function:

   Monitor not just what the unit is doing, but *why*. This includes analyzing interaction patterns, usage distribution, and exception rates for signals that the unit's Purpose may no longer be appropriately interpreted or fulfilled.

Implementation Considerations:

- Observe which interfaces or features are most used—and whether they align with Purpose.
- Flag significant deviations in usage that suggest consumer interpretation is drifting.
- Analyze failed coordination attempts as potential signals of Purpose misalignment.

6. **Designated Purpose Stewards or Structural Owners**

   Function:

   Assign explicit accountability for monitoring, defending, and adapting the unit's Purpose. This role is not managerial—it is architectural.

   Implementation Considerations:

   - Stewardship embedded in architectural review boards or system design authorities
   - Stewards maintain version control, structural rationale, and interface alignment records
   - Act as escalation points for ambiguity or contradiction in structural decisions

Sustaining Coherence While Enabling Change

Continuous alignment mechanisms ensure that Purpose remains structurally actionable, even as the system adapts and evolves. Without them, changes may proceed unchecked—driven by performance optimizations, external demands, or opportunistic scope creep. These changes, however well-intentioned, can erode the unit's architectural identity if not evaluated against Purpose.

By embedding alignment mechanisms into the design and operation of the unit, Purpose becomes not just a statement—but a continuous reference. It remains visible, interpretable, and structurally reinforced. This transforms adaptation from drift into *architected evolution.*

3.7.3. Versioning and Governance

In adaptive systems, architectural constructs must support both continuity and change. Purpose, as the orienting construct of a unit, must remain stable enough to guide coherent design, yet modifiable enough to remain relevant. Achieving this balance requires explicit **versioning and governance**—mechanisms that ensure Purpose can evolve without compromising structural integrity or violating system-level agreements.

Versioning and governance do not formalize bureaucracy; they establish the *conditions under which Purpose may change*, the *implications of such change*, and the *mechanisms by which alignment is restored.*

Purpose As a Versioned Architectural Object

Treating Purpose as versioned acknowledges that it may need to be clarified, narrowed, or repositioned over time. Each version should reflect a discrete, justified structural state, complete with

- *Version identifier* (e.g., semantic or chronological)
- *Change rationale* (what triggered the revision and why)
- *Structural impact* (constructs or surfaces affected)
- *Approval metadata* (review authority, date, context)

This record enables traceability, post-change evaluation, and informed interaction with external stakeholders.

Triggers for Governance Activation

Purpose revision should not occur reactively or informally. The following conditions should activate governance protocols:

- *Systemic reconfiguration* (e.g., redistribution of responsibilities across units)
- *Repeated misalignment* with Function, Process, or external interfaces
- *Emergence of new stakeholders* requiring role expansion or contraction
- *Environmental disruption* (e.g., legal, regulatory, or market shifts)

Each trigger initiates structured review to determine whether Purpose revision is necessary and whether adjacent constructs must be updated accordingly.

Governance Responsibilities

Governance must be assigned to accountable roles with the authority and contextual awareness to evaluate structural impact. Responsibilities include

- *Initiating Purpose review* in response to internal or external signals
- *Validating alignment* across all constructs before approving a new version
- *Coordinating communication* of Purpose change to affected stakeholders
- *Maintaining the audit trail* of all Purpose declarations and their rationale

Governance does not guarantee correctness, but it guarantees that change is deliberate, interpretable, and structurally recoverable.

Coordination with Adjacent Units

In federated or multi-unit systems, Purpose revisions must not occur in isolation. Governance must include

- *Dependency analysis* to identify upstream and downstream effects
- *Boundary re-evaluation* to ensure no role duplication or overlap emerges
- *System-level notification protocols* so other units can assess their own alignment

When coordination is neglected, units risk violating systemic fit and undermining composability.

Publishing and Communicating Purpose Versions

Each version of Purpose must be made available in a manner consistent with the unit's digital and coordination surfaces. This includes

- Updating all identity and onboarding materials to reflect the new Purpose
- Ensuring that interface expectations align with revised scope or commitments
- Communicating to partners, stakeholders, and system registries that a new structural commitment has been declared

Transparent communication is not optional—it is the means by which legitimacy and trust are maintained.

Enabling Change Without Eroding Structure

Versioning and governance transform Purpose from a static declaration into a *governed constraint*—a durable yet adaptable anchor that can evolve without compromising the unit's architectural role. They make it possible to *change with continuity*, enabling the unit to remain coherent internally, composable systemically, and legitimate externally—even as its structural context shifts. Without these mechanisms, Purpose becomes either rigid and obsolete, or fluid and incoherent. With them, it remains both stable and viable.

3.8. Summary: Purpose As a Structural First Principle

Purpose is not an aspirational or symbolic element—it is a *structural first principle*. In Purpose Driven Design, it serves as the originating constraint that defines why a sociotechnical unit exists, for whom it delivers Value, and under what conditions its presence is justified. Unlike goals, strategies, or performance targets, Purpose operates architecturally: it constrains design decisions, aligns internal constructs, situates the unit within a broader system, and anchors its legitimacy in relation to the external environment.

Throughout this chapter, Purpose has been examined not as a philosophical abstraction, but as an operational structure that governs the unit's viability across time and scale:

- *Conceptually*, Purpose provides directional clarity and boundary definition.

- *Functionally,* it stabilizes internal design and enables external engagement.
- *Systemically,* it ensures composability and role differentiation.
- *Digitally,* it is made visible through identity, interface, and coordination surfaces.
- *Socially and operationally,* it is realized through roles, norms, and decision frameworks.
- *Dynamically,* it is maintained through refinement, continuous alignment, and governed versioning.

Purpose integrates with all other constructs—Function, Process, Structure, Order and Culture, Memory, and Storage—not as a background context but as their organizing rationale. When misapplied, neglected, or superficially defined, Purpose ceases to constrain and instead becomes a liability. Misalignment across constructs leads to incoherence, erosion of trust, and eventual system fragility.

Conversely, when Purpose is properly defined, structurally embedded, and continuously maintained, it enables units to act with integrity, adapt with accountability, and coordinate with minimal overhead. It allows complex systems to evolve without losing identity and supports composable architectures that remain legible and governable.

In the design of sociotechnical systems, Purpose is not optional. It is the first architectural decision—and the one that all others must serve.

CHAPTER 4

Value

In the 1950s, the idea of a theme park was not new. Fairs, carnivals, and amusement parks had long provided mechanical entertainment and food stalls in open fields or along waterfronts. But in 1955, Walt Disney introduced something fundamentally different—Disneyland, a fully designed experience in which every path, building, performance, and interaction was orchestrated to produce a cohesive form of delight.

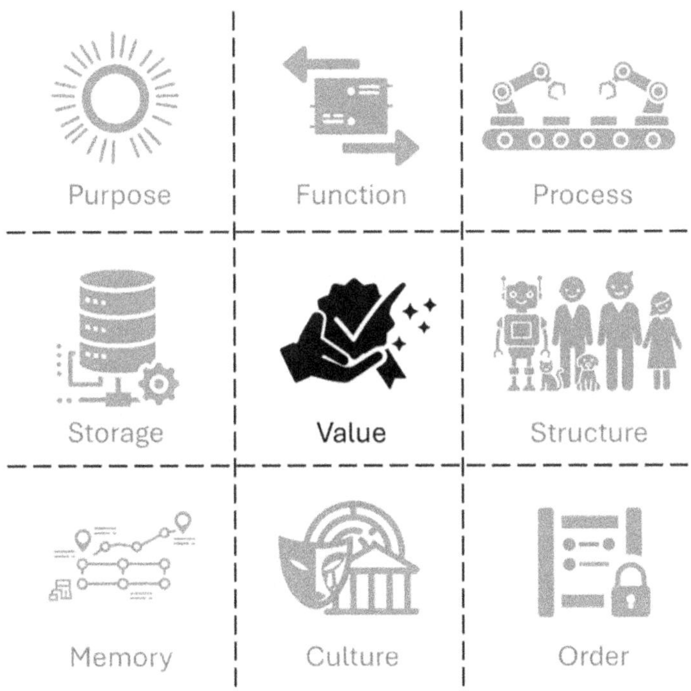

Disney understood that visitors were not purchasing tickets to ride machines. They were entering a narrative space—one that promised wonder, control, safety, immersion, and memory. From the moment a guest walked through the gates, they were no longer navigating a park; they were participating in a constructed reality where even the trash bins matched the architecture of each themed zone.

A. Milchman, *The Purpose Driven Design Framework for Modern Enterprises*,
https://doi.org/10.1007/979-8-8688-2223-0_4

This was not merely entertainment—it was Value delivered through design. Disney engineered every aspect of the environment to produce emotional, relational, and aesthetic outcomes. Unlike traditional amusement parks, which assembled independent rides and food stalls around logistical convenience, Disneyland was conceived as a coherent world—where architecture, sound, and story formed a continuous experience of designed delight. The rides were only one part. The layout, sound design, uniforms, scripted character behavior, and even the queuing systems were meticulously crafted to ensure that guests remained within the designed experience, rather than being periodically interrupted by reminders of the ordinary world.

The most remarkable aspect was not the scale or spectacle but the internal coherence. Disneyland was not a loose collection of features—it was a system. And within that system, Value was not an abstract promise; it was something guests could point to, remember, and feel.

Over time, this architectural approach to Value shaped industries far beyond entertainment. The rise of experience design in retail, hospitality, software, and education can all be traced to the insight that Value is not what a system offers—it is what a participant receives, experiences, and retains. As Joseph Pine and James Gilmore later observed in *The Experience Economy*, this same principle animates environments such as Starbucks cafés, Apple Stores, and immersive hotels, where design choreographs emotional and sensory continuity rather than discrete transactions. Disney showed that products and services are only as effective as the total system that surrounds them. And the most successful systems are those that make Value feel inevitable, embedded, and continuous.

Today, this design philosophy can be found in the layout of an Apple Store, the onboarding flow of a mobile app, the structure of an immersive museum, or the interface of a high-trust digital platform. In each case, the system is not organized around what it does—it is organized around what the participant experiences as valuable.

Walt Disney did not invent theme parks. He invented a new way of thinking about how systems create meaning. That shift—from provision to perception, from offering to reception—is what marks the evolution of Value from a transaction to a design construct.

Function opens the door to interaction, but Value gives that interaction meaning. It is through Value that a unit affirms its relevance, maintains its legitimacy, and sustains its existence. The relationship among Purpose, Function, and Value is summarized in Table 4-1.

Table 4-1. *Relationship Among Purpose, Function, and Value in Purpose Driven Design*

Construct	Core question	Role in design	Relationship to the whole
Purpose	Why does the unit exist?	Defines intent and orientation.	Provides direction for all design decisions.
Function	How does the unit interact?	Exposes interfaces through which Purpose is made accessible.	Translates Purpose into usable form.
Value	What does the interaction produce?	Represents the outcome experienced or received by others.	Validates Purpose through experience and contribution.

Value is not inherent in what the unit produces; it emerges from how that production is perceived, received, and retained by others. The same offering can carry different Value depending on who it serves—for instance, a financial report may represent transparency and accountability to regulators, strategic insight to investors, and operational guidance to internal teams. In this chapter, Value is treated as a structural design element—one that must be defined, delivered, contextualized, and preserved across time. This section explores how units can design for Value explicitly, ensuring that what they do produces outcomes others consider meaningful, desirable, or beneficial.

4.1. Foundations of Value As a Design Construct

In Purpose Driven Design, *Value* is not an economic abstraction or a subjective benefit—it is a structural construct. It defines what a unit delivers into the world that others can see, use, and respond to. A compliance unit, for example, delivers verified assurance reports to regulators and executives—artifacts that make its Purpose externally testable. Value is the first outward-facing artifact of design: the moment when internal Purpose becomes externally verifiable. It is how a unit proves its relevance—not through intentions but through offerings that are discoverable, interpretable, and composable by others. We need a conceptual and architectural foundation for understanding Value not as a by-product of good design but as an essential surface of it. We will define what Value is, explore why it must exist as a construct, identify the systemic risks that emerge in its absence, and trace its historical transformations from intrinsic worth to modular interface. In doing so, we position Value as the structural bridge between Purpose and Function—the anchor that ensures that what a unit *is for* results in something others can *actually use*.

4.1.1. Defining Value Architecturally

In Purpose Driven Design, *Value is not a subjective benefit or market abstraction*. It is a structurally expressed outcome—a tangible or intangible offering that a unit delivers in exchange for interaction. Value represents the external realization of Purpose: it is how Purpose enters the world, becomes testable, and is validated by others. Architecturally, Value is not defined by intent but by effect. It is not what the unit *claims* to offer but what others *receive*, use, or benefit from.

Value is always *relational*, even when no monetary exchange is involved. It defines what the unit *offers* and what others *experience*. These offerings are structurally exposed as products, services, experiences, or by-products—each representing a distinct surface through which Value becomes accessible and usable. The key criterion is that they are legible and meaningful to the unit's environment—designed not for internal metrics, but for external consumption, reuse, or response. In commercial systems, Value may appear as a completed insurance claim, a delivered product, or an executed financial transaction—outcomes that others can recognize and benefit from. In civic or social systems, it may take the form of an issued passport, an approved grant, or an enrolled student. In each case, Value represents a tangible result that fulfills Purpose through structured design.

From an architectural perspective, *Value is a surface-level artifact with systemic consequence*. It is not merely the by-product of Function and Process—it is their directional outcome. It is what a unit is accountable for producing, delivering, and preserving. Value can be shaped, stored, versioned, evaluated, or even withdrawn, but it must always be traceable to the unit's Purpose and realized through coherent design.

To serve as a meaningful construct, Value must be

- *Externally interpretable*: Value cannot be self-referential. It must make sense to consumers, partners, or other units. What the unit deems valuable must resonate with what the environment considers useful, usable, or desirable.

- *Structurally expressed*: Value is not a performance metric or an evaluative label—it manifests as specific offerings encoded in channels, catalogs, dashboards, interfaces, and delivery pipelines. These surfaces are not optional—they are the architectural expressions of what the unit exists to provide.
- *Purpose-constrained*: Not all outputs constitute Value. A unit may produce side effects, generate data, or support workflows that are operationally useful but structurally irrelevant. Value, in this framework, is restricted to those artifacts that directly fulfill the Purpose of the unit.
- *Platform-ready*: In distributed architectures, Value must be composable. It should be possible to treat Value artifacts as modules, endpoints, or reusable elements—capable of integration, reuse, or extension by other parts of the system.

These criteria can be formalized as shown in Table 4-2.

Table 4-2. *Structural Criteria for Defining Value in Purpose Driven Design*

Criterion	Description
Externally interpretable	Value cannot be self-referential. It must make sense to consumers, partners, or other units. What the unit deems valuable must resonate with what the environment considers useful, usable, or desirable.
Structurally expressed	Value must manifest as tangible offerings—products, services, experiences, or by-products—encoded in channels, catalogs, dashboards, or delivery pipelines.
Purpose-constrained	Only outputs that directly fulfill the unit's Purpose qualify as Value; side effects or internal utilities do not.
Platform-ready	In distributed architectures, Value must be composable and reusable, enabling integration across systems and contexts.

The *architectural function of Value* is to define the terms of engagement between a unit and its environment. It makes the unit legible to others and creates the basis for trust, coordination, and system-wide composition. Without Value, Purpose remains unfulfilled, and Function lacks justification.

Importantly, *Value is not static.* It may evolve as consumer needs change, environments shift, or capabilities expand. But at any given point, it provides the most concrete architectural test of a unit's alignment. If Purpose is the reason for existence, Value is the proof of relevance.

In this framework, *Value is not measured in economic terms but in structural fidelity:* Does the unit deliver what it promises? Can others rely on its outputs? Are those outputs discoverable, accessible, and usable? These are not philosophical questions—they are architectural diagnostics.

Thus, in Purpose Driven Design, Value becomes a first-class construct—*not a vague benefit but a contractually expressed outcome,* testable through real interfaces and accountable within the system's architecture. It is how the system makes its Purpose legible, and how Purpose becomes real.

4.1.2. Why Value Exists

Value exists because without it, Purpose remains inert and Function becomes directionless. In sociotechnical design, Value is not an optional output or a post hoc metric—it is the *necessary consequence* of Purpose fulfilled through Function and Process. If a unit does not produce Value, it cannot be evaluated, trusted, or justified within its system. Without Value, there is no external referent by which the unit can be judged, engaged, or composed. Therefore, Value is not simply a benefit—it is an architectural requirement.

In this framework, Value is the important point at which design becomes externally legible. It allows others—users, systems, regulators, or partners—to perceive what the unit *is for* in operational terms. It creates a bridge between internal intent and external reality. When this bridge is missing, systems lose their capacity for coordination, discoverability, and trust. Conversely, when Value is made explicit, structured, and constrained by Purpose, it becomes the basis for integration, reuse, and systemic alignment.

Value solves more than a packaging problem—it solves a visibility problem of purposeful action in a distributed system. In architectures composed of autonomous but accountable units, each with its own direction and boundary, Value is the way a unit declares its *contribution.* It externalizes what the unit offers—intentionally and reliably—so that others can engage without needing to inspect or interpret internal workings.

Whether expressed through products, services, experiences, or by-products, Value becomes a *promise surface*: a visible signal of capability and commitment, designed to invite interaction, reduce friction, and enable scalable collaboration without structural entanglement.

Moreover, Value serves as a moral and strategic constraint. In the absence of structurally defined Value, units may focus on internal efficiency, performance metrics, or tactical wins that create no actual benefit for those they are meant to serve. When Value is structurally required, these distortions are harder to justify. The unit must demonstrate not only activity but *contribution*. It must produce outcomes that are externally relevant and contextually appropriate. This constraint acts as a form of architectural discipline—ensuring that design remains not just operationally viable, but meaningful to those it aims to serve.

Finally, Value exists to provide a reference for adaptation. As environments change and feedback accumulates, systems must evolve. But evolution is directionless unless guided by a stable criterion. Value provides that criterion. It enables a unit to ask: Is the Value we produce still needed? Is it discoverable, usable, and trusted? Has its form or surface become misaligned with our Purpose? Without these questions, change becomes arbitrary. With them, it becomes intentional and anchored.

In sum, Value exists to make Purpose *observable*, Function *justifiable*, and design *legible*. It is the test of a unit's relevance and the condition of its composability. It exists not to declare worth but to *deliver* it—in a form that others can recognize, engage with, and build upon.

4.1.3. What Problem Value Solves

Value solves the core architectural problem of ***unmeasurable relevance***. In the absence of Value, a unit may operate, process, or communicate—but there is no meaningful way to determine whether it matters. Without structurally expressed Value, a unit may produce artifacts that are opaque, untrusted, or irrelevant to the environment it is meant to serve. Over time, this leads to a breakdown in legitimacy, coordination, and adaptation. The unit becomes a black box—functioning internally but disconnected from the system it inhabits.

The first and most critical failure that arises in the absence of Value is *invisible output*. Units may perform elaborate internal processes—generate reports, execute jobs, store data—but if these outputs are not packaged into recognizable, consumable Value, they offer no reliable interface to external actors. This disconnect produces what might be called *output theater*: work is performed, but no one can use it. Systems degrade into clusters of internal motion with no shared surface of contribution.

Second, Value solves the problem of *internal-external misalignment.* Without a clearly defined Value surface, units tend to optimize for what they *can do*, not what others *need*. Functions expand without constraint. Processes grow brittle under internal complexity. And what the unit produces becomes increasingly detached from what the system or environment expects. This leads to a condition of *structural narcissism*: units serve themselves, not their purpose.

Third, the absence of Value produces *coordination entropy*. In distributed architectures, units must interact without centralized orchestration. For that to happen, each must expose clear, trustworthy Value. If outputs are ambiguous, hidden, or mismatched, units cannot align, compose, or substitute. The result is fragile glue code, duplicated efforts, and constant negotiation over what is actually being offered. Value solves this by making offerings explicit, durable, and interpretable.

Fourth, Value addresses the problem of *non-accountability*. When no clear Value is declared or exposed, there is no basis for asking: Did the unit deliver what it promised? Did the system receive what it needed? Without Value, responsibility becomes diffused. Blame circulates, but accountability is lost. In contrast, when Value is treated as a structural construct, expectations can be stated, tested, and improved—turning outputs into commitments, and commitments into trust.

Finally, Value solves the *evolution trap*: the inability to adapt intelligently. In fast-changing systems, the question is not just whether a unit functions, but whether its output remains relevant. Without a visible, structured expression of Value, it is impossible to know what needs to change, or how to interpret feedback. Adaptation becomes reactive or misguided. Value provides a reference surface against which learning can occur and recalibration can be applied.

In sum, Value solves the problem of *disconnected relevance*. It ensures that units do not merely produce—they contribute. It makes Purpose visible, Function testable, and interaction meaningful. Without Value, systems can still operate—but they cannot evolve intelligently, scale coherently, or justify themselves to those they serve. Value is what prevents design from becoming noise. It is the condition that makes systems legible, relationships reciprocal, and architecture accountable.

4.1.4. Historical Patterns and Structural Evolution

The concept of Value has always existed—but not always as a design construct. In early systems, Value was embedded, assumed, or inferred. What mattered was *doing the work*—harvesting, building, administering, deciding. The worth of that work was often taken for granted within closed social or organizational settings. But as systems grew in complexity, scale, and modularity, it became necessary to articulate not just what was done, but *what was delivered*, to *whom*, and *why it mattered*. This shift—from embedded to expressed, from assumed to designed—is what makes Value a structural concern in modern sociotechnical systems.

Historically, systems have expressed Value in different forms. These forms reflect not only changing technologies and economies, but also evolving assumptions about *what counts* as Value and *how it should be delivered*. In the context of Purpose Driven Design, we can trace four dominant architectural expressions of Value:

1. **Product—Value As a Finished Artifact**

   Products are the oldest and most stable form of expressed Value. They are discrete, durable, and often transferable outcomes—tools, documents, datasets, dashboards, books, blueprints. Products allow Value to be stored, distributed, and reused independent of the producer. In early societies, this meant physical goods. In modern systems, it includes digital artifacts: APIs, models, templates, packaged reports.

   Product Value is legible, ownable, and often asynchronous. It represents *what the unit gives the world that can be used again.*

2. **Service—Value As a Capability-on-Demand**

   With industrialization and professionalization came the rise of services: *performances triggered by a request and shaped by the requestor's identity, needs, or input.* A product—like a manufactured tool—exists independently once created, while a service—such as legal advice or medical consultation—comes into being only when requested and vanishes when complete. Unlike products, services require not only coordination in time, but also *a specification*—whether explicit (a form) or embodied (a person in the chair). They are ephemeral but purposeful—like technical support, data transformation, recommendation generation, or real-time analytics.

Service Value is dynamic, interactive, and responsive. It represents what the unit *does for others*, but only in response to engagement. The Value is not prepackaged—it is realized through execution.

3. **Experience—Value As Perception and Engagement**

   As Pine and Gilmore argued in *The Experience Economy* (1998), a new layer of Value emerged when organizations began *designing the way people feel* during interactions. Experience is not a separate kind of Value but an overlay that shapes how Product and Service Value are received. In digital systems, this includes interface design, responsiveness, emotional tone, and trust signals.

   Experience Value is *modulatory*. It enhances or undermines the usability and resonance of other forms of Value.

   When ignored, poor experience can obscure real Value. When mastered, it can transform simple offerings into meaningful engagements.

4. **By-Product—Value As Emergent Output**

   Not all Value is designed. Some arises unintentionally—as a consequence of Function or Process. Log files, behavior traces, content metadata, and failed outputs can become valuable when reframed, structured, and shared. Much of the modern data economy is built not on primary Value, but on by-products. However, not every by-product is inherently valuable—or safe—to expose. Without curation and ethical oversight, such data may mislead, violate privacy, or create unintended dependencies. Purpose Driven Design treats by-product Value as conditional: it becomes legitimate only when responsibly contextualized and governed.

By-Product Value is opportunistic. It represents *what the unit unintentionally generates that others find useful.* But until it is made legible and reusable, it remains *latent*, not architectural.

From Implicit to Explicit: The Structural Shift

Across these categories, a broader shift can be observed:

- From *embedded* to *exposed* Value
- From *function-implied* to *design-declared* Value
- From *closed delivery* to *open legibility*
- From *static artifact* to *dynamic engagement*

This shift reflects deeper changes in how systems are structured. In monolithic organizations, Value could remain implicit—everyone knew the outputs. But in distributed, composable architectures, *Value must be surfaced.* It must have an interface, a name, a structure. It must be usable by agents and systems that neither created it nor fully understand its internals.

In this context, Value evolves from a side effect of work to a *structured surface of commitment*—not in the legal sense, but in the architectural one. Others engage with the unit because of the Value it reliably exposes, not the internal activities it performs.

Designing for the Future

As systems continue to scale and fragment, the need for structured, intentional Value only intensifies. Whether through reusable products, responsive services, engaging experiences, or mined by-products, the challenge remains the same:

Can others discover, trust, and use what your unit produces?

If the answer is yes, then Value exists—and architecture is working.

If not, then no amount of Purpose, Function, or Process will make the unit meaningful to the world around it.

4.2. Role of Value in Sociotechnical Systems

In a well-designed system, Value is never confined to a single domain. It radiates outward—anchoring the unit internally, aligning it with the broader system, and making it intelligible to the world beyond. It acts as both an internal stabilizer and an external interface, defining how the unit holds its shape and how it enters into relationships. The presence or absence of Value determines whether a unit can coordinate without confusion, adapt without chaos, and be trusted without inspection. Understanding how Value operates across these layers is key to designing systems that are not only efficient but also legible, modular, and resilient.

4.2.1. Internal Role: Stabilizing the Unit

Value plays a critical stabilizing role inside the boundaries of a unit. It provides the internal system with a clear orientation: *This is what we exist to deliver.* Metrics, KPIs, and internal dashboards may track performance, but they derive meaning only when aligned with the unit's defined Value—what it actually contributes to others. Without that alignment, measurement becomes activity without direction. Unlike Purpose, which offers directional meaning, or Function, which provides operational capability, Value gives the unit a *concrete anchor*. It defines the end condition that internal efforts must converge toward. Without this anchor, units may remain active but incoherent—performing tasks without clarity on whether those tasks produce anything others can use, want, or rely upon.

Internally, Value serves three primary stabilizing functions:

1. **Alignment Anchor**

   When clearly defined, Value aligns the unit's internal components—Purpose, Function, Process, Storage, and Structure—toward a shared outcome. It ensures that what the unit does, how it does it, and how it preserves and exposes its outputs are all in service of a coherent contribution. In addition, it constrains expansion: not every capability needs to be pursued, not every idea needs to be realized. The question becomes: *Does this contribute to the Value we exist to produce?* This framing acts as a governor on overreach and fragmentation. It helps teams say no to distractions and yes to intentional design.

2. **Justification Surface**

   Value enables self-evaluation. It provides a way for internal agents to ask: *Is our work producing something meaningful?* In practice, this means testing Value through direct questions: Does what we produce still serve our defined Purpose? Is it recognized as valuable by those we serve? Does it justify the effort and resources it consumes? This is especially important in complex environments where units may become overloaded with internal KPIs, ticket queues, or platform responsibilities. When Value is treated as a construct, the team can distinguish between *activity and contribution*, and redirect effort accordingly.

   A unit that knows its Value is better able to defend its time, define its scope, and justify its decisions.

3. **Boundary Clarifier**

   Perhaps most importantly, Value helps define what *is* and *is not* within the unit's mandate. It prevents the unit from becoming a catch-all or a dependency sponge. For example, when a data-governance unit begins accepting ad hoc analytics requests from every department, it ceases to enforce standards and becomes an informal service desk—eroding both focus and accountability. By making its Value explicit, a unit can state: *This is what we offer. Everything else is someone else's domain.* This clarity reduces ambiguity, protects team autonomy, and simplifies interfacing. It encourages *structural composability* by allowing others to build on what the unit reliably delivers, without having to entangle themselves in its inner workings.

   When these stabilizing functions are absent, units often drift. They grow in scope without purpose, accumulate tools without outcomes, or become reactive to every request. Internally, they suffer from fatigue. Externally, they become illegible. The result is organizational noise: busy systems that produce little of value. But when Value is embedded as an internal design constraint, the unit gains shape, clarity, and direction. It becomes not just a set of functions or a cluster of people but a coherent contributor to the system—a node with a purpose and a surface others can trust.

4.2.2. Systemic Role: Integration into System-Wide Coordination

Value is what allows a system to coordinate without collapsing into centralization. In a well-structured sociotechnical architecture, coordination happens not through command and control, but through the *disciplined exchange of Value.* Each unit contributes to the system not by reporting upward but by exposing clearly defined Value surfaces—artifacts that others can interpret, depend on, and integrate into their own work.

System-wide coherence depends on three primary relational patterns, each of which relies on structurally expressed Value: *organizational, collaborative,* and *cooperative.*

1. **Organizational Role—Direction and Contribution Between Differentiated Units**

   In system design, not all units play the same role. Some are *integrative units*—responsible for setting direction, defining constraints, and preserving alignment with broader goals. Others are *operational units*—focused on delivering targeted outcomes within bounded scopes of responsibility.

   The integrative units generate Value in the form of

   - Strategic objectives and intent
   - Risk and compliance boundaries
   - Resource and performance expectations
   - Architectural and policy guardrails

   This Value is not passive documentation—it is an *active, structured surface* that orients the system. It gives operational units the directional clarity they need to act with autonomy while remaining aligned with the whole.

   Operational units, in turn, generate Value in the form of

   - Products, services, or transformations
   - Reports, feedback, and signals
   - Event traces or traceable outcomes
   - Constraint-aware delivery mechanisms

This return flow is equally structural. It allows integrative units to assess impact, surface friction, and refine direction. The relationship becomes *reciprocal*: direction flows outward as contextual Value; contribution flows inward as verified Value.

When these roles and exchanges are clearly defined, the system stabilizes without top-down control. Integration happens through Value alignment, not through structural subordination. Each unit, whether integrative or operational, becomes accountable for what it exposes and interprets from others.

2. **Collaborative Role—Joint Construction of Shared Outcomes**

   In this role, the unit participates in a structured effort to achieve a common goal. It coordinates its Value production with others via an explicit choreography or process context—such as a transaction ID, a shared case, or a defined interaction protocol. Value must be timely, relevant, and aligned to the ongoing flow of work.

   This is not just information-sharing—it's synchronized contribution within a known coordination frame.

3. **Cooperative Role—Shared Environment Without Shared Context**

   Here, the unit contributes Value without direct interaction or alignment with other units. Its outputs are published, exposed, or made accessible in shared spaces (catalogs, logs, reports), and may be consumed by others *without coordination*. The unit does not know how, when, or whether its outputs will be used—but must design them for interpretability and reliability.

   This is not orchestration—it's environmental contribution, where others can depend on what's available, but not on when or why it's offered.

Systems Coordinate Through Value, Not Authority

These three modes—organizational, collaborative, and cooperative—differ in structure but share a common dependency: *the explicit, constrained, and reusable expression of Value.*

In systems that treat Value as a structural surface, coordination emerges naturally. Units align because their outputs interlock. Units adapt because they can interpret changing inputs. Units scale because their roles are not hardcoded but *legible through what they produce.* These distinctions become clearer when examined through the three primary modes of Value exchange shown in Table 4-3.

Table 4-3. *Comparison of Organizational, Collaborative, and Cooperative Modes of Value Exchange*

Mode	Primary relationship	Coordination mechanism	Form of value exchange
Organizational	Between differentiated units (e.g., direction from integrative to operational)	Structured through hierarchy, governance, and accountability boundaries	Contextual Value: objectives, constraints, and expectations flow downward; verified results flow upward
Collaborative	Between units jointly pursuing a shared outcome	Managed through defined processes, schedules, or interaction protocols	Shared Value: co-created outcomes within a coordinated workflow
Cooperative	Between independent units sharing an environment	Enabled by exposure of discoverable, reusable offerings	Ambient Value: each unit contributes to and draws from a shared environment without direct coordination

When Value is unclear, systems compensate with bureaucracy. When it is clear, systems evolve with grace.

4.2.3. Role in Environmental Interaction

A unit does not exist in isolation. It is always embedded within an environment—an external world filled with stakeholders, authorities, observers, allies, and adversaries. Each of these actors forms a different kind of relationship with the unit, and each expects or extracts some form of Value. These relationships are often asymmetric, indirect, or slow to develop—but they are real, and they shape the unit's trajectory. The ability to sustain these relationships depends not on persuasion but on design. *A unit must express Value that is meaningful to its environment, even if that environment does not reciprocate in kind.*

Value is what makes a unit *visible* and *trustable* to the outside world. It allows the environment to interpret what the unit is, assess what it stands for, and decide whether it is worth engaging, approving, funding, joining, or opposing. Without clearly structured Value, the unit remains opaque—a black box that may function internally but exerts no meaningful presence beyond its boundaries.

Across the landscape of environmental interaction, we can distinguish at least six major relational patterns—each shaped by different expectations, and each requiring a different kind of Value expression:

1. **Transactional Relations**

   These are the most familiar: Interactions with *customers, clients, suppliers,* and *distributors.* Here, Value is exchanged directly—through products, services, subscriptions, deliveries, or licensing agreements. The expectations are clear: *give me something I can use or consume.* These are structured relationships, often governed by contracts or APIs, and measured by satisfaction, fulfillment, or renewal. If Value is not visible or reliable, the relationship collapses.

2. **Compliance Relations**

   With *regulators, authorities,* and *standards bodies,* the relationship is not transactional, but *conformational.* The unit must expose Value in the form of *demonstrable adherence:* audit trails, certifications, documented controls, disclosures, or testing outcomes. These artifacts are not consumed, but *evaluated.* The Value here is legitimacy—the structural permission to exist and operate within a governed space. In this context, silence is not safety; absence of declared Value may be interpreted as non-compliance.

3. **Investor and Oversight Relations**

   Investors, boards, lenders, and *auditors* interact with the unit through a lens of performance, accountability, and risk. Their expectation is *transparent Value* in the form of strategic clarity, financial outcomes, forward signals, and governance maturity. They do not ask what the unit sells; they ask whether it is viable, ethical, and likely to succeed. Here, Value is exposed through *artifacts of trajectory*—reports, roadmaps, KPIs, and architectural evidence of scalability.

4. **Talent and Employer Brand Relations**

   Some units are in relation to those who might *join, support,* or *build* with it: potential employees, contractors, interns, or research collaborators. These actors assess Value not as consumers, but as potential co-creators. They look for signs of purpose, integrity, direction, culture, and technical elegance. Value here is reputational, and often indirect. A unit's ability to attract and retain talent depends not just on what it does, but *how well it signals what it stands for.*

5. **Ecosystem Relations**

   Selected units may contribute to and benefit from shared standards, platforms, or communities. Their Value is designed for openness and interoperability—exposed through APIs, public artifacts, or contributions to joint initiatives. The goal is mutual enablement and legitimacy. Value must be accessible, well-structured, and aligned with external expectations, but without committing the unit to dependencies it cannot manage. The posture is cooperative, but architecturally bounded.

6. **Competitor Relations**

   Certain units are responsible for managing market-facing interactions where visibility is political. The Value they expose is designed to shape perception, establish presence, or misdirect competitors—while withholding strategic detail. This requires tight control, selective disclosure, and sometimes intentional ambiguity. The design goal is to remain legible without becoming transparent. What's not shown is as important as what is.

7. **Community and Societal Relations**

   Finally, units operate within *public and civil environments.* This includes communities, advocacy groups, local populations, and society at large. In these relations, Value is expressed through *impact, ethics, and transparency*. A unit may not seek this relationship, but it cannot avoid it. Environmental groups assess sustainability; civil society monitors fairness; journalists investigate inconsistencies. The Value expressed—or withheld—by a unit affects not only its public perception but its social license to operate.

Visibility As Design, Not Afterthought

What unites all these external relations is this: *the environment cannot engage with what it cannot see*. And what it sees is not Function or Purpose—it sees *Value*. Not just as marketing language or corporate narrative, but as *structured, observable, composable artifacts* that express what the unit actually offers the world.

In this context, Value becomes more than a mechanism of coordination—it becomes a mechanism of *existence*. If a unit cannot express Value in a way the environment can recognize, interpret, and use, it does not simply fail to thrive. It begins to disappear—organizationally, economically, and socially.

4.3. Design Principles and Heuristics

4.3.1. Principles of Value-Oriented Design

Value-oriented design focuses on ensuring that what a unit offers to others—whether a product, service, experience, or by-product—is intelligible, usable, and relevant in context.

Value-oriented design ensures that the offerings produced by a unit are structurally sound, externally meaningful, and contextually appropriate. Whether the Value takes the form of a product, service, experience, or by-product, the way it is constructed, exposed, and recognized must reflect architectural intent—not happenstance or effort alone.

1. **Value must be constructed for external use, not internal completion**

   An offering is not valuable simply because it is finished. It must be designed for consumption—by someone else, in a context the provider does not fully control. Internal polish or satisfaction means little if the output is unusable from the outside.

Illustration (Accounting): The accounting unit produces a detailed cost model for internal validation, but department heads need concise cost allocations they can apply in budgeting. The Value isn't the model—it's the usable financial insight.

Implication: Value exists only when others can recognize, access, and apply it independently.

2. **Value should anchor in the most common and consequential use**

   Designing Value begins with the most frequent or impactful scenario. This primary use case should guide structure, constraints, and delivery. Designing for edge cases too early leads to brittle or unfocused offerings.

   Illustration (Sales): Sales operations deliver complex performance reports that overwhelm sales teams. Reps only need clear weekly summaries showing target progress. The intended Value is actionable insight, not analytical complexity.

   Implication: When the main path is clear and trusted, peripheral uses can be managed or eliminated.

3. **Value must be appropriate in scope, not maximal in scale**

   Over-delivering can obscure clarity and waste resources. Value must meet the need precisely—no more, no less. Architectural elegance favors sufficiency, not excess.

   Illustration (IT): The IT unit builds a multi-layer analytics product integrating dozens of feeds, but business users only need daily metrics on customer activity. The Value—the reliable daily view—is obscured by unnecessary expansion.

   Implication: Fit-for-purpose offerings are easier to adopt, sustain, and trust.

4. **Value must declare its form: product, service, experience, or by-product**

   Each Value form creates different expectations. A product implies repeatability; a service, responsiveness; an experience, memorability; a by-product, opportunistic utility. Recipients interpret offerings based on form—even if the provider is unclear.

   Illustration (HR): HR publishes leadership toolkits, runs mentoring programs, and hosts cultural events under one banner. Employees can't tell which deliverables are products and which are experiences. The Value itself becomes indistinct.

   Implication: When form is ambiguous, trust and adoption erode.

5. **Value must be discoverable before it is consumed**

   Design doesn't end with the offering—it must include signals, interfaces, and cues that allow others to locate, interpret, and trust the Value surface.

   Illustration (Finance): Finance provides automated budget forecasts but hides them deep within an internal drive. The deliverable—the forecast itself—has Value only if discoverable.

   Implication: Undiscovered Value is indistinguishable from nonexistent Value.

6. **Value must carry weight even in the absence of monetary exchange**

   Free offerings—dashboards, summaries, feedback tools—can still be architecturally critical. Structural Value often has no price tag but significant impact.

   Illustration (Legal): Legal provides a free self-assessment service that helps partners identify risk exposures early. Although no payment is involved, the Value—risk reduction—is tangible and measurable.

   Implication: Design for utility, not price, and measure impact, not transaction.

7. **Value must trace back to Purpose**

   If the Value exposed by a unit cannot be understood as a projection of its Purpose, it will eventually confuse stakeholders or duplicate effort. Traceability keeps Value from drifting into contradiction.

   Illustration (Marketing): Marketing invests in awareness campaigns that generate engagement but no qualified leads. The activities are energetic, but the Value—lead generation—is absent.

   Implication: Purpose-aligned Value strengthens system coherence and unit identity.

8. **Value must be surfaced in a form that conveys meaning and trust**

   While interfaces belong to Function, the visible form of Value must still communicate scope, status, and intention. Recipients should not have to decode or infer what an offering does.

   Illustration (Customer Support): The support unit closes issues with automated confirmations, but customers need clear documentation of resolution. The Value isn't the closure—it's the verified outcome.

   Implication: Value must be communicated in a way that inspires confidence.

9. **Value must adapt through feedback**

   Value is not static. It should evolve based on usage, environment, and input from those who consume it. Feedback is not a favor—it is a structural requirement.

   Illustration (Product): The product team offers an onboarding program that fails to adjust to user feedback about confusion points. The Value—effective onboarding—deteriorates without adaptation.

   Implication: Static Value loses relevance; adaptive Value retains trust.

10. **Value must respect the cost of attention**

    Even a useful offering becomes burdensome if it is too noisy, complex, or demanding. The pathway to Value must be smooth, not exhausting.

    Illustration (Internal Communications): The communications unit issues frequent newsletters with overlapping content. Employees stop reading, missing critical updates.

    Implication: Attention is the gateway to Value—waste it, and the Value will disappear.

11. **Value must strengthen the system, not just the task**

    Some offerings—like standards, audit trails, or shared artifacts—enable others to perform better, even if they are not task-specific. Their Value lies in infrastructural support.

    Illustration (Procurement): Procurement standardizes vendor contracts, enabling smoother renewals across departments. The Value is not the contract itself but the reduction in systemic friction.

    Implication: Some forms of Value are infrastructural—they strengthen the system as a whole.

4.3.2. Architectural Heuristics for Value

While principles define what Value must represent, heuristics provide practical guidance for how to design, deliver, and evolve it under real-world conditions. These are not universal rules—they are tested design shortcuts that help reduce ambiguity, accelerate iteration, and prevent common failure modes. They are especially useful when working under constraint: limited attention, ambiguous demand, uncertain timing, or partial visibility into how Value is used.

Below are architectural heuristics to support the design of discoverable, interpretable, and purpose-aligned Value surfaces:

1. **Design the offering as an interface, not a destination**

   Value should initiate or extend a relationship—not end it. Even when the offering is declarative or stand-alone, it must invite interpretation, response, or action from the recipient.

   Rationale: If a Value surface acts as a dead end, it limits reusability, weakens feedback, and reduces the likelihood of systemic integration.

2. **Separate creation from exposure**

   The mechanism that generates Value should not constrain the way it is delivered. Offerings should be exposed through flexible interfaces, adapted for different contexts, without replicating or reinventing the internal logic.

   Rationale: Tightly coupling creation and delivery makes it harder to scale, evolve, or adapt offerings for new recipients or channels.

3. **Package Value around outcomes, not components**

   Group outputs by what they help someone achieve—not by how they were produced. Offerings should reflect the goals, actions, or decisions they support.

   Rationale: Value becomes legible and attractive when presented in terms of usefulness, not architecture.

4. **Deliver from Storage when reliability matters; from Memory when speed or informality suffices**

   Determine whether the Value being offered must persist beyond the moment, survive transfer across units, or meet compliance expectations. Use Memory for exploratory, low-risk exchanges. Use Storage for durable, dependable, repeatable offerings.

   Rationale: Value served from Memory is fragile and often unverifiable. Storage ensures that what was offered can be proven, reused, and trusted—especially in external or asynchronous contexts.

5. **Signal what the offering assumes**

   Every Value surface encodes assumptions—about inputs, timing, environment, or consumer behavior. These should be stated, surfaced, or implied through strong affordances.

   Rationale: When assumptions are hidden, recipients encounter silent failures or misuse offerings unintentionally.

6. **Make Value usable at a glance**

   Design the offering so its utility is immediately evident. Reduce the cognitive load required to interpret, test, or adopt it.

   Rationale: Attention is expensive; unclear or overcomplicated offerings get ignored, regardless of their actual usefulness.

7. **Design for reusability, not just delivery**

   A well-structured offering may be consumed in unintended ways or reused in other contexts. Anticipate flexible application—but protect core intent.

   Rationale: Offerings that are too brittle or too tightly scoped become disposable or duplicated elsewhere.

8. **Align perceived Value with actual contribution**

   Ensure the surface appearance of the offering reflects its impact and relevance. Avoid under-signaling high-impact Value or over-designing trivial outputs.

   Rationale: Misalignment between surface and substance reduces trust, uptake, and correct usage.

9. **Fail safely and visibly**

   When Value cannot be delivered as intended, expose failure in a way that preserves trust and invites response. Don't hide degradation.

   Rationale: Silent failure undermines reliability. Clear failure creates pathways for correction, feedback, and redesign.

4.4. Realization Across Implementation Planes

While Value is defined at the design level, it becomes real only through implementation—when it manifests in tangible or observable form across the social, physical, and digital planes of the system. In practice, this means offerings must be expressed not only as internal constructs but as externally accessible surfaces: through roles and relationships, physical products and services, software interfaces, analytical feedback loops, and interaction channels. Understanding where and how Value appears—who carries it, what formats encode it, and what surfaces expose it—is critical for ensuring its legitimacy, trustworthiness, and continued relevance within a sociotechnical environment.

4.4.1. Manifestation in the Social and Physical Planes

Value has always been shaped—and often constrained—by the nature of its delivery mechanisms. Before the digital layer, Value was realized primarily through *people, artifacts, and environments.* What a unit offered was expressed through physical products, face-to-face services, human roles, and material resources. These expressions remain vital in modern sociotechnical systems, even as they become increasingly entangled with digital delivery.

In the *social plane,* Value manifests through *roles, relationships, and behaviors*. A teacher, a technician, a concierge, or a compliance officer are not simply placeholders—they are human Value surfaces. Each role exposes the unit's offering to the outside world, interprets requests, applies judgment, and modulates delivery. The reliability of Value in this plane depends on *clear mandates, embedded expertise*, and *performative alignment with Purpose.* When Value is expected but the role is misaligned—or absent—the result is perceived failure, regardless of internal capacity.

In the **physical plane**, Value is delivered through **material expressions** of products, services, experiences, and by-products. A product may be embodied as a physical device, printed output, or packaged good. A service might rely on tools, instruments, or co-located environments to perform responsively. Experiences are shaped by the design of physical spaces, tactile interactions, and environmental cues. Even by-products—such as leftover materials or autogenerated reports—may take tangible form. Each of these expressions must be designed with care, ensuring they convey the right expectations and support intended use.

Across both planes, *symbolic clarity* and *operational reliability* are essential. A product that looks beautiful but performs inconsistently, or a person whose title suggests authority but lacks agency, results in Value mismatch. Physical and social Value delivery must therefore be *intentionally constructed, aligned with the unit's Purpose,* and *interpretable by its intended audience.*

4.4.2. Value in the Digital Plane

The digital plane offers the broadest stage for Value realization—where offerings can be stored, accessed, delivered, recomposed, and scaled with unprecedented precision and reach. Unlike the physical or social planes, where delivery is shaped by proximity and presence, digital Value is defined by *structure, accessibility, observability,* and *interpretability.* A successful offering must be equally legible to humans and machines, whether it is a static artifact or a dynamic interaction.

Each Value form—*product, service, experience, by-product*—has recognizable expressions in digital environments:

- *Products* include downloadable or reusable digital goods: Images and photographs, music and audio files, video content (films, tutorials, animations, livestreams), fonts and graphics, e-books, research reports, online courses, and software applications. These artifacts are designed for repeated access with minimal dependency on time or context.
- *Services* appear as digital offerings that require contextual input from the consumer—whether data, parameters, preferences, or presence. Examples include custom website design, API-driven data transformations, software integration, copywriting, or machine-learning model tuning. These services adapt to the needs or identity of the requester and are delivered through interfaces, workflows, or human-in-the-loop platforms.
- *Experiences* are immersive and choreographed interactions that leave behind memory, insight, or transformation. These include VR games, interactive storytelling platforms, live virtual concerts, role-playing simulations for medical or business training, gamified

shopping journeys, virtual wellness programs, or metaverse-based gatherings. Unlike products, experiences unfold over time and are co-constructed with the user.

- *By-products* consist of system outputs that were not the primary design objective but hold downstream Value: Autogenerated reports, metadata trails, analytics logs, intermediate data snapshots, prompt histories, and derived content artifacts. In well-designed systems, these are captured, packaged, and exposed as reusable Value.

In all these forms, digital Value must be *structured, documented, and externally interpretable*. That means APIs must be contract-bound; dashboards must carry legible signals; downloadable files must follow naming conventions; and digital services must present meaningful defaults and actionable errors. The boundary of the offering is not just its output—but also the way it declares assumptions, invites interaction, and supports observability.

Finally, not all digital Value requires the same design properties—but the more embedded, dynamic, or systemic the offering, the greater the architectural responsibility. Offerings such as services and interactive experiences must be *traceable, composable,* and *resilient* to asynchronous use. By contrast, stand-alone products like stock photos or ebooks may prioritize *clarity, portability,* and *metadata integrity*. The design of Value must reflect not only what is offered but how it will be used and relied upon in practice.

4.4.3. Implementation Surfaces and Sociotechnical Contexts

Value is not only what is offered, but *where and how it is encountered*. An offering may be well-designed internally yet fail to deliver impact if it is exposed in the wrong place, to the wrong audience, or in the wrong format. To be meaningful, Value must appear at the *right surface*, in the *right context*, with the *right framing*.

In digital systems, implementation surfaces include *catalogs, marketplaces, APIs, front-end dashboards, portals, search interfaces, notification layers,* and *chat-based agents*. These are not just delivery mechanisms—they are interpretive spaces where Value is *framed, filtered,* and *perceived*. A high-quality offering buried three clicks deep in a poorly indexed interface may go unnoticed, while a simpler one surfaced through search or alerting may drive meaningful outcomes.

But surfaces alone are not sufficient. The *sociotechnical context* matters equally: what expectations users bring, what roles they occupy, what organizational culture surrounds them, and what processes govern access and response. A dashboard delivered to a non-analytical audience, or a service API exposed without change management, may erode trust even if technically sound.

Designing Value at this level requires *mapping not just channels but intentions and capabilities*. What is the user or recipient trying to achieve? How do they currently access similar offerings? Who decides what counts as Value in this context? Whether the surface is a Slack notification, a Jira ticket comment, a ServiceNow catalog item, or a mobile app screen—each implementation must align with both *technical affordances* and *social rhythms*.

Value also travels across domains and systems. A report generated by a compliance team may later appear as a heatmap in a risk dashboard; an ML model trained by one unit may fuel recommendations in another. These cross-domain reuses require *deliberate packaging, contextual cues*, and *usage boundaries* to preserve interpretability.

Ultimately, Value is not just designed—it is *staged*. Where it appears, who encounters it, and what meaning they can make from it determines whether it succeeds architecturally. Designing the offering is only half the task; *placing it well within the system is the rest*.

4.5. Strategic Impact of Value Interfaces

Value interfaces are not neutral—they shape how systems evolve, how units interact, and how strategy becomes visible in execution. Every offering—whether a product, service, experience, or by-product—acts as a *point of contact* between intent and outcome, between internal capability and external relevance. The design of these interfaces carries strategic consequences: it affects alignment, modularity, legitimacy, and leverage. Some effects are subtle, unfolding over time; others are immediate, measurable, and structural. Understanding the *strategic impact of Value* means recognizing that offerings do more than deliver utility—they define how the enterprise is seen, used, and ultimately judged.

4.5.1. Nuances of Impact

The impact of Value cannot be measured solely in transactions or consumption rates. Once exposed, a Value offering becomes more than just an object of utility—it becomes a *signal, a constraint, a memory, a symbol*, and sometimes even a *structural force*. These effects may not be immediate or easily quantified, but they are architecturally significant. Understanding the nuances of impact means recognizing what Value *does* to the system once it becomes accessible.

Perception of Purpose

What a unit chooses to offer reveals what it believes it exists to do. The framing of Value—whether a diagnostic model, a curated dataset, or a patient guidance experience—makes Purpose concrete. If Purpose is claimed but not offered in any usable form, the unit becomes performative. Conversely, well-framed Value reinforces Purpose not through assertion but through presence.

Emotional and Aesthetic Resonance

Some offerings carry affective weight: a design system that feels elegant, a wellness experience that cultivates calm, or a compliance warning that evokes seriousness. These qualities are not peripheral—they *are* part of the Value. Aesthetic coherence, emotional tone, and symbolic clarity shape how offerings are remembered, reused, or rejected. The most trusted services are often the ones that feel quietly competent.

Strategic Framing and Legibility

How Value is presented changes how the unit is perceived—by internal stakeholders, partners, regulators, or the public. A curated insight platform communicates maturity. A raw data dump suggests indifference. The same Value can be exposed as empowerment or as obligation. Legibility—what others can infer about the Value and its intent—is a form of power.

Creation of Architectural Boundaries

Once Value is made available, it defines a *boundary of responsibility*. A published risk score becomes something others depend on. A generated artifact becomes part of someone else's workflow. These offerings form *operational interfaces*, even when not technically wrapped in Function. Their presence conditions what others do next—and what they no longer need to build themselves.

Cultural Shaping Through Offering Structure

Offerings train users. A Value delivered as self-service empowers autonomy; one that requires repeated requests promotes gatekeeping. The level of polish, consistency, and clarity shapes expectations—not just of the offering but of the unit behind it. Over time, repeated patterns of Value exposure teach others how to work, how to wait, and how to trust.

Non-Functional Characteristics As Strategic Differentiators

Many offerings compete not on core functionality, but on *non-functional attributes*: safety, security, trustworthiness, emotional alignment, aesthetic clarity, ethical framing, or even ambient tone. A product that feels honest may outperform a faster one. A service that inspires confidence may outlast one with better specs. These qualities become the source of long-term strategic advantage—and must be considered part of the Value itself.

4.5.2. Measurable Benefits and Architectural Payoffs

Well-designed Value offerings do more than satisfy a need—they clarify the unit's role, increase the system's usability, and create architectural leverage. They reduce friction, accelerate integration, and establish visible trust boundaries across the enterprise. The benefits of such offerings are not abstract—they can be measured, observed, and amplified. The architectural payoff lies in their ability to align distributed actors while preserving autonomy and integrity.

A. **Discoverability Reduces Duplication and Reinvention**

When Value is clearly defined and made findable—via catalogs, APIs, dashboards, or public repositories—other teams can reuse rather than rebuild. This increases leverage and reduces redundancy.

- Higher reuse of existing products, services, and data
- Shorter lead times for dependent initiatives
- Lower costs from eliminating shadow systems

Payoff: Discoverable Value enables horizontal efficiency. What exists can be confidently reused rather than re-invented.

B. **Clarity of Offering Reduces Negotiation Overhead**

When it's clear what is offered, under what terms, and to whom, cross-team collaboration becomes lighter and faster. Meetings are no longer needed just to ask "what do you provide?"

- Fewer clarifying meetings or informal escalations
- Quicker alignment between producer and consumer
- Fewer cycles wasted on assumption mismatches

Payoff: Clear Value accelerates cooperation by reducing transactional drag.

C. **Boundary Exposure Supports Modularity**

Exposed Value creates modular seams in the system—enabling teams to innovate internally without affecting others, and to depend on others without internalizing their complexity.

- Easier integration across units
- Better isolation of changes and updates
- Lower coupling between systems and domains

Payoff: Value exposure builds clean, stable boundaries for composable enterprise design.

D. **Visibility Generates Feedback and Evolution**

When Value is exposed, it's also observable. This creates a feedback loop—offering data on what's useful, ignored, misused, or misunderstood.

- Analytics on usage, performance, and relevance
- Informed decisions about refinement or retirement
- Quicker discovery of broken assumptions

Payoff: Observable Value enables continuous refinement and architectural maturity.

E. **Exposure Converts Capability into Legitimacy**

An internal capability only gains strategic weight when it produces something visible, usable, and consumed. Offered Value becomes the traceable evidence of contribution.

- Demonstrable outputs from support and enabling units
- Tangible ROI from platform, data, and design teams
- Stronger justification for resourcing and investment

Payoff: Offering Value transforms abstract potential into recognizable impact.

4.6. Implementation Patterns

Designing Value is one part of the task; *delivering it repeatedly, accessibly, and coherently* is another. Implementation patterns translate architectural intent into usable form. They reflect the real-world strategies units use to expose, scale, or evolve Value—sometimes deliberately, sometimes retroactively. These patterns are not universal prescriptions, but recurring moves that bring Value to life across diverse contexts. Each comes with its own structural implications and design considerations.

4.6.1. Productization of Reusable Outputs

When a unit repeatedly generates artifacts that others rely on—datasets, insights, models, components—it often transitions from ad hoc delivery to formal *productization*. This involves naming the output, assigning ownership, versioning releases, and wrapping it with documentation and support. The offering becomes stable, referenceable, and eligible for reuse at scale.

Productization creates architectural anchors. It signals readiness, defines responsibility, and invites integration.

4.6.2. Service Wrapping of Internal Capability

Some Value is inherently contextual—it depends on requests, inputs, or presence. Rather than publishing fixed outputs, the unit exposes a *service* that responds dynamically. This can be a human-in-the-loop process (e.g., legal review), a technical API (e.g., personalization engine), or a hybrid workflow. Framing the Value as a service allows it to be invoked on demand without requiring internal access to the producing unit.

Wrapping a capability as a service enables reuse without entanglement, but demands clear contracts and guardrails.

4.6.3. Packaging of Experiences for Target Outcomes

Experiential Value must be staged—not just produced. Whether digital, physical, or hybrid, an experience involves narrative flow, temporal coordination, and emotional tone. Value is not just delivered—it is *lived*. When experiences are intentionally packaged (e.g., onboarding journeys, virtual learning arcs, immersive diagnostics), they guide perception and memory toward specific outcomes.

Experience packaging requires cross-functional choreography and intentional framing of transformation.

4.6.4. Harvesting of By-Products for Secondary Use

Not all Value is designed—some of it is *captured*. Units often generate outputs in the course of other work: logs, drafts, usage traces, recordings, errors, feedback, raw data. When these by-products are stabilized and reused, they become second-order Value—supporting observability, recommendations, or continuous learning.

By-product harvesting turns exhaust into asset. It rewards reflection and instrumentation.

4.6.5. Externalization via Catalogs and Marketplaces

Even well-designed offerings can fail if they're invisible. Externalization involves surfacing the Value through shared discovery channels—internal portals, team catalogs, developer platforms, or commercial marketplaces. This step formalizes availability and makes offerings visible to broader audiences.

Externalization expands reach and demand but requires maturity in positioning, labeling, and support.

4.6.6. Progressive Framing Through Iterative Exposure

Not all Value is ready at once. In fast-moving or uncertain environments, teams may expose early or partial versions—previews, betas, experiments—and refine based on usage. This allows offerings to evolve organically while maintaining architectural coherence.

Iterative exposure builds legitimacy, aligns with real demand, and surfaces blind spots early.

4.7. Interlocking Dependencies and Alignment

Value does not emerge in isolation. Within the Purpose Driven Design framework, it is a *structurally interdependent construct*—manifested as an offering, but always shaped by deeper design elements. The coherence, relevance, and sustainability of any Value depends on how well it aligns with the unit's Purpose, is enacted through Function, delivered via Process, enabled by Structure, shaped by Order and Culture, accumulated in Memory, and surfaced from Storage.

When these dependencies are neglected, Value becomes brittle, performative, or disjointed—useful in the moment, perhaps, but untrustworthy in the long term. What appears to be an offering may instead be a leak, a misrepresentation, or an accidental liability. Architectural alignment ensures that Value is *not only deliverable, but dependable*—and that it reinforces the system's integrity over time.

4.7.1. Alignment with Purpose: Justification and Relevance

Purpose clarifies *why* the unit exists and *whom* it serves. Value is the outward expression of that commitment. If a Value offering cannot be traced to Purpose, it may reflect misplaced energy or symbolic compliance. Conversely, if Purpose exists but produces no external Value, the unit risks irrelevance. Alignment ensures that Value is not just an output but a promise fulfilled.

4.7.2. Alignment with Function: Access and Framing

Function makes Value *usable*—it exposes offerings through defined interfaces, protocols, and channels. Misaligned Function can obscure or distort Value, while excessive Function may expose offerings without context or constraint. Proper alignment ensures that what is offered is *appropriately accessible*—neither buried nor overexposed.

4.7.3. Alignment with Process: Delivery and Timing

Process governs *how* Value is created, updated, and maintained. Without a clear link to Value, processes become rituals or back-office churn. Without supporting processes, Value becomes episodic or decays over time. Well-aligned Value is not just present—it is deliverable on time, at quality, and in rhythm with the system it serves.

4.7.4. Alignment with Structure: Ownership and Accountability

Structure assigns agents to roles—defining who is responsible for producing, refining, approving, or retiring Value. Without structural alignment, Value offerings become orphaned or contested. Aligned units have clear boundaries of responsibility and visibility: someone owns the outcome, not just the mechanics.

4.7.5. Alignment with Order and Culture: Framing and Trust

Order sets rules; Culture sets tone. These shape how Value is *perceived, accepted, and reused.* A Value offering that violates organizational norms or bypasses expected flows may be rejected, even if technically sound. Alignment ensures that Value shows up *with the right framing*—legible, trustworthy, and culturally resonant.

4.7.6. Alignment with Memory: Learning and Evolution

Memory retains what has been offered before—past iterations, usage patterns, regrets, and informal insights. When Value aligns with Memory, teams can evolve their offerings thoughtfully, drawing from lived experience rather than repeating mistakes. In many cooperative or collaborative scenarios—especially between trusted units—Value may be delivered directly from Memory rather than formalized in Storage. This enables fast, context-rich exchange without the overhead of packaging. However, overreliance on Memory can lead to drift or loss if team members change or scale increases. Alignment ensures that Value is grounded in lived knowledge, but not limited by it.

4.7.7. Alignment with Storage: Durability and Transferability

Storage determines how Value is *preserved and served.* Without it, Value is ephemeral—only available when memory is fresh or a person is present. In inter-unit or external interactions, Value must be formalized in Storage to enable delegation, traceability, and asynchronous use. Alignment ensures that Value is *not only created but sustained.*

4.8. Failure Modes and Antipatterns

Value may be present, praised, or heavily consumed—yet still fail architecturally. These failures often go unnoticed at the interface or usage level, but their impact is systemic. They originate from misalignments across Purpose, Function, Process, Structure, or Memory—or from poor judgment about what should be made available, to whom, and how.

This section identifies recurring *antipatterns in Value design* that obscure, distort, or hollow out architectural integrity. They offer a vocabulary for diagnosing contribution-related dysfunction across modular, composable systems.

1. **Phantom Value**
   - *Definition*: Value is claimed or assumed, but no actual offering is delivered.
   - *Symptoms*: Stakeholders refer to a dataset, insight, or service that cannot be found or accessed; usage inquiries result in vague reassurances.
   - *Consequence*: Frustration, workarounds, erosion of trust, informal replication.
   - *Architectural Fault*: Misalignment with Function (nothing exposed) or Purpose (offering never properly defined).
   - *Example:* A "data insights service" is announced on an internal portal, but the promised dashboards were never built. Requests arrive, but nothing can be fulfilled—eroding trust and credibility.
2. **Output Without Purpose**
   - *Definition*: Something is created and delivered, but no longer serves a meaningful goal.
   - *Symptoms*: Teams continue to maintain dashboards, reports, or APIs out of habit or contractual inertia.
   - *Consequence*: Waste, distraction, signal pollution.
   - *Architectural Fault*: Misalignment with Purpose (offering no longer aligned with why the unit exists).
   - *Example:* A finance unit continues publishing a weekly risk report years after automation eliminated the underlying risk. The report consumes effort but provides no actionable Value.

3. **Overstated or Performed Value**

    - *Definition*: The appearance of offering something is mistaken for actual delivery.
    - *Symptoms*: Polished documentation or interfaces hide thin, unstable, or superficial capabilities.
    - *Consequence*: Strategic credibility loss, user distrust, unmet expectations.
    - *Architectural Fault*: Misalignment with Process or Structure—form without function or sustained delivery.
    - *Example:* A "feedback improvement" campaign launches with branded emails and slogans, but no mechanism exists to collect or act on user input. Engagement looks high, but nothing changes.

4. **Accidental Value Leakage**

    - *Definition*: Internally useful artifacts are adopted as offerings without being stabilized or secured.
    - *Symptoms*: Consumers informally rely on draft dashboards, experimental models, or unapproved APIs.
    - *Consequence*: Fragile dependencies, legal exposure, lack of accountability.
    - *Architectural Fault*: Misalignment with Storage (not productionized) and Structure (no declared ownership or boundaries).
    - *Example:* Engineers share an internal monitoring tool with partners before it's hardened for external use. It later exposes sensitive configuration data—turning an informal tool into a governance failure.

5. **Ambiguous or Absent Recipient**

    - *Definition*: An offering exists, but no one is clearly defined as its user, beneficiary, or audience.
    - *Symptoms*: Conflicting usage assumptions; low engagement despite effort; Value tailored to no one in particular.
    - *Consequence*: Wasted potential, misuse, lack of adoption.
    - *Architectural Fault*: Misalignment with Purpose or Order—Value not designed with clear stakeholder framing.
    - *Example:* A compliance dashboard is created "for transparency," but no specific audience owns or uses it. It drifts into neglect, yet remains part of the system's declared Value set

6. **Channel Proliferation Without Intent**

    - *Definition*: Value is offered simultaneously through too many inconsistent or duplicative channels.
    - *Symptoms*: Multiple dashboards, APIs, or documents all claiming to represent the same thing.
    - *Consequence*: Confusion, contradictory usage metrics, interface fatigue.
    - *Architectural Fault*: Misalignment with Structure and Memory—no single source of truth or offer governance.
    - *Example:* HR publishes policy updates by email, chat, intranet, and newsletters, each with slightly different wording. Employees stop trusting any version, creating systemic confusion.

7. **One-Time Artifact Masquerading As Offering**

    - *Definition*: A custom, situational output is mistaken for a repeatable or maintainable product.
    - *Symptoms*: Teams treat a one-off deck, notebook, or query as a reusable product; consumers are misled about its scope or freshness.

- *Consequence*: Fragile reuse, failed automation, unmet expectations.
- *Architectural Fault*: Misalignment with Process and Storage—no sustainable delivery mechanism.
- *Example:* A data science team builds a one-off executive presentation model and leadership markets it as a permanent "analytics product." Maintenance fails within months.

8. **Instrumentalized Metrics or Proxy Offerings**
    - *Definition*: Metrics, traces, or proxies are offered in place of real Value—often due to measurement bias or KPI incentives.
    - *Symptoms*: Dashboards full of activity metrics instead of outcomes; service responsiveness masquerading as effectiveness.
    - *Consequence*: Misaligned optimization, misleading decisions, outcome blindness.
    - *Architectural Fault*: Misalignment with Memory and Purpose—Value distorted by what's easy to measure.
    - *Example:* A support unit reports "average ticket resolution time" as its Value, even when quick resolutions come from premature closures rather than customer satisfaction.
9. **Legacy Artifact Persistence**
    - *Definition*: Outdated offerings remain publicly listed or technically available long after they are useful.
    - *Symptoms*: Broken links, stale models, misleading versions continue to circulate or appear in catalogs.
    - *Consequence*: Accidental misuse, system bloat, analytical contamination.
    - *Architectural Fault*: Misalignment with Memory and Process—no deprecation or retirement signal.
    - *Example:* An obsolete API stays online to "avoid disruption," even though no current systems depend on it. Its lingering presence confuses new developers and expands the attack surface.

4.9. Summary: Value As a Design Construct

Within Purpose Driven Design, Value is the *system's surface of contribution*—the explicit manifestation of what each unit offers to others in order to fulfill its Purpose. It defines not only what is produced but also what is received, interpreted, and relied upon across the system's boundaries. In applied contexts, these same criteria serve as a self-assessment lens, helping teams test whether what they deliver remains interpretable, purposeful, and systemically aligned. Value transforms Purpose into observable contribution, anchoring Function and providing the foundation for coherent interaction. It is how the system produces relevance, earns trust, and maintains directional cohesion across internal and external interactions. Unlike constructs that govern internal logic or behavioral flow, Value defines *what is made available*—as product, service, experience, or by-product—and *why it matters.*

Value resolves the fundamental tension between autonomy and integration in composable systems. It enables distributed units to operate independently while still contributing meaningfully to the success of the whole. This contribution is not theoretical; it is realized through specific interfaces, formats, and engagements—designed, exposed, and sustained as part of the system's architecture.

Realized effectively, Value functions as

- *A signal of Purpose,* expressing why the unit exists and what others can expect from it.
- *A boundary surface,* enabling coordination without entanglement by making offerings explicit, dependable, and reusable.
- *A coordination accelerant,* reducing ambiguity and negotiation overhead by clarifying what is available, in what form, and under what terms.
- *A feedback channel,* capturing usage, trust, and impact—allowing the unit to evolve its offerings in response to need and context.
- *A legitimacy mechanism,* converting potential and capability into demonstrable contribution and traceable outcomes.

The architectural treatment of Value reveals that it is not a business term or delivery artifact. It is a structural commitment to relevance, designed into the very interface between units and their environment. When neglected, Value becomes performative, ambiguous, or misaligned—leading to drift, duplication, or systemic opacity. When designed deliberately, Value enables clarity of role, coherence of system, and acceleration of purposeful change.

In the age of modular, AI-augmented, and interdependent systems, Value is not just a deliverable. It is the primary mechanism of architectural accountability—the visible, usable proof that each unit is not just active, but meaningfully aligned.

CHAPTER 5

Function

In 1916, a grocer from Memphis, Tennessee, reimagined the way people bought food. His name was Clarence Saunders, and he was frustrated. Traditional grocery stores were chaotic. Customers handed lists to clerks, who gathered items behind counters. Errors were frequent. Bottlenecks formed. Theft was common. But more than anything, Saunders saw lost opportunity. Customers were disconnected from the products they consumed. The transaction was functional, but the experience was fragmented.

Saunders' idea was radical: let the customer do the work.

He opened the first Piggly Wiggly store on September 6, 1916. Unlike any store before it, Piggly Wiggly allowed shoppers to pick up a basket, walk through organized aisles, inspect goods, compare prices, and bring their selections to a cashier. It was a physical interface—a functional structure—designed to shift agency. Instead of relying on clerks, customers now engaged directly with the inventory.

A. Milchman, *The Purpose Driven Design Framework for Modern Enterprises*,
https://doi.org/10.1007/979-8-8688-2223-0_5

The concept was not merely about efficiency. It reshaped behavior. Shoppers lingered longer, discovered new products, and made spontaneous purchases. Manufacturers adjusted packaging to appeal visually. Prices became transparent. Saunders had introduced a new Function—one that placed interaction, exploration, and transaction into the hands of the consumer.

The result was explosive. By 1923, there were more than 1,200 Piggly Wiggly stores across the United States. The self-service model quickly spread, transforming not just grocery retail but the entire landscape of consumer commerce. The core insight was simple: *designing Function as an intentional interface for interaction unlocks latent value.*

Saunders' legacy extends far beyond groceries. His invention foreshadowed the ATM, the self-checkout machine, and the e-commerce storefront. Every modern app, website, or physical space that puts control in the hands of the user owes something to Piggly Wiggly. Even the structure of cloud-based services, where APIs replace clerks and user dashboards replace order forms, can be traced back to that same principle.

But Saunders also revealed the stakes of Function. His patent battles, financial overreach, and eventual ouster from the company he founded illustrate a core design lesson: *Function must align with underlying structure, governance, and economic logic.* A beautiful interface without operational integrity can scale chaos instead of value.

Still, the Function Saunders introduced endures. Its success stemmed not just from clever mechanics, but from an architectural shift: a unit designed around empowered interaction. The shopper became an active participant, not a passive recipient. And in doing so, the unit—the store—became more than a channel. It became a system with purpose, process, and structure aligned.

Purpose defines why a unit exists—but it is through Function that this purpose becomes accessible and experienced. Function is the structured point of engagement, the boundary surface through which the unit interacts with its environment. It transforms internal intent into external availability, allowing others to search, request, navigate, transact, or respond. A well-designed Function does not simply expose internal capability; it structures interaction in a way that is coherent, purposeful, and extensible. In this chapter, Function is treated not as an implementation detail but as a first-class design construct—defined, structured, and aligned with other architectural elements. What follows is a technical and conceptual analysis of Function as a means of system participation.

5.1. Foundations of Function As a Design Construct

Function is the first outward expression of a unit's Purpose. It defines how the unit's capabilities are made accessible to others—much like a television's remote control, which translates the device's internal complexity into a clear, usable interface. It is through Function that Purpose becomes navigable, actionable, and experientially real. In Purpose Driven Design, Function is not an implementation detail but an architectural construct—it defines how a unit reveals itself to the world, structures engagement, and enables the exchange of Value. It is the interactive surface of a unit, shaping the means by which users browse, search, compare, configure, submit, and confirm. These transactional flows are not auxiliary features; they are the very architecture of interaction. Whether a unit operates through a digital portal, physical interface, voice command, or automated API, its Function determines the path by which Value is accessed, evaluated, and received. This section defines Function as a structural construct—its conceptual role, design necessity, and evolving patterns across sociotechnical systems.

5.1.1. Defining Function Architecturally

In Purpose Driven Design, Function is the structural construct that exposes a unit's Purpose to its environment through pathways of interaction. It is not merely what the unit does, but how it makes itself available—how it becomes operable, accessible, and intelligible to consumers, collaborators, and systems. If Purpose provides the reason for a unit's existence, Function provides the interface through which that reason is enacted and validated.

Architecturally, Function manifests as the surfaces and flows through which Value is requested, navigated, configured, and received. These may include user interfaces, APIs, portals, service calls, kiosks, or voice assistants—any medium through which a unit's offerings become usable by others. Function is thus the connective tissue between internal capability and external utility.

In sociotechnical units, Function frequently takes the form of transactional flows—discrete, legible steps that guide engagement and define utility. Browsing, searching, comparing, configuring, submitting, confirming, receiving, and rating are all functional acts. Together, they establish a logic of engagement that consumers learn, navigate, and rely on. These flows are not ancillary—they are structurally significant. They shape not only how Value is delivered, but also how it is interpreted, evaluated, and trusted.

What distinguishes Function in this framework is its architectural intentionality. It is not the accidental result of product design, nor the sum of ad hoc endpoints. Function is deliberately composed to express Purpose through structured interaction. As such, its design carries implications for usability, accessibility, legitimacy, and coherence.

A well-formed Function exhibits the following architectural properties:

- *Externally operable*: Function must be legible and actionable from outside the unit. It provides discoverable points of entry into the system's capabilities and services.
- *Structurally expressive*: Function should reflect Purpose and Value in its configuration. The interface logic must signal what the unit is for, what it offers, and how to use it appropriately.
- *Composable and modular*: Function must support structured integration—able to be composed into broader workflows, ecosystems, or federated architectures without friction or duplication.
- *Governable*: Function is not merely a pathway of use—it is a surface of accountability. It must support authentication, authorization, observability, and compliance mechanisms where required.
- *Aligned with process and structure*: Behind every function lies a process; behind every process, a structure of roles and agents. Function must align with these internal constructs to avoid interaction fragility or operational mismatch.

Function is not reducible to an interface. It is an architectural construct that frames how interaction becomes meaningfully possible. Its presence determines whether a unit is engaged or ignored, trusted or bypassed, intelligible or opaque. It is where design meets use, and where structural logic becomes lived experience.

5.1.2. Why Function Exists

Function exists because interaction must be made possible, repeatable, and meaningful. In a sociotechnical unit, Purpose provides the reason for existence—but without Function, that Purpose remains inert. No matter how well-conceived a unit's internal capabilities may be, if there is no structured way to access and engage them, the unit cannot contribute, coordinate, or create Value.

Architecturally, Function is the bridge between *intent* and *use*. It creates the terms of engagement between a unit and its environment—terms that must be legible, navigable, and trustworthy. Without Function, offerings cannot be discovered, understood, or invoked. Without Function, Process cannot be initiated, Structure remains idle, and Value cannot be realized.

In most sociotechnical units, Function is not a single operation but a structured pattern of engagement—a designed flow through which users interact with the unit's capabilities and extract Value. These flows often take the form of transactional steps: filtering results, adding items to a cart, scheduling deliveries, making payments, uploading documents, reviewing status updates, or responding to system prompts. Each step serves as a gateway to the unit's offerings, shaping how interaction unfolds and how utility is perceived. Whether configuring a policy, onboarding a client, submitting a request, or accessing analytics, the unit becomes intelligible and usable through these functional pathways.

Function also establishes interactional constraint. Not every capability should be exposed. Function determines what is accessible, to whom, and under what conditions. It governs what can be invoked directly, what must be orchestrated, and what remains encapsulated. In this way, Function serves not only usability but also *security, control, and responsibility*.

Furthermore, Function enables systemic operability. In federated architectures, composable platforms, and decentralized ecosystems, units must not only be visible—they must be usable. Purpose may surface a unit's identity through static elements such as landing pages, service descriptions, or narrative framings. But Function is what makes that identity actionable. It defines how a unit behaves when engaged—what it accepts, what it returns, how it responds, and under what rules. A unit with only Purpose is legible but inert. A unit with well-formed Function becomes operable within a system of systems—discoverable, invocable, and governable.

Thus, Function exists to fulfill four design necessities:

1. *Interaction*: Translating Purpose into structured, repeatable patterns of engagement
2. *Access*: Making capabilities usable under defined rules, roles, and conditions

3. *Integration*: Enabling the unit to participate in workflows, systems, and platform-wide coordination

4. *Operability*: Ensuring the unit is not just visible but capable of being invoked, configured, and governed meaningfully

Without Function, a unit may present itself through Purpose—but it remains inert. Function is what renders the unit *active*, giving it an interactive form that supports real-world use, participation, and system-wide contribution.

5.1.3. What Problem Function Solves

Function solves the architectural problem of *inert capability*—the condition in which a unit possesses internal purpose and potential but lacks any structured means for that potential to be realized through engagement. Without Function, a unit may be well-formed internally but remains *interactionally unavailable*. Its services, processes, and offerings cannot be discovered, invoked, configured, or coordinated. The result is a breakdown in the core mechanism of sociotechnical viability: meaningful, repeatable interaction.

In modern systems, this problem manifests in several ways:

- *Isolated capability*: Units are built with processes and roles, but lack usable interfaces. Consumers and collaborators cannot engage with them meaningfully—either because entry points are missing, poorly designed, or inconsistent.

- *Interaction debt*: Functional flows accumulate over time through fragmented patches—forms bolted onto portals, disconnected APIs, undocumented triggers. This leads to brittle, non-reusable, and often duplicative interaction surfaces.

- *Opaque behavior*: Systems expose interfaces without clarifying what will happen. Users are left guessing what inputs are allowed, what outputs to expect, or what conditions govern interaction. This erodes trust, prevents automation, and inhibits integration.

- *Non-composability*: In the absence of structured Function, units cannot be orchestrated into broader workflows. Even if a unit serves a clear purpose, its lack of a usable functional surface makes it unfit for inclusion in system-wide processes.

Function addresses these failures by introducing structured interactionality into the architecture. It creates designed surfaces that mediate between users (human or machine) and internal capability. These surfaces are not incidental—they are what render the unit operable. Function converts potential into participation.

Specifically, Function solves four persistent structural deficiencies:

1. *Inaccessibility*: When units are technically available but practically unreachable.
2. *Ambiguity of use*: When it is unclear what can be done, how, and with what outcome.
3. *Non-repeatability*: When interactions depend on tribal knowledge, manual workarounds, or context-specific hacks rather than consistent design.
4. *Coordination failure*: When systems must interact but cannot do so reliably because functional contracts are missing or unstable.

By solving these, Function ensures that the unit does not merely *exist*—it *engages*. It becomes a structural participant in its environment, not through symbolic presence but through designed interaction. And it does so in a way that is interpretable, governable, and usable by the systems and actors around it.

5.1.4. Historical Patterns and Structural Evolution

The role of Function in sociotechnical systems has evolved from implicit interaction to explicit interface—a shift that mirrors broader transitions from physical co-presence to digital disconnection, from centralized orchestration to federated participation. Early systems relied on proximity, routine, and fixed roles to coordinate behavior. Function was enacted socially—through desks, counters, procedures, and shared habits. It did not need to be formally designed because the surrounding structure was the interface.

As systems scaled and abstracted, especially with the rise of software and networks, Function could no longer remain implicit. It had to be externalized and encoded—first as buttons and forms, then as APIs, SDKs, microservices, and event streams. This evolution introduced powerful capabilities—but also persistent risks: fragmentation, inconsistency, opacity, and interactional overload.

Three structural eras mark the changing nature of Function:

1. **Embodied Function (Pre-Digital Systems)**

   In early institutions—banks, libraries, factories—Function was physically embodied. You walked to the desk, handed over a form, or joined a line. The Function was inseparable from the physical setting, the staff role, and the process step. There was no need to define "submit" or "confirm"—you could see what to do. Interaction was self-evident, but non-scalable.

2. **Interface-Centric Function (First Digital Systems)**

   With digitization, Function shifted into screens. Interfaces replaced clerks. Buttons replaced verbal commands. These systems were built around defined interactions—search, apply, register, pay. Function was now visible and reusable, but often rigid and siloed. Coordination across systems required integration projects, custom bridges, and point-to-point mappings. The surface was legible, but the structure underneath was fragile.

3. **Composable Function (Contemporary Systems)**

   In modern platforms, Function is increasingly modular, programmable, and decoupled. It manifests as APIs, composable workflows, automation surfaces, or even autonomous agents. This offers extreme flexibility—but also introduces new challenges: proliferation of endpoints, versioning chaos, inconsistent behavior, and functional drift. Without clear architecture, systems become interactionally incoherent despite appearing technically integrated.

   This structural history reveals a paradox: as Function becomes more powerful, it becomes *harder to govern*. What was once visible and embodied is now virtualized and distributed. Function can be triggered invisibly, composed dynamically, and deployed across contexts. The risk is not absence of interaction—but *loss of semantic control* over what those interactions mean, do, or require.

Purpose Driven Design responds to this by treating Function as an architectural construct—not a UI pattern or service hook but a structural commitment. Function must be *defined, constrained,* and *made intelligible*—not only to end users but to systems, processes, and teams. In this view, the evolution of Function is not complete—it is ongoing. The challenge is no longer to create Function, but to architect it deliberately.

5.2. Role of Function in Sociotechnical Systems

Function is not merely a surface for user interaction—it is a structural participant in how sociotechnical systems operate, coordinate, and scale. It mediates between internal capability and external demand, translating Purpose into usable form while shaping how systems behave in context. Within the unit, Function stabilizes operations by defining how roles, processes, and resources are engaged. Across systems and into the environment, Function plays an even more nuanced role: it *implements the protocols of interaction that reflect distinct relational patterns between units and between a unit and entities in its surrounding environment.*

These *relational patterns*—such as *exchange, organizational, collaborative, cooperative, compliance, investment,* and *employment*—describe not just who interacts, but *how* and *why*. Function gives these relations operable form. It defines the triggers, inputs, constraints, and flows appropriate to each type of engagement. Whether a Function processes a purchase, shares intelligence, executes a command, or submits a report to a regulator, its structure reflects the logic of the relationship it serves.

These patterns are described in more detail in the *Unit Oriented Enterprise Architecture* book, where they form the basis for understanding how systems within a Large Sociotechnical System (LSS) interact and align. Here, we focus on how Function expresses these patterns operationally—at the level of design and implementation.

Treating Function as the *interface of relationship* allows us to design more coherent units and more governable systems. In the sections that follow, we examine how Function stabilizes the unit (2.1), enables system-wide coordination (2.2), and structures engagement with external actors and entities (2.3), all while respecting the logic of the relational contract in play.

5.2.1. Internal Role: Stabilizing the Unit

Within the boundaries of a unit, Function provides structural clarity and behavioral regularity. It establishes how agents—human or machine—interact with the unit's internal capabilities in a way that is predictable, governable, and aligned with Purpose. Without Function, even the most well-defined roles and processes remain inert. Function turns potential into action by framing what can be done, by whom, and how it should be initiated and completed.

Internally, Function stabilizes the unit in three critical ways:

1. **Enabling Role Execution**

   Each role within a unit requires access to specific capabilities. Function defines how those capabilities are surfaced—whether through tools, screens, forms, APIs, or scripts. A role cannot fulfill its duties without a functional interface that makes the relevant actions accessible and usable. Well-designed Function ensures that agents encounter only the actions they are authorized and expected to perform—nothing more, nothing less.

2. **Concretizing Process Initiation and Handoffs**

   While Process defines the logic of flow, Function provides the trigger points. It defines how a process starts—via a button, a request, a schedule, a webhook—and how it moves between agents through defined interaction points. Without these triggers and transitions, the process remains conceptual. Function stabilizes execution by making it actionable.

3. **Reducing Internal Friction and Error**

   Function reduces the cognitive load and ambiguity that can lead to mistakes. It offers structure, guardrails, and feedback loops that help agents understand what actions are available, what information is required, and what the system will do in response. This minimizes coordination overhead and reduces the risk of breakdowns.

When Function is poorly formed, internal behaviors become brittle. Roles must rely on undocumented steps or workaround tools. Processes must be manually shepherded through unclear stages. Conflicts arise over timing, responsibility, and authority. In such cases, the unit may still operate—but with drag, opacity, and instability.

Function solves this by providing a stable interaction model within the unit. It does not replace Process or Structure—it activates them. It gives shape to work and enables each agent to act within a coherent, interpretable frame. This internal stability is the foundation on which external coordination can be built.

5.2.2. Systemic Role: Integration into System-Wide Coordination

In a sociotechnical system composed of multiple interacting units, Function serves as the connective layer through which coordination becomes operable. It is not enough for units to exist and hold Purpose—*they must function in ways that allow them to participate in larger workflows, federations, or platform ecosystems.* The system cannot rely on informal channels or shared memory alone; it requires *structured, interpretable, and enforceable functional surfaces* that allow units to work together without ambiguity or friction.

These surfaces are not neutral. Every functional interaction expresses a particular *relational pattern*—a kind of systemic posture that defines how a unit engages with others across boundaries. The three most prevalent relational patterns in system-wide coordination are *organizational, collaborative,* and *cooperative*:

1. *Organizational relations* involve *direction-setting and execution.* Integrative units (e.g., governance, policy, coordination units) issue functional directives—approvals, schedules, constraints, templates—while operational units implement them. Function here takes the form of *controlled interfaces*: submission portals, workflow triggers, status updates, and feedback loops that formalize top-down alignment.

2. *Collaborative relations* emerge when *multiple units jointly contribute to a shared goal.* These relations require synchronized action, mutual state awareness, and often negotiated decisions.

Function here must support *shared workspaces, co-authored objects, version control, and reconciliation mechanisms.* Coordination through collaboration depends not on hierarchy but on *functional reciprocity.*

3. *Cooperative relations* focus on *situational awareness and information sharing,* even in the absence of shared goals. One unit may surface early signals, operational metrics, or event logs to others—not to direct action, but to enhance systemic responsiveness. Function in this case supports *notifications, subscriptions, dashboards,* and *informational APIs*—tools that increase visibility without asserting control.

In each of these patterns, Function becomes the *structural expression of how coordination is expected to work.* Without it, units may intend to collaborate or comply, but find themselves blocked by incompatible formats, missing triggers, or implicit dependencies. Poorly defined Function leads to coordination breakdowns—misaligned timing, redundant processing, invisible failures, and systemic drift.

By treating Function as a contract *bound to the logic of a specific relational pattern,* we move beyond superficial integration. We begin to architect systems where interaction is not just technically enabled but structurally appropriate—each functional interface carrying not just input and output, but *relational intent.*

5.2.3. Environmental Role: Interface with External Entities

Every sociotechnical unit exists within an environment—composed of consumers, suppliers, regulators, investors, talent pools, and the broader public. Function determines how the unit becomes *legible and operable* to these external entities. It is through Function that a unit reveals not only *what it offers,* but *how it expects to interact*—what the entry points are, what the rules of engagement require, and what each party can expect in return.

Just as coordination within systems follows structural patterns, so too do external interactions. In the environment, Function expresses relational patterns such as *exchange, compliance, investment,* and *employment.* Each of these implies a different mode of exposure, a different behavioral contract, and a different set of expectations:

1. *Exchange relations* involve *transactional engagement with consumers and suppliers*. Here, Function structures offerings into flows—*browsing, filtering, selecting, configuring, submitting, paying, tracking, receiving*. These are often self-service and must prioritize clarity, trust, responsiveness, and feedback. The Function is designed not just for access but for *conversion*.

2. *Compliance relations* involve *formal, often mandatory, reporting to regulators and authorities*. Function in this case must be *constrained, auditable, and non-optional*. Interfaces for compliance—such as regulatory reports, attestations, disclosures, and automated submissions—must enforce correctness, timing, and traceability. Failure here affects legitimacy and legal standing.

3. *Investment relations* focus on *external confidence and capital flow*. Function provides access to performance dashboards, strategic roadmaps, and disclosure APIs. While these interactions may not trigger operations directly, they influence *strategic alignment and resource continuity*. Function here supports *trust-building through structured visibility*.

4. *Employment relations* govern how the unit *interacts with the labor market*. From talent portals to onboarding flows, from performance feedback interfaces to internal mobility tools, Function defines how the unit attracts, integrates, and retains agents—human or otherwise. The design of these Functions reflects the unit's *identity as an employer*.

Each of these environmental relations carries distinct *relational intent*. Treating Function as a mere access point ignores the subtleties of trust, compliance, responsibility, and long-term engagement. *What is offered, how it is requested, what metadata is attached, what responses are expected, and what follow-up is triggered*—these all vary depending on the nature of the relationship.

This is why well-formed Functions should not only be technically correct but also *relationally annotated*. When a message leaves a unit—whether a request, a report, or a response—it should carry with it an *indication of the relationship it expresses*. This allows surrounding systems to interpret it correctly, prioritize it accordingly, and respond in structurally aligned ways.

In this sense, Function is not just an interface—it is *an act of declared intent.* It is how the unit shows up in the world, not simply as a provider of capability, but as *a participant in ongoing relationships* with entities that matter.

5.3. Design Principles and Heuristics

Having explored what Function is, what roles it plays, and how it reflects structural and relational logic, we now shift to the question of *how to design it well.* This section outlines a set of principles and heuristics that support the creation of Function with architectural clarity and strategic intent. The principles focus on embedding Function meaningfully within a unit's design—ensuring it expresses Purpose, respects relational patterns, and supports operational reliability. The heuristics provide practical shortcuts, trade-offs, and design moves to reduce the risk of ambiguity, misalignment, or unnecessary complexity. Together, they guide designers toward Functions that are not only usable but structurally appropriate and evolution-ready.

5.3.1. Principles of Function-Oriented Design

Function-oriented design approaches Function not as a mere interface or implementation detail, but as *a structural mechanism for mediating interaction, expressing relational intent, and preserving compositional integrity.* A Function does not stand alone—it exists in context, governed by Purpose, shaped by the unit's role, and constrained by the kind of relationships it supports. Designing Function well means making it *structurally appropriate, intelligible,* and *governable* from the start.

The following principles support the architectural quality and operational reliability of Function:

1. **Function must express Purpose behaviorally**

   A Function should make Purpose *performative.* It transforms static intent into dynamic interaction—enabling users and systems to engage with what the unit exists to do. A unit designed to recommend should not only store or display data—it must provide a Function to *rank, match, filter,* or *propose.*

   Implication: If Purpose cannot be invoked, tested, or experienced through Function, it remains abstract.

2. **Function must communicate Value as offering**

   Function is not just a gateway—it is a *wrapper around Value*. It determines what is made accessible, under what terms, and with what constraints. Poorly designed Functions may expose data but obscure utility.

   Implication: Value should be experienced as a service or capability—not merely as access.

3. **Function must be designed as a coherent interface, not a fragmented shell**

   A unit may expose multiple entry points, but Function must present a *cohesive and recognizable surface*. Fragmented flows, ambiguous endpoints, or inconsistent naming lead to functional drift.

   Implication: Coherence makes Function composable, governable, and understandable across teams.

4. **Function must serve both human and machine agents**

   In blended sociotechnical environments, Functions are accessed by people, bots, and services. Design must account for *multiple modes of invocation*: graphical, scripted, programmatic, voice-based.

   Implication: A Function that serves only one type of agent undermines usability, automation, or both.

5. **Function must reflect relational intent**

   Each Function operates within a *specific relational pattern*—such as exchange, collaboration, or compliance. These patterns determine behavioral expectations: who initiates, what protocols apply, what trust or control is needed.

   Implication: Every Function should be relation-aware—designed to support the logic of its relational type.

6. **Function must support governance, traceability, and observability**

   As Functions shape behavior and decision-making, they become sites of accountability. They must allow for auditing, policy enforcement, and monitoring. Functions that can't be traced can't be trusted.

   Implication: Every Function is a potential compliance surface—design it accordingly.

7. **Function must evolve without breaking contracts**

   Functions change—usage expands, inputs grow, contexts shift. To prevent systemic fragility, Functions must evolve in *versioned, backward-compatible, and discoverable* ways.

   Implication: Change should extend Function, not rupture it.

8. **Function should expose only what is minimal and necessary**

   Overexposure invites misuse, overwhelms users, and complicates security. Good Functions are intentionally scoped, surfacing only the interactions that reflect the unit's Purpose and relationships.

   Implication: Design Function as an economy of interaction, not an exhaustive menu.

Together, these principles guide designers to treat Function not as a bolt-on access point, but as *a formal commitment to meaningful, structured, and sustainable interaction.* In doing so, Function becomes not only usable—but *architecturally sound.*

5.3.2. Architectural Heuristics for Function

While principles clarify what Function *should* achieve, heuristics offer practical shortcuts and patterns for how to *design it effectively.* These heuristics are not universal rules, but tested design moves that help reduce ambiguity, accelerate implementation, and avoid common pitfalls. They are especially useful when operating under real-world constraints—limited time, evolving requirements, legacy dependencies, or mixed-agent environments.

Below are eight architectural heuristics to support the design of clear, usable, and structurally appropriate Functions:

1. **Design the most typical interaction flow first**
   Start with the primary user journey or system flow—the one that occurs most often or delivers the most value. Get this path working clearly and reliably before designing edge cases or exceptions.

   Rationale: Anchoring Function in its dominant use pattern stabilizes design and exposes true complexity early.

2. **Treat each Function as a relation-bound contract**

   Clarify the relational pattern—*exchange, collaboration, compliance*, etc.—before defining the Function's inputs, outputs, triggers, and response patterns.

   Rationale: Misaligned Functions emerge when relational logic is treated as an afterthought.

3. **Design for testability, not just usability**

   Functions should be easy to simulate, mock, validate, and monitor—not just easy to click or invoke. Expose predictable inputs, observable outputs, and meaningful error states.

   Rationale: A Function that can't be tested can't be trusted.

4. **Minimize cross-functional entanglement**

   Avoid Functions that span too many responsibilities or depend on unstable external state. Split Functions that serve different relational patterns (e.g., user submission vs. system integration).

   Rationale: Entangled Functions are brittle, hard to reuse, and difficult to govern.

5. **Surface failure states deliberately**

   Design Functions to signal not only success but *structured failure*: rejections, violations, incompletions, and timeouts. Use explicit status codes or domain-specific feedback.

   Rationale: Silent or vague failures erode trust and block automation.

6. **Use relation-specific scaffolds and templates**

   Establish standard design frames for common relational patterns—for example, "submit-confirm-track" for exchange; "propose-review-align" for collaboration.

   Rationale: Structural reuse improves consistency across units and reduces onboarding time for new agents.

7. **Bundle interface with metadata**

   Expose not only the Function but also its *description, relational type, version, constraints,* and *available actions*. Include this metadata in API docs, UIs, and registries.

   Rationale: Self-describing Functions are easier to integrate, evolve, and govern.

8. **Default to narrow interfaces, then widen as needed**

   Start with the smallest viable Function—what's necessary to expose one piece of Value—and expand only as relational context demands.

   Rationale: Narrow Functions reduce coupling, lower risk, and make testing and governance easier.

These heuristics are not replacements for judgment, but tools to sharpen it. They help designers avoid accidental complexity and ensure that Function serves not just interaction—but *architecture, evolution, and coordination*.

5.4. Realization Across Implementation Planes

Function is not a purely digital phenomenon—it is a *cross-plane construct* that manifests differently in the social, physical, and digital layers of a sociotechnical system. While its contemporary expression is overwhelmingly digital, its conceptual roots and behavioral implications span multiple dimensions of design and execution. Function links Purpose to action by shaping how agents—human or nonhuman—engage with offerings, make decisions, and coordinate activities. To understand its full architectural role, we must examine how Function has historically manifested in physical and social forms, how it is now implemented and governed in the digital plane, and how it surfaces across interfaces, APIs, dashboards, and control layers that connect teams to systems. Each of these planes reveals a different aspect of Function's structure, scope, and sociotechnical entanglement.

5.4.1. Manifestation in the Social and Physical Planes

Before the advent of digital systems, Function was embodied in *rituals, routines, tools,* and *formalized roles*. In social terms, Function manifested as job responsibilities, scripts, instructions, or the tacit expectations embedded in institutions. A librarian checking out a book, a banker processing a loan, a conductor orchestrating transit schedules—each performed Functions that made a system operable. These Functions were often distributed, physically enacted, and socially negotiated.

Physically, Function took form in *machines, forms, windows, booths,* and *objects built for use*. A slot in a wall labeled "inquiries," a lever that operates a gate, a call bell at a hotel desk—each represented a bounded interface with clear expectations for use and response. The *design of space, placement of signage,* and *material configuration* all communicated Function, often without requiring digital mediation.

Even in purely analog systems, relational patterns were evident. Exchange Functions were facilitated through storefronts and counters; compliance through formal reporting structures and oversight visits; collaboration through in-person meetings and co-authored ledgers. These Functions were not less real for lacking code—they were simply enacted through *presence, ritual, and trained behavior.*

In such environments, Function was *concrete, visible,* and often *non-portable.* It relied heavily on shared norms, social context, and physical infrastructure. Its governance was informal but enforced through repetition, observation, and social expectation. Understanding these historical patterns helps us appreciate which aspects of Function are *inherent to structure,* and which are *products of digitization.*

5.4.2. Function in the Digital Plane

In the digital plane, Function becomes programmable, replicable, and distributed. It is no longer bound to a location, a role, or a ritual—it is embedded in *interfaces, APIs, workflows, bots,* and *service endpoints*. What was once physically enacted is now codified in interaction patterns: click, tap, submit, respond, trigger, sync, retry. Each digital Function is an encoded surface that makes a unit's Purpose operable at scale.

This shift brings enormous flexibility—but also demands *precision*. Unlike in social contexts, digital Functions cannot rely on tacit knowledge or environmental cues. Every interaction must be explicitly structured: inputs validated, responses handled, exceptions accounted for. In digital environments, *Function is not just possibility—it is protocol*. And that protocol must be clear enough to be interpreted not just by humans but also by machines.

In this plane, Function *binds the unit to its external relations*. Exchange Functions appear as checkout flows, payment APIs, product configurators. Collaborative Functions are implemented through shared documents, task boards, code repositories, and coordination tools. Compliance Functions surface as regulatory APIs, mandatory submissions, and automated audit logs. The structure of the Function encodes the *relational logic*—who may interact, under what conditions, with what guarantees.

Digital Functions are also *governable and observable*. They can be versioned, monitored, throttled, deprecated, and composed. They support telemetry, access control, and policy enforcement. These characteristics make them suitable for *platformization*, allowing other systems—or other units—to build upon them. The digital plane turns Function into *infrastructure*, enabling not just interaction, but systemic composability.

Yet, this composability comes at a cost: *Function must now align with both internal architecture and external expectations*. Misalignment—between what the Function exposes and what the Purpose requires, or between what the Function permits and what the relational context demands—leads to confusion, drift, or failure. The architectural challenge, then, is not merely to digitize Function, but to *digitize it faithfully*—in a way that preserves its structural role and supports meaningful interaction.

5.4.3. Implementation Surfaces and Sociotechnical Contexts

To be usable, a Function must be implemented as a surface—*an interface where an agent can invoke, engage with, or observe* what the Function enables. These surfaces are not just technical; they are sociotechnical. They mediate between structured capability and the agent's context, role, and expectations. A Function becomes actionable only when it is meaningfully exposed to those who need it—whether they are humans, machines, or hybrid agents.

Typical Function surfaces include

- *APIs*, which expose Function to machines, agents, and automation layers. These support scripting, integration, chaining, and orchestration—but only if they are well-defined, versioned, and constrained. APIs serve developers, platform teams, automation pipelines, and compliance tooling.
- *User Interfaces (UIs)*, which expose Function to human agents. Portals, submission forms, visual editors, and task panels frame Function through layouts, labels, and constraints. UIs are shaped by usability, accessibility, and domain fit—and serve users across frontline roles, analytical teams, and internal services.
- *Conversational interfaces*, such as chatbots or voice assistants, which expose Function in *natural language form*. These require semantic precision, intent recognition, and fallback design. They serve not only users but also support agents, advisors, and service coordinators.
- *IoT and physical interfaces*, including sensors, devices, and service robots, which engage Functions based on *environmental stimuli or programmed routines*. These may invoke Functions via machine-to-machine protocols, actuator signals, or embedded workflows—and serve nonhuman agents operating autonomously or cooperatively.
- *Catalogs and registries*, which make Functions *discoverable and interpretable*. Whether embedded in data catalogs, service registries, or marketplace-style platforms, these help agents understand what a Function does, when to use it, and how it connects to Purpose.

Each of these surfaces serves a different class of agent—but all require architectural care. A Function must expose only what is structurally appropriate, reflect its relational logic, and avoid creating ambiguity or overreach. Interfaces should not flatten complexity or conceal intent—they should *clarify scope, usage, and limits*.

While some Functions may also emit metrics or logs that are captured by *shared observability systems*, these are *not Functions themselves*—they are *reflections of Function behavior*. Exposure to dashboards or telemetry platforms must be designed intentionally, without violating the unit's autonomy or enabling external control. Visibility is not governance; it is perspective.

Above all, Function surfaces must tie back to *roles, teams, and processes.* A Function that is technically sound but misaligned with the workflow of its intended users fails to deliver value. A Function that is used successfully by one team but misused by another introduces operational risk. Implementation surfaces must therefore respect both the architecture of the system *and* the social structure around it.

Function is not confined to a single surface, syntax, or mode of interaction. It travels across planes—embodied in physical action, encoded in digital interfaces, and interpreted through social roles. Each implementation surface makes Function legible to a different kind of agent. The architectural challenge is not just to implement Function, but to ensure it *aligns with Purpose, respects relational logic, and fits into the broader sociotechnical system* in which it operates.

5.5. Strategic Impact of Function Interfaces

5.5.1. Nuances of Interaction

Not all Function interfaces are created equal. Even when exposing similar capabilities, their structure, surface, and relational context can vary dramatically—shaping how they are used, trusted, and evolved. These nuances are not superficial. They reflect underlying decisions about *autonomy, coupling, relational purpose,* and *organizational structure.* In this section, we examine the key architectural dimensions that give Function interfaces their distinct strategic shape.

Interface Shape and Function Type

A Function may manifest as a *product,* a *service,* or a *self-service interface*—each implying different expectations:

- *Product interfaces* offer packaged, stable outputs. They are typically stateless and expectation-driven—for example, a currency conversion tool or a risk rating model. Consumers do not need to specify how the Function works; it encapsulates expertise or logic in a ready-made form.
- *Service interfaces* require the consumer to participate or specify. This may take the form of providing parameters (e.g., onboarding preferences, constraints) or simply making oneself available as the

subject of the service (e.g., haircut, consultation, diagnostic exam). The Function adapts based on this engagement—often involving interaction over time, role-awareness, and contextual cues.

- Self-service interfaces aim to reduce reliance on others. They empower agents to initiate, modify, or retrieve without direct coordination. Their success depends on the *clarity* of boundaries and the *predictability* of outcomes.

Each type demands a different balance of affordances, constraints, and failure modes. Designing a self-service interface as if it were a service can introduce ambiguity and dependency. Designing a product interface without clarity on inputs and outputs risks misuse.

Relational Dynamics and Function Intent

Function interfaces are shaped by the *type of relation* they enable:

- *Exchange Functions* involve give-and-take: Submitting requests, receiving outputs, making payments, accessing services.
- *Collaborative Functions* support *joint action*: Co-creation, mutual validation, or shared decision-making. These require alignment, identity awareness, and sequencing.
- *Cooperative Functions* support *situational alignment without shared goals*. Data sharing, alerts, or observational telemetry are common examples. These require *interpretability* and *decoupled intent*.
- *Compliance Functions* formalize obligations. Their interfaces are highly structured, with strict validation, audit trails, and predictable schedules.

The relational logic determines not just *what* the Function does but *how it must behave*. A Function meant for exchange but designed like a compliance submission can create undue friction. A cooperative Function disguised as collaboration can lead to misaligned expectations.

Organizational Embedding: Integrative vs. Operational Interfaces

Functions are also shaped by the organizational layer they serve:

- *Integrative units* define direction, policy, and boundaries. Their Functions expose constraint-setting capabilities—configurations, limits, rule definitions. These interfaces *govern behavior.*
- *Operational units* perform actions and report outcomes. Their Functions involve execution, measurement, exception handling, and feedback. These interfaces *reflect outcomes.*

When these roles are blurred—for example, when an operational unit designs its own constraints without integrative oversight—Functions may become brittle or misaligned. Conversely, when integrative Functions demand excessive coupling, they may restrict autonomy.

Functions should express not only *technical role,* but also *organizational posture.*

Patterns of Interaction: Timing, Trust, and Surface Behavior

Other key nuances include

- *Symmetry*: Is the Function uni-directional (submit and done) or bi-directional (negotiate, confirm, adjust)?
- *Timing*: Does it require synchronous response or operate asynchronously?
- *Friction*: Is the Function designed for smooth flow or intentional pause (e.g., confirmation prompts)?
- *Trust cues*: Does it signal success and failure clearly? Are constraints visible or implicit?

These subtle choices shape how users engage. A Function that fails silently or responds unpredictably undermines trust. A Function that overloads optionality or delays feedback introduces unnecessary cognitive burden.

By-Products and Strategic Reuse

Most Functions emit *by-products*: logs, audit trails, intermediate states, analytic signals. These are not always the Function's purpose—but they become *strategic assets* when

- Shared with other units to support *cooperation*
- Exposed as *data products* to external consumers
- Used to support *governance, observability*, or *insight generation*

A Function that emits clean, interpretable by-products multiplies its architectural value. One that generates opaque or inaccessible trails loses opportunities for reuse, alignment, and evolution.

Function interfaces are never just endpoints. They are *statements of structure, intent, and relation.* The subtle choices embedded in their design—about direction, timing, audience, and by-products—shape not just how they operate but how they are perceived, trusted, and reused. Architectural awareness of these nuances is critical for any system that aims to scale its Functions across agents, domains, and evolutionary phases.

5.5.2. Measurable Benefits and Architectural Payoffs

Well-designed Functions do more than perform—they clarify, align, and scale. They reduce ambiguity, enable autonomy, and make systems more accountable to their Purpose. The benefits of such Functions are not vague; they can be *measured, observed,* and *accumulated* across the system. Their architectural payoff lies in both *local clarity* and *systemic coherence.*

A. **Functional Clarity Reduces Coordination Overhead**

   When a Function is clearly exposed—through an interface that makes its Purpose, inputs, and expected outcomes unambiguous—fewer meetings are needed, fewer misunderstandings arise, and fewer intermediaries are required. This leads to

   - Faster onboarding of new users and agents
   - Lower friction in cross-unit collaboration
   - Fewer escalations and mismatches

Payoff: Clarity in Function reduces the need for meta-coordination. Agents can engage directly, confidently, and correctly.

B. **Functional Autonomy Enables Parallel Progress**

Functions that are *self-contained and discoverable* allow units and agents to move independently. Teams can build on one another's Functions without waiting for internal approvals or manual handoffs.

- Product development accelerates through composability.
- Operations scale horizontally without entanglement.
- New offerings emerge from reusing existing Functions.

Payoff: Clear Functions become reusable building blocks—internally and externally.

C. **Functional Observability Supports Feedback and Trust**

Functions that expose their outcomes and emit meaningful by-products create a *feedback loop*—supporting improvement, accountability, and trust.

- Analysts can observe Function usage and quality.
- Errors can be traced to specific invocations or patterns.
- Governance teams can verify compliance with minimal intrusion.

Payoff: Functions that "report themselves" reduce manual auditing, enable observability, and make platforms more trustworthy.

D. **Functional Encapsulation Protects Structural Integrity**

Well-bounded Functions make internal changes possible without breaking external behavior. Teams can evolve logic, refactor internals, or optimize performance—*as long as the Function contract remains stable.*

- Enables technical debt management without service disruption
- Encourages domain ownership and architectural resilience
- Prevents accidental coupling across layers

Payoff: Encapsulated Functions enable change without chaos.

E. **Functional Surfaces Reflect Organizational Maturity**

A system's Function interfaces reflect its structural alignment. If Functions are unclear, inconsistent, or overly entangled, it often signals

- Poorly defined Purpose
- Unclear accountability
- Lack of relational awareness

Conversely, a system with coherent, role-aligned, composable Functions reflects

- Intentional architecture
- Healthy sociotechnical interaction patterns
- Design maturity

Payoff: Function becomes a diagnostic surface. You can assess the system's health by observing how well it exposes what it does.

The architectural value of Function lies not only in what it enables, but in *how cleanly, clearly, and coherently* it is exposed. Measurable benefits include reduced coordination costs, accelerated development, increased trust, greater autonomy, and enhanced resilience. When Function is treated as a first-class design concern—not just as a backend detail or front-end affordance—it becomes a strategic lever for building systems that scale *structurally, socially,* and *evolutionarily.*

5.6. Implementation Patterns

Functions become real when they are made available for interaction. From an architectural standpoint, this availability is expressed through a set of implementation patterns—recurrent ways in which Functions are surfaced, constrained, and maintained in relation to external actors and systems. These patterns do not describe internal behavior; they define how Functions are *encountered, engaged with, and interpreted* by the outside world.

While these patterns may be implemented via technical protocols, they are fundamentally structural. Each one implies a particular kind of expectation, dependency, and relational context.

5.6.1. Connection Surfaces

Some Functions are exposed as stable, standardized surfaces meant for establishing interaction. These connection surfaces serve as the formal points of contact—typically well-documented, version-controlled, and explicitly bound to a protocol or access mechanism. They are discoverable by design and are structured to manage compatibility, visibility, and trust. Connection surfaces are often implemented through APIs, secure upload endpoints, or integration connectors, and they define the boundary across which others may formally engage the unit's Function.

5.6.2. Interaction Signals

Other Functions manifest as signals: they do not wait to be invoked but instead emit status, intent, or readiness. These one-way communications support cooperation, observability, and asynchronous coordination. They allow other systems or agents to infer the current behavior or condition of the unit without direct interaction. Implementation forms include event broadcasts, heartbeat indicators, or state change messages. Though passive in nature, interaction signals are *architecturally active*—they shape how other units synchronize or defer action.

5.6.3. Conditional Interfaces

Not all Functions are universally available. Some are exposed only under specific conditions: to particular roles, during defined time windows, or after prerequisite interactions. Conditional interfaces enforce access policies at the surface level, shaping not just what can be done but when and by whom. These may be implemented via authentication gates, role-based workflows, or contextual feature flags. Conditionality makes Functions safer, but also more complex—requiring architectural clarity around state, authority, and scope.

5.6.4. Exchange Channels

Many Functions exist to facilitate structured exchange—of data, payments, services, or decisions. These exchange channels are interaction patterns that establish submission, processing, delivery, and acknowledgment flows. They often require clear contracts, robust error handling, and traceability. Whether realized through batch pipelines, form submissions, transactional APIs, or payment interfaces, exchange channels must balance structure and flexibility: they define not only *what* is transferred but *how reliably and reciprocally* it occurs.

5.6.5. Delivery Points

Some Functions are not about submission or exchange, but direct consumption. These are the surfaces where agents—human or machine—receive the output of a Function or engage with its offering in real time. Delivery points include user interfaces, dashboards, mobile apps, and voice interfaces. Their design is less about integration logic and more about *cognitive clarity* and *task completion*. Good delivery points hide complexity and make Function frictionless. Poor ones obscure capability and degrade trust.

5.6.6. Feedback Interfaces

An often-overlooked Function pattern is the ability to receive input *after* the main interaction occurs. Feedback interfaces allow consumers, partners, or internal agents to influence future behavior—by reporting satisfaction, surfacing anomalies, or suggesting changes. These inputs may come via explicit channels (e.g., ratings, surveys) or implicit ones (e.g., telemetry, usage traces). Architecturally, the feedback path must be *connected* to the Function's life cycle—if a Function cannot learn or adapt, feedback becomes noise.

5.6.7. Resilience Protocols

Finally, some Functions are designed not just for ordinary use but for failure states. Resilience protocols expose surfaces for recovery, redirection, or degradation when things go wrong. These interfaces handle exception conditions—providing fallback options, triggering compensation, or rerouting responsibility. They may be implemented as retriable endpoints, error queues, or contingency APIs. Their presence makes Functions trustworthy not just when things work but when they don't.

Each of these patterns reflects a strategic decision: *how to make the Function interactable in a way that matches its Purpose, aligns with its Structure, and respects its relational context.* Recognizing and designing for these patterns ensures that Functions are not only implemented, but *made meaningful*—visible, trustworthy, and usable in the sociotechnical landscape they inhabit.

5.7. Interlocking Dependencies and Alignment

Function does not stand alone. In the Purpose Driven Design framework, it is a structurally interdependent construct—activated at the surface of the unit but always tethered to deeper architectural commitments. The effectiveness of any Function depends not merely on its technical correctness or usability but on its structural *alignment* with adjacent design elements: Purpose, Value, Process, Structure, Order and Culture, Memory, and Storage.

Each of these constructs informs or constrains what the Function can—and should—do. Without these anchor points, Function risks becoming decoupled from the unit's identity, drifting toward incoherence, misuse, or overexposure.

5.7.1. Alignment with Purpose: Justification and Boundaries

Purpose defines why the unit exists. Function exists to make that Purpose accessible—translating intent into action. If a Function cannot be traced back to Purpose, it constitutes architectural drift. The Function may appear legitimate at the interface level, but it lacks a structural mandate. Purpose determines what Functions are *allowable*—not in terms of syntax or code, but in terms of *teleological legitimacy*.

5.7.2. Alignment with Value: Meaningful Exposure

Function is the gateway to Value. It must expose Value in legible, bounded, and consumable ways. A Function that presents capability without Value becomes hollow—technically operational but semantically void. This misalignment often leads to interfaces that are overcomplicated, misunderstood, or misused. The Value construct ensures that Function carries something worthwhile across the interaction boundary.

5.7.3. Alignment with Process: Behavioral Grounding

Function is not execution itself—it is *entry into* execution. That execution is defined by Process: the transformation of inputs into outcomes. If Function lacks a corresponding Process, it will fail at invocation. Conversely, if Process changes but the Function remains unadjusted, the contract is broken. The two must evolve in synchrony. Together, they ensure that what is exposed can be enacted.

5.7.4. Alignment with Structure: Responsibility and Operability

No Function can be sustained without agents responsible for it. Structure provides that accountability: who owns the Function, who maintains it, who is called into action when it misbehaves. If a Function is designed without alignment to Structure—whether human teams or automated systems—then responsibility is diffuse or absent, and reliability is undermined.

5.7.5. Alignment with Order and Culture: Normative Framing

Even the most technically sound Function is not free from social constraint. Order and Culture define how Functions are invoked, who may access them, what priorities are honored, and what behaviors are acceptable. Escalation paths, compliance requirements, fairness guarantees—all of these live in the normative framing. Function must operate within this space, not above it.

5.7.6. Alignment with Memory: Traceability and Learning

A Function that cannot be remembered cannot be governed. Memory provides logs, outcomes, decisions, and failure records—connecting each invocation to a larger pattern. Through Memory, Function becomes learnable and improvable. Without it, the Function exists only in the present tense, unable to support reflection, accountability, or adaptation.

5.7.7. Alignment with Storage: Durability and Continuity

Some Functions interact with persistent assets—receiving, creating, or modifying data or Value that must endure. Storage ensures that these interactions are governed by appropriate logic, constraints, and access rules. Misalignment here leads to data fragmentation, accidental exposure, or breakdowns in the continuity of Value delivery.

In sum, Function is not just an exposed capability—it is a structurally integrated expression of the unit's identity. It is where Purpose is made actionable, Value is delivered, Process is triggered, and Structure is held accountable—within the bounds of Order, enriched by Memory, and anchored by Storage.

A Function that fails to respect these dependencies may still be technically operable—but it will not be architecturally valid.

5.8. Failure Modes and Antipatterns

Function may be well-intended, even technically sound—yet still fail architecturally. These failures rarely originate at the level of code or interface alone; they are often consequences of structural misalignment. When Purpose is unclear, Value is obscured, Process is unstable, or Structure is absent, Function becomes brittle, misleading, or unusable.

This section identifies and defines recurring *antipatterns* in Function design, offering a diagnostic lens grounded in architectural thinking.

1. **Latent Function**

   - *Definition*: A Function exists within the system but is not externally discoverable or consumable.
   - *Symptoms*: The Function is implemented but inaccessible; usage is near zero; consumers replicate it unknowingly.
   - *Consequence*: Duplicated effort, untapped potential, architectural drift.
   - *Architectural Fault*: Misalignment with Structure (no responsible surface for exposure), or with Purpose (Function exists without audience clarity).

2. **Overexposed Function**

    - *Definition*: A Function is too widely available or underspecified, leading to unpredictable or excessive usage.
    - *Symptoms*: Downstream overload, accidental misuse, security gaps, resistance to change due to external dependency.
    - *Consequence*: Fragility, coupling, ungovernable surface area.
    - *Architectural Fault*: Weak Order constraints, insufficient Structure, and lack of scoping through Purpose.

3. **Misaligned Function**

    - *Definition*: The Function operates technically, but diverges from the unit's Purpose or the expectations of consumers.
    - *Symptoms*: Stakeholder confusion, manual compensations, Value not delivered despite usage.
    - *Consequence*: Erosion of trust, hidden workarounds, mounting design debt.
    - *Architectural Fault*: Conflict between Purpose, Process, and Value. Often introduced under pressure.

4. **Ambiguous Function**

    - *Definition*: The Function's behavior, inputs, or expected outputs are unclear to the consumer.
    - *Symptoms*: Hesitation to use, frequent questions, incorrect assumptions, support escalations.
    - *Consequence*: Underuse, misuse, or distrust.
    - *Architectural Fault*: Failure to align with Order and Culture (missing naming conventions, affordances, documentation norms).

5. **Duplicated Function**

    - *Definition*: The same or similar Function is implemented independently in multiple units.
    - *Symptoms*: Inconsistent behavior, confused consumers, fragmented feedback, unclear ownership.
    - *Consequence*: Redundancy, incoherence, elevated maintenance overhead.
    - *Architectural Fault*: Breakdown in Structure and lack of shared Memory (e.g., no canonical Function registry).

6. **Phantom Function**

    - *Definition*: A Function is documented or exposed but does not exist or fails silently.
    - *Symptoms*: Catalog entries with 404s, references to deprecated endpoints, "zombie" UI elements.
    - *Consequence*: Frustration, false trust, disillusionment.
    - *Architectural Fault*: Misalignment between Memory (historical state) and current Storage or deployment reality.

7. **Incoherent Function**

    - *Definition*: A Function is implemented without clear ties to any design construct—it exists for reasons no longer understood.
    - *Symptoms*: Nobody can explain why it exists; rarely used, but politically sensitive to remove.
    - *Consequence*: Design stagnation, institutional inertia.
    - *Architectural Fault*: Detachment from Purpose and Structure, likely a legacy artifact.

8. **Overcoupled Function**

    - *Definition*: The Function is tightly bound to other Functions, processes, or services, making it difficult to evolve independently.
    - *Symptoms*: Fear of breaking changes, blocked refactors, unscalable behavior under stress.

- *Consequence*: Paralysis, system brittleness.
- *Architectural Fault*: Poor Process modularity, weak Structure boundaries, or excessive implicit dependencies.

9. **Unbounded Function**
   - *Definition*: The Function tries to handle too many roles, serve too many consumers, or adapt to too many use cases.
   - *Symptoms*: Sprawling interfaces, tangled code paths, inconsistent behavior across inputs.
   - *Consequence*: Maintenance complexity, unclear user experience, escalating risk.
   - *Architectural Fault*: Ambiguous Purpose, unresolved stakeholder conflicts, avoidance of structural decisions.

Each of these antipatterns reveals not just a weak Function but a deeper failure of alignment. Function is not simply what a unit *does*—it is how the unit *engages* the world. When engagement becomes incoherent, excessive, or invisible, the Function fails not as an implementation, but as a design construct.

Resisting these antipatterns requires more than cleanup. It requires reconnecting Function to its architectural foundations—reasserting Purpose, simplifying Process, clarifying Structure, and reestablishing the legitimacy of its presence in the system.

5.9. Summary: Function As a Design Construct

Within Purpose Driven Design, *Function* is the system's exposed mode of engagement—its interface to the world. While Purpose defines *why* a unit exists, Function determines *how* that Purpose is made tangible to others. It is the system's actionable presence: the structured, governable surface through which agents and systems interact with the Value the unit produces.

Function is not merely capability or implementation. It is the designed *expression* of intent, Value, and coordination—constrained by structure, contextualized by process, and legitimized by its alignment with Purpose. It forms the contact boundary where promises are made, transactions initiated, and relationships enacted.

Realized effectively, Function operates as

- *A structural interface,* shaping how external agents perceive, access, and engage with the system's capabilities
- *A behavioral contract,* ensuring that each invocation reflects internal intent and delivers expected Value with coherence and traceability
- *A coordination entry point,* where systems initiate interaction and align behavior—via protocols, roles, and interface surfaces
- *A mechanism of trust,* reinforcing the system's reliability, learnability, and continuity across diverse contexts and consumers
- *A link between constructs,* activating Process, exposing Value, enabling Memory, and requiring Structure for ownership and evolution

Function is where design becomes legible. It manifests architecture to the outside world—not in diagrams or blueprints but in action. A well-formed Function expresses Purpose; a broken Function betrays it. When neglected, Function becomes incoherent, unsafe, or invisible. When treated as a first-class construct—bounded, aligned, and situated in its sociotechnical context—it enables systems to deliver on their promises with clarity, integrity, and adaptability.

In an age of dynamic interfaces, autonomous agents, and persistent interaction, Function is no longer a passive endpoint of design. It is the *forefront*—the structure that mediates intention and reality, behavior and accountability, offering and reception. A unit's Function is its most public truth.

CHAPTER 6

Process

In postwar Japan, the Toyota Motor Corporation faced a problem that could not be solved with speed or volume. The company was undercapitalized, operating in a recovering economy, and could not afford the luxury of waste. Every mistake—every defect, delay, or inefficiency—carried disproportionate consequences. What Toyota needed was not just productivity. It needed *precision, adaptability, and responsiveness.* And from this constraint emerged one of the most influential ideas in modern production: *the Andon cord.*

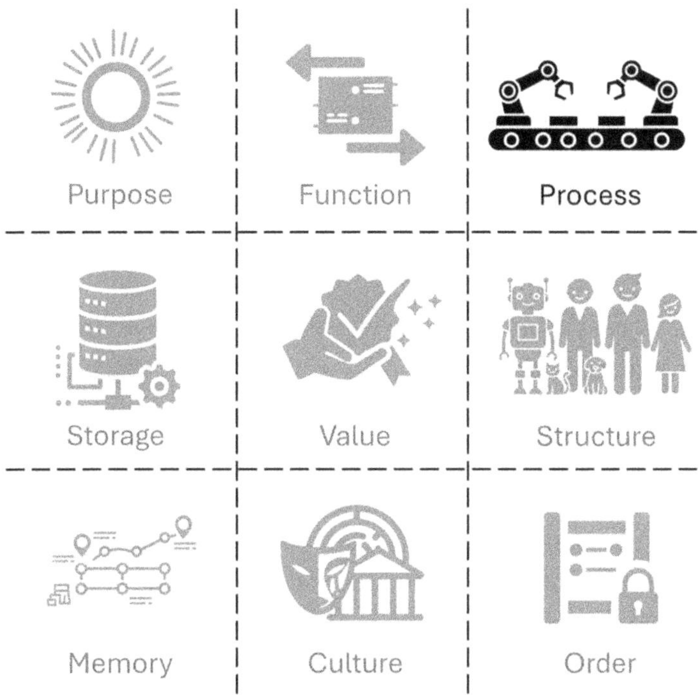

A. Milchman, *The Purpose Driven Design Framework for Modern Enterprises*,
https://doi.org/10.1007/979-8-8688-2223-0_6

Introduced in the 1950s, the Andon cord was a simple overhead pull-line that allowed any worker on the assembly line to stop the entire production process. At first glance, the idea seemed reckless. Why give line workers the power to halt the entire factory? But the purpose was not to slow production—it was to make *quality and awareness everyone's responsibility*. If a worker noticed a misaligned component, a tool malfunction, or any inconsistency, they were not expected to wait for inspection. They were empowered to act.

The Andon cord was part of a larger system that came to be known as the Toyota Production System (TPS)—a structured, principle-based approach to Process that emphasized *flow, visibility, and intervention*. Every task was part of a sequence. Every sequence had a reason. And every reason could be questioned, refined, or improved. What mattered was not just that the process worked, but that it could *reveal its own failures* and adapt in real time.

In other domains, the same structural logic applies. In a hospital, a nurse may pause a procedure to flag a safety concern. In customer service, an agent may escalate a flawed script that confuses clients. In software, a release engineer may halt a rollout after detecting unexpected user behavior. Each is a human-triggered intervention—a modern Andon moment—designed to expose issues early, invite accountability, and restore flow through correction rather than concealment. The principle is universal: when systems make deviations visible and empower participants to respond, Process becomes adaptive rather than fragile.

This philosophy transformed how organizations understood Process. No longer just a series of steps, Process became an *embedded system of feedback and alignment*. Every element—timing, materials, roles, decisions—was part of a designed structure. And critically, the process was not "owned" by management or automation systems. It was *distributed*. Each person participating in the process had both the knowledge and the agency to influence it.

The long-term impact of this approach is visible far beyond automotive manufacturing. The principles of *lean production, continuous improvement, and systems thinking* all trace their lineage to the Andon cord. In digital systems, the equivalent is the error log, the retry loop, the escalation alert. In healthcare, it's the surgical checklist or the nurse's power to escalate. In software, it's CI/CD pipelines that stop deployment when a test fails. In each case, the Process is not just a pipeline—it is a *responsive, accountable structure* that can detect deviation and support correction without hierarchy or delay.

Toyota's insight was not to make processes more efficient—it was to make them *more intelligent*. That intelligence was not centralized; it was designed into the process itself. And it began with one deceptively simple gesture: the pull of a cord.

Process is how Function is sustained over time. It defines not just what is done, but how, when, by whom, and under what conditions. A unit without a defined Process may still act, but it cannot adapt, coordinate, or scale. Process transforms Function into reproducible behavior and makes purposeful activity observable, governable, and improvable. In this chapter, Process is treated as a structured, evolvable construct—designed not just for flow but for interruption, feedback, and learning.

6.1. Foundations of Process As a Design Construct

Process is where architecture becomes motion. If Purpose gives direction and Function opens the doorway of interaction, Process provides the path that carries work forward. It is the operational logic of a unit—the flows, rules, and sequences that transform intention into repeatable outcomes. Unlike Function, which is outward-facing, Process is primarily inward-facing, concerned less with how a unit is discovered than with how it sustains continuity once engaged. This relationship is illustrated in Figure 6-1.

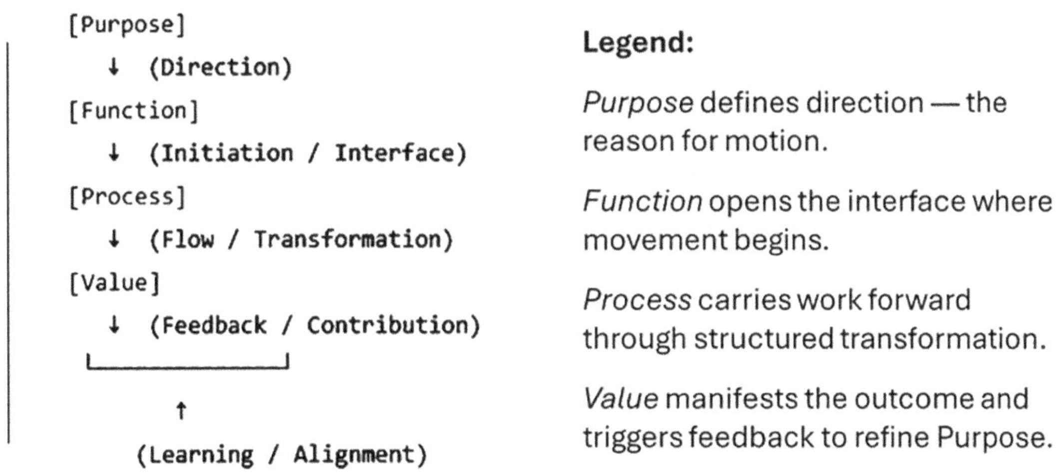

Figure 6-1. *The architecture of motion—showing how Purpose defines direction, Function initiates interaction, Process carries transformation, and Value closes the loop through feedback and alignment*

To design Process is to decide not only what happens but in what order, under what conditions, and with what safeguards. It is the structural choreography of execution: visible in workflows, embedded in routines, and encoded in systems. When well-formed, Process transforms potential into reliability; when neglected, it leaves units brittle, inconsistent, and opaque. The discussion that follows clarifies the conceptual and structural rationale for Process, treating it not as a by-product of tools or habits but as a design construct in its own right—an architecture of flow that makes Purpose durable and Function trustworthy.

6.1.1. Defining Process Architecturally

In Purpose Driven Design, Process is not an incidental workflow or a collection of operational steps. It is an architectonic construct: the structured logic by which a unit transforms inputs into outcomes, ensuring that Purpose is not only declared and exposed through Function but enacted consistently over time. Process provides the repeatable flow, decision logic, and sequencing that make Value creation durable. In a broken process, decisions depend on memory and goodwill, producing inconsistency and delay; in a well-architected process, rules, triggers, and roles create a self-sustaining flow where work moves predictably from intent to outcome.

Process is not reducible to automation scripts or checklists. It is the architecture of movement—what makes execution coherent, governable, and traceable.

Architecturally, Process defines how a unit *does its work*. It clarifies

- *What must occur, in what order, and under what conditions.*
- *Who or what is responsible for each stage of execution.*
- *How exceptions, variations, and adaptations are handled without collapse.*
- *Where accountability is preserved, and how the flow can be inspected, audited, or improved.*

While Function opens the door to interaction, Process carries the request across a coherent pathway. A unit without Process may still expose an attractive Function, but beneath the surface it will struggle—failing to deliver reliably, scaling inconsistently, or producing outcomes that cannot be traced back to intent.

A well-formed Process exhibits the following architectural properties:

- *Internally coherent*: Activities follow a deliberate order, minimizing ambiguity and dependency sprawl. *Example*: In onboarding, tasks progress logically from identity verification to access provisioning, avoiding circular approvals.
- *Repeatable and governable*: The same request produces predictable outcomes, with defined guardrails for variation. *Example:* In procurement, identical purchase requests yield consistent approvals regardless of requester or department.
- *Transparent and traceable*: Flow can be inspected, audited, and learned from—errors and decisions leave structural evidence. *Example:* In financial reporting, each adjustment links to its origin and justification, enabling review without speculation.
- *Resilient under pressure*: Designed with recovery, fallback, and exception handling that preserve continuity. *Example:* In order fulfillment, delayed shipments automatically trigger substitute routing rather than halting operations.
- *Aligned with Purpose*: Every stage exists to fulfill the unit's declared reason for being—superfluous or contradictory steps are excluded. *Example:* In compliance, reviews focus solely on risk and regulation, not convenience or preference.
- *Supportive of Function and Structure*: Processes enact what Functions expose and allocate responsibilities in line with Structure. *Example:* In customer support, escalation follows clear tiers that mirror the organizational structure and defined Functions.

In this sense, Process is the *choreography of the unit.* It is what converts Purpose into lived execution and what ensures that Function is more than a facade. When designed architecturally, Process does not simply describe how work is done—it ensures that work remains legible, accountable, and evolution-ready.

6.1.2. Why Process Exists

Process exists because work, once initiated, must flow. Purpose defines why a unit exists, and Function makes that existence available to others—but without Process, these constructs remain inert. A unit cannot sustain itself on entry points alone; it must carry requests through structured execution. Process ensures that Value is not delivered by chance but by design: repeatable, governable, and resilient to variation.

Architecturally, Process addresses three design necessities:

- *Continuity*: It links actions together, creating a reliable path from initiation to completion.
- *Consistency*: It stabilizes outcomes, reducing the risk of ad hoc interpretation, dependency fragility, or one-off workarounds.
- *Accountability*: It encodes decision points and handoffs so they can be traced, governed, and improved.

Without Process, a unit becomes fragile. Activities may occur, but they do so irregularly, depending on memory, improvisation, or informal negotiation. This leads to hidden dependencies, silent failures, and outcomes that cannot be reproduced. In sociotechnical terms, the absence of Process produces opacity: agents cannot see how work moves, stakeholders cannot trust that work will finish, and systems cannot adapt because no structural logic exists to calibrate change. The organizational symptoms are visible—missed deadlines, unclear ownership, repeated errors—but they are merely surface manifestations of a deeper architectural disorder.

Process also exists to *absorb complexity*. Real environments are variable—inputs arrive late, conditions shift, exceptions arise. Process provides the architectural mechanism by which such variation is handled without undermining continuity. It does not eliminate uncertainty but channels it into designed points of adaptation, ensuring that deviation does not escalate into systemic collapse.

Finally, Process enables *learning*. Because it creates observable flows, it generates the history that feeds Memory and informs structural improvement. At Toyota, this principle evolved into kaizen—continuous refinement driven by data from visible work, not intuition. A unit with well-formed Process can look back, diagnose patterns, and evolve responsibly. A unit without Process can only react—adapting through improvisation rather than design.

In this sense, Process exists not simply to "get things done" but to make execution sustainable. It bridges Purpose and Function with a structural pathway that is repeatable, inspectable, and aligned. Where Function makes a promise, Process fulfills it.

6.1.3. What Problem Process Solves

Process solves the problem of incoherent execution in a world where little can be accomplished by one agent. Modern work requires division of responsibilities, but division introduces dependency. Dependency demands coordination. Coordination requires communication. And communication depends on reliable connectivity. Without a designed Process, this chain frays—leaving activity scattered, accountability unclear, and outcomes unpredictable.

Architecturally, the absence of Process produces four systemic risks:

- *Fragmentation*: Tasks drift into local improvisation. Each agent executes differently, making outcomes inconsistent and often incompatible.
- *Coordination failure*: Without defined flows and handoffs, dependencies are missed. Work piles up at unclear boundaries, creating bottlenecks, duplication, or outright collapse.
- *Opacity*: When steps are undocumented or untraceable, errors cannot be located, and accountability dissipates. Stakeholders lose trust because outcomes appear accidental rather than intentional.
- *Reactive adaptation*: Lacking a baseline of structured flow, change becomes improvisational. Every adjustment is disruptive because no one knows what the "normal" path is supposed to be.

In practice, this means that units without Process cannot scale beyond small, tightly coupled groups. Trust depends on personal memory rather than structural traceability. Communication compensates for missing design—endless status meetings, escalation emails, or side-channel negotiations take the place of stable pathways.

Process introduces the structural logic that prevents these failures. It aligns divided responsibilities into coordinated flow, ensuring that each role knows not only *what* to do but *when, with whom*, and *under what conditions*. It provides the choreography of connectivity, turning communication from a source of fragility into a backbone of continuity.

In this sense, Process solves the deepest coordination problem of sociotechnical systems: *how to move together without constant supervision.* It is the architecture that allows distributed agents—human, digital, or hybrid—to act independently while still producing a coherent outcome.

6.1.4. Historical Patterns and Structural Evolution

The history of Process is not a straight line but a dialectic between two enduring modes of coordination: command-and-control and stigmergy. Each has shaped human systems, often in tension, sometimes in combination.

Command-and-control excels where reliability, safety, or legal accountability dominate—assembly lines, aviation, and clinical procedures. Stigmergy, by contrast, emerges where coordination is unplanned and signals accumulate through shared environments—visible in open-source communities, emergency response improvisation, or even the unstructured flow of email. Yet unlike social species that use stigmergy as a primary mode of coordination, most human enterprises experience it only incidentally. In modern systems, stigmergy is most valuable when new or unknown processes are forming: it reveals natural pathways of coordination that, once understood, can be visualized, stabilized, and formalized into deliberate Process.

The command-and-control lineage has dominated formal organizations for centuries. Borrowed from military hierarchy, it promises discipline, predictability, and visible accountability. Its industrial incarnation was the assembly line: a process so precisely sequenced that it could transform labor into throughput with astonishing efficiency. Production workflows, aviation checklists, pharmaceutical protocols, and surgical procedures all carry this heritage. In these contexts, the strength of command lies in reducing error under high stakes. Reliability is purchased through standardization, and deviation is treated as risk. Where safety and volume are paramount, the model works.

Alongside this, however, has always been the stigmergic lineage. Social animals build, forage, and adapt without central command, leaving signals in their environment—pheromone trails, movement cues, structural traces—that guide collective behavior. Francis Heylighen calls stigmergy a *Universal Coordination Mechanism*, and its resilience is evident across species. Humans, too, follow this path: desire lines cut across manicured campuses, email threads bypass rigid workflow engines, open-source projects grow through decentralized contributions. Coordination emerges not because someone designed the flow, but because participants left interpretable traces that others followed.

Why then, given the evolutionary success of stigmergy, did humans institutionalize command? Path dependence provides the answer. Early states, armies, and bureaucracies required visible hierarchies to levy taxes, wage wars, and enforce laws. Once codified, this model became self-reinforcing—embedded in law, management science, and organizational culture. Even as environments shifted, the design instinct remained: build processes as rulebooks, codify them in workflow engines, enforce them through BPM systems. These efforts succeed only where tasks are rule-bound and conditions stable. Elsewhere, people revert to stigmergy—falling back on email, chat, or improvisation to coordinate.

Modern processes, then, are neither purely designed nor purely emergent. They are hybrids, combining the rigidity of rulebooks with the fluidity of desire paths. Medicine offers a telling example: surgical teams follow strict checklists, yet also rely on informal cues and emergent collaboration to respond when cases deviate from protocol. The same duality appears in logistics, emergency response, and creative industries: processes exist on paper, but real coordination is lived through signals, improvisations, and adaptive flows.

The architectural lesson is that Process has never evolved linearly. It oscillates between the designed and the emergent, between the imposed and the discovered. This dialectic makes Process the most complex and sophisticated construct of Purpose Driven Design. Unlike Purpose or Function, it cannot be reduced to a single lineage or model. Its history shows that Process is not simply about codifying flow—it is about negotiating the boundary between what must be structured and what must be allowed to emerge. Designing Process, therefore, is less about choosing one tradition over the other and more about orchestrating their coexistence. In that coexistence lies both the difficulty and the promise of Process as an architectural construct.

6.2. Role of Process in Sociotechnical Systems

Process holds a paradoxical place in sociotechnical systems. It is the choreography that makes execution possible, yet it is often hidden from view. Most of the time, consumers, regulators, and investors never see a unit's Process directly; they encounter only its Value and its Function. In product and by-product scenarios, Process remains encapsulated, shaping outcomes without presenting itself as an interface. Yet in services and experiences—where delivery unfolds in time, often with the customer present—Process becomes visible, even constitutive of what is being offered. A haircut, a hospital admission, a technical support call, or a museum visit is experienced not just as an outcome but as a process lived alongside the provider. This duality complicates the role of Process. It is at once internal—stabilizing coordination within the unit—and sometimes external, exposed as a surface of engagement. Understanding this dual role is key: Process is the least outward-facing of the design constructs, yet in certain contexts it becomes the very fabric of the interaction. Its role must therefore be interpreted dialectically: mostly hidden, sometimes revealed, always decisive. To see this dialectic in practice, Table 6-1 contrasts internal and inter-unit expressions of Process.

Table 6-1. *Intra-Unit vs. Inter-Unit Processes: Contrasting Internal Discipline with Boundary-Spanning Coordination*

Dimension	Intra-unit process	Inter-unit process
Primary role	Maintains internal discipline, efficiency, and accountability within a unit	Enables coordination, alignment, and shared adaptation across units
Dominant mode	Command-and-control: clearly defined roles, rules, and checkpoints	Stigmergic: coordination emerges through shared signals, artifacts, or environments
Focus	Execution, control, and optimization of known work	Negotiation, synchronization, and learning across boundaries
Visibility	Mostly internal; observed through performance metrics or exceptions	Externally visible to multiple parties through shared tools, tickets, or communication threads
Governance	Defined by internal procedures, hierarchies, or schedules	Defined by mutual standards, service protocols, or informal signaling
Examples	Code review, cost reconciliation, documentation upkeep	Onboarding workflow, incident response, order processing

6.2.1. Internal Role: Stabilizing the Unit

Within a unit, Process is the silent architecture of *continuity*. It aligns divided responsibilities into a coherent flow, ensuring that what begins with Purpose and becomes accessible through Function does not disintegrate in execution. Without Process, work fragments into improvisation; outcomes become unreliable; accountability dissolves. With it, the unit gains rhythm, repeatability, and trust in its own operations.

The stabilizing role of Process can be seen in three interlocking ways:

- *Coherence*: Process integrates roles, activities, and resources into an intelligible sequence. Each actor—human or artificial—knows when to act, with what input, and to what effect. Coherence prevents duplication, omission, and drift.
- *Continuity*: Process creates the expectation that outcomes will be delivered in the same way today as yesterday, regardless of which agent performs the task. This continuity anchors the unit, even as people rotate, technologies evolve, or conditions fluctuate.
- *Accountability*: Process encodes decision points and handoffs, producing traces that can be inspected, audited, and improved. It makes the invisible visible—not to external consumers but to the unit itself.

The dialectical nature of Process is evident here: while its logic is designed, its stability often depends on emergent adaptation. People improvise when exceptions arise, technologies evolve to automate steps, and informal shortcuts—desire paths—appear alongside formal flows. Far from undermining Process, these adaptations often reinforce it, providing evidence of where rigidity fails and where flexibility must be introduced.

In this sense, the internal role of Process is not to eliminate improvisation but to hold it within bounds—to ensure that emergence feeds back into structure rather than eroding it. By stabilizing the unit through coherence, continuity, and accountability, Process makes it possible for Purpose to be realized consistently and for Function to remain trustworthy.

6.2.2. Systemic Role: Integration into System-Wide Coordination

In system-wide coordination, Process recedes further into the background. Units primarily interact through Purpose, Function, and Value; their internal flows remain encapsulated. Unlike Function, which must expose an interface, or Value, which must surface an offering, Process does its work behind the scenes. To the broader system, it is almost always invisible.

This invisibility is by design. Processes stabilize execution within a unit but rarely serve as mechanisms of integration between units. When one unit requests something from another, the expectation is framed through Function and justified through Value. What happens inside—how the work is sequenced, routed, or resolved—is immaterial to the requesting unit, so long as the outcome arrives as promised.

There are, however, limited exceptions. When services cross organizational boundaries, or when experiences unfold across multiple units, elements of Process must align to avoid visible friction. A patient admitted through an emergency room, a logistics shipment handed from carrier to carrier, or a financial transaction passing through clearing houses all require that internal processes interlock just enough to preserve continuity across boundaries. In these cases, systemic coordination depends on partial exposure of Process—interfaces that synchronize timing, validate transitions, or confirm completion.

Still, these are the minority. In most contexts, the systemic role of Process is not to integrate directly but to remain encapsulated, enabling each unit to coordinate externally without revealing its internal machinery. This encapsulation is a strength: it preserves autonomy, reduces coupling, and allows units to evolve their processes without destabilizing the wider system.

The dialectic, however, persists. Where environments demand seamless continuity, processes are pressured to interlock. Where autonomy and resilience matter more, they remain opaque. The architectural role of Process at the systemic level is therefore limited but not negligible: it is the silent negotiator of continuity across boundaries, revealed only when exposure is unavoidable.

6.2.3. Environmental Role: Interface with External Entities

If Process is usually encapsulated in system-to-system coordination, it is far more visible in the environment. Customers and consumers often encounter not just the outcome of a unit but the flow by which that outcome is delivered. Services and experiences are structured as lived processes—engagements in which the consumer participates or observes execution directly. A haircut, a medical appointment, a classroom session, or a customer support interaction is not merely about the end result; it is defined by the sequence, timing, and quality of the process itself.

This exposure makes Process a critical site of meaning and trust. A poorly designed product process may remain hidden if the final object is sound, but a poorly designed service process is felt immediately. Long wait times, confusing handoffs, or opaque sequencing undermines trust just as effectively as defects in the final outcome. Conversely, a well-crafted process—transparent, efficient, and participatory—can itself become part of the Value, differentiating an otherwise commoditized offering.

Regulators and authorities also press Process into the open. For compliance in healthcare, finance, aviation, or pharmaceuticals, it is not enough to demonstrate that a product works; units must show that the process by which it was created, tested, or delivered meets defined standards. Here, exposure of Process is not for consumer experience but for legitimacy and accountability.

The environmental role of Process, then, is paradoxical. For products and by-products, it remains concealed, a silent guarantor of reliability. But for services, experiences, and regulated domains, Process becomes the Value itself or the condition of trust. No other construct of Purpose Driven Design is so double-edged: mostly hidden, yet sometimes inseparable from what the environment perceives. This duality makes Process uniquely challenging to design—because its visibility is not optional but context-dependent.

6.3. Design Principles and Heuristics

Designing Process is less about inventing steps than about architecting flow. The challenge is not simply to write procedures or map workflows but to articulate the structural principles that make any process coherent, resilient, and aligned with Purpose. Because Process oscillates between the designed and the emergent, its principles must balance structure with adaptability, rule with responsiveness, choreography with improvisation. Heuristics, in turn, provide practical guidance—shortcuts and tested patterns that help reduce error without extinguishing flexibility. Together, principles and heuristics shape Process not as a static diagram but as a living design surface: a way of channeling effort so that units remain legible, accountable, and capable of moving in rhythm with their environment.

6.3.1. Principles of Process-Oriented Design

Process-oriented design treats flows not as incidental sequences but as structural commitments. Every step in a process is an architectural decision: about how work is initiated, how it moves, how it adapts, and how it concludes. Poorly designed processes create opacity, fragility, and mistrust; well-designed ones generate continuity, accountability, and trust in outcomes. Because Process oscillates between designed structure and emergent adaptation, its principles must embrace both—embedding enough order to sustain coherence, while leaving space for adjustment and evolution.

The following principles guide the architectural quality of Process:

1. **Process must convert Purpose into continuity**

   A Process is the bridge from intent to sustained execution. It transforms Purpose from a guiding statement into a repeatable sequence of action.

   Implication: If Purpose cannot be enacted through a coherent Process, it remains aspirational rather than operational.

2. **Process must preserve accountability across flow**

   Every process should encode who or what acts, in what order, and under what conditions. Without such clarity, responsibility diffuses.

   Implication: If decisions and handoffs cannot be traced, outcomes cannot be governed or improved.

3. **Process must coordinate divided responsibilities**

Modern work depends on role separation. Process provides the logic that binds division into cooperation—sequencing steps, managing dependencies, and creating rhythm.

Implication: Without Process, coordination collapses into endless meetings, improvisation, or failure at the boundaries.

4. **Process must balance designed order with emergent adaptation**

   Rigid flows fracture under changing conditions, while unstructured improvisation collapses into chaos. Processes must define stable pathways while leaving room for adjustment.

   Implication: A process that cannot adapt is brittle; one that cannot hold structure is incoherent.

   Example: Telehealth triage emerged informally during the pandemic as clinicians improvised remote assessments, then evolved into standardized care pathways—turning spontaneous coordination into durable Process.

5. **Process must integrate human and technical agents**

   Work today is carried out by blended actors—people, machines, and AI agents. Processes must allocate roles across them coherently, respecting constraints and strengths.

   Implication: Processes designed for only one class of agent will misalign, creating fragility at integration points.

6. **Process must create legibility for learning**

   Flows generate signals—timestamps, logs, outcomes—that feed Memory. A well-designed process ensures these traces are observable and usable for reflection and redesign.

   Implication: If a process leaves no structural evidence of its execution, it cannot evolve responsibly.

7. **Process must minimize hidden dependencies**

Unseen linkages between steps or roles create systemic fragility. Processes should modularize flow so that one failure does not cascade uncontrollably.

Implication: A process riddled with invisible ties undermines both reliability and trust.

8. **Process must evolve without disrupting continuity**

   Change is inevitable: Steps will be automated, reordered, or removed. Process must accommodate revision while still ensuring stable delivery of Value.

   Implication: If evolution breaks continuity, the unit cannot remain coherent over time.

6.3.2. Architectural Heuristics for Process

Whereas principles articulate what constitutes sound process-oriented design, architectural heuristics offer practical, situation-sensitive guidance for structuring processes under real-world constraints. These heuristics reflect accumulated design insight—tendencies that, while not universally prescriptive, generally yield clarity, resilience, and maintainability when applied appropriately. The following heuristics support the implementation of process structures that are modular, traceable, and fit for dynamic sociotechnical environments.

1. **Anchor the process in its dominant flow**

   Stabilize the most common path before adding exceptions or edge cases.

   Rationale: Predictability in the typical flow reduces fragility when complexity increases.

2. **Adapt design to the nature of the process**

   Human-centric flows need clarity of roles and feedback; integration-centric flows need robust interfaces; case management thrives on branching flexibility; straight-through processing (STP) depends on automation with exception channels.

Rationale: One-size-fits-all processes create brittleness when applied across different modes.

3. **Separate coordination logic from transformation logic**

   Process flow should decide what happens and in what order; individual steps should define how.

   Rationale: Separation simplifies testing, reuse, and reasoning about flow.

4. **Prefer declarative structures over imperative scripts**

   Model processes using DAGs, state machines, or workflow DSLs where possible.

   Rationale: Declarative models improve transparency, traceability, and reasoning about dependencies.

5. **Design for idempotence and retryability**

   Steps should tolerate re-execution without harm.

   Rationale: Distributed environments rarely guarantee "exactly once"; resilience requires safe retries.

6. **Decompose long-running flows into durable segments**

   Insert checkpoints and well-defined handoffs.

   Rationale: Segmentation improves recovery, reduces risk of cascading failure, and supports incremental optimization.

7. **Make handoffs explicit and observable**

   Every transfer should be visible, timestamped, and owned.

   Rationale: Silent handoffs cause ownership confusion and lost accountability.

8. **Minimize temporal and structural coupling**

   Favor event-based triggers, abstraction, and delegation over fixed timing or hard-coded assumptions.

   Rationale: Decoupling improves flexibility and reduces systemic fragility.

9. **Design escalation as a first-class construct**

   Define thresholds, alternate flows, and human interventions before failure occurs.

   Rationale: Processes collapse when escalation is improvised instead of architected.

10. **Surface feedback early and often**

    Expose metrics, errors, and progress signals near their origin.

    Rationale: Timely feedback reduces latent failures and supports trust in Function.

11. **Continuously align formal flow with desire paths**

    Observe how people actually coordinate (email, chat, side channels) and absorb viable patterns into formalized processes.

    Rationale: Ignoring emergent stigmergy creates brittle structures; integrating it strengthens adoption.

12. **Start informal, stabilize through formalization**

    Use stigmergic or lightweight coordination when uncertainty is high, then codify patterns once behavior stabilizes.

    Rationale: Premature rigidity locks in fragility; formalization should follow proven practice.

Principles define the architectural necessities of Process—what must hold true for flows to remain coherent, accountable, and aligned with Purpose. Heuristics provide the practical moves that make those necessities real under constraint: separating logic, designing for retries, modeling declaratively, absorbing desire paths, and formalizing only when stability emerges. Together they frame Process not as a static diagram but as a living system of coordination, always negotiating between designed order and emergent adaptation. This dual perspective is essential: without principles, Process devolves into ad hoc improvisation; without heuristics, it becomes rigid, brittle, and detached from practice. The strength of Process in sociotechnical design lies precisely in its ability to hold both—an architecture of flow that is at once principled and pragmatic, structured yet evolution-ready.

6.4. Realization Across Implementation Planes

Process is never abstract; it must be enacted. How it is realized depends on the plane in which it operates. In the social plane, processes appear as rituals, routines, and role enactments. In the physical plane, they are embodied in production lines, logistics chains, and procedural manuals. In the digital plane, they take the form of workflows, orchestration engines, state machines, and automation scripts. Each plane reveals different strengths and vulnerabilities: social processes adapt quickly but may be fragile; physical processes anchor reliability but resist change; digital processes scale rapidly but risk opacity and overcomplexity. These distinctions become clearer when examined across the three planes in which Process manifests, as shown in Table 6-2.

Table 6-2. *Manifestations of Process Across the Social, Physical, and Digital Planes*

Plane	Manifestation of process	Characteristics	Examples
Social plane	Rituals, routines, and role enactments	Human-centered; adaptive through judgment and shared norms	Daily stand-ups, review meetings, emergency drills
Physical plane	Production lines, logistics chains, procedural manuals	Tangible; sequenced through material flow and spatial arrangement	Assembly line, hospital triage, airport ground operations
Digital plane	Workflows, orchestration engines, automation scripts	Programmable; scalable and repeatable through logic and code	CI/CD pipelines, data-processing workflows, CRM automation

To design Process architecturally requires seeing across these planes—recognizing that every flow in practice is sociotechnical, and that coherence depends on balancing the affordances and constraints of each domain.

6.4.1. Manifestation in the Social and Physical Planes

Long before workflows were drawn in BPM software, processes were lived socially and enacted physically. The earliest forms of Process appeared as rituals and routines—shared understandings of "what comes next" embedded in human interaction. A courtroom hearing, a wedding ceremony, or a religious rite each unfolds through an agreed sequence of steps, collectively upheld and recognizable across time. In these cases, Process was not documented in diagrams but carried in memory, reinforced by repetition, and stabilized through shared legitimacy.

In the physical plane, Process manifested as production workflows, transport routines, and operational checklists. The industrial assembly line remains the most iconic example: a sequence of steps physically arranged to transform raw material into finished product with speed and consistency. Factories, warehouses, and logistics chains all extend this pattern—relying on physical ordering of tasks, tools, and people to minimize error and maximize throughput. Aviation, medicine, and nuclear operations similarly use checklists and procedural manuals to guarantee that critical flows are followed step by step, especially when safety is paramount.

The strength of these manifestations lies in their tangibility. Steps are visible, roles embodied, and outcomes traceable. Yet they can also become rigid, resistant to adaptation when conditions shift. Over time, informal shortcuts emerge—what urbanists call *desire paths*. Workers create more efficient routines, pilots develop cockpit shorthand, and medical teams improvise responses to unexpected complications. These adaptations show that even in the most rule-bound domains, social and physical processes are never purely scripted; they are negotiated in practice.

For architects, the lesson is twofold: processes in the social and physical planes demonstrate both the power of sequencing to produce reliability and the inevitability of emergence when reality deviates from design. They remind us that Process is not only a digital construct but an embodied choreography of human and material action—anchoring continuity in ways that remain indispensable even in a digital age.

6.4.2. Process in the Digital Plane

With digitization, Process left the factory floor and entered the domain of code. What was once embodied in rituals or embedded in assembly lines became formalized as workflows, orchestration engines, and automation scripts. The digital plane made processes programmable, composable, and scalable—able to span geographies, operate continuously, and integrate diverse agents. Yet this shift also introduced new risks: what was once visible and embodied could now become opaque, fragmented, and brittle.

Digital workflows offer significant strengths. They can be versioned, monitored, and replayed; their handoffs are encoded in APIs rather than physical spaces; and their steps can be automated for speed and cost reduction. Enterprise workflow systems, BPM engines, and cloud orchestration platforms allow units to connect processes across services and teams with unprecedented reach. The promise of "straight-through processing" in finance or automated fulfillment in logistics depends on this digital capacity for coordination at scale.

But the risks are equally clear. As processes proliferate in scripts, APIs, and workflow engines, visibility often erodes. Steps become hidden in configuration files, exceptions are buried in logs, and dependencies multiply across services. The result is fragility: a change in one system cascades unpredictably across others. Digitalization does not eliminate the need for Process design; it amplifies both its importance and its difficulty.

Here the dialectic between designed and emergent coordination resurfaces. Much work in digital environments still flows through stigmergic channels—emails, chats, ticket systems, informal signals—because formal workflows rarely capture the full range of lived coordination. But unlike in the physical world, digital environments leave traces: logs of conversations, histories of workarounds, patterns of repeated improvisation. These traces open the possibility of *Directed Emergence*: the systematic observation of desire paths and their translation into formalized, resilient processes.

Directed Emergence closes the loop between emergent practice and designed structure. Instead of treating informal coordination as noise, digital platforms can detect, validate, and institutionalize it—turning repeated improvisations into stable workflows. In this way, the digital plane not only scales Process but also enables its continuous evolution, balancing rigidity with adaptability. It is here that Process reveals its true sophistication: not as a frozen diagram of flow but as a living choreography that learns from its own use.

6.4.3. Implementation Surfaces and Sociotechnical Contexts

Processes only become usable when they are surfaced—exposed in a way that agents can engage, monitor, or intervene. These surfaces are not neutral interfaces; they are sociotechnical touchpoints where design decisions meet human and machine activity. The form a process takes at its point of exposure strongly shapes how it is understood, trusted, and enacted.

Implementation surfaces vary widely. Workflow engines provide dashboards where tasks are sequenced, deadlines are shown, and exceptions are raised. Runbooks and procedural manuals capture operational steps in human-readable form, often for environments where automation is incomplete or infeasible. Ticketing systems and collaboration tools embed processes into the rhythms of team communication, making status visible and handoffs explicit. APIs and orchestration layers expose process logic to machines, enabling automation, chaining, and integration at scale. Even physical artifacts—kanban boards, hospital charts, or assembly floor signals—serve as process surfaces, situating coordination in a form that agents can read and act upon.

Each surface carries trade-offs. Dashboards improve visibility but can overwhelm with signals. Runbooks preserve institutional knowledge but risk becoming stale. APIs scale automation but conceal decision logic. Informal tools like email or chat lower friction but undermine accountability. The architecture of Process is therefore inseparable from the sociotechnical context of its surface: who must use it, under what constraints, and with what expectations.

For architects, the key is to design surfaces that make Process both legible and governable. A process that is invisible cannot be trusted; a process that is overexposed becomes noise. The balance lies in ensuring that what is surfaced enables the right agents—human or machine—to understand their role, act appropriately, and leave behind traces that feed back into accountability and learning. When this balance fails—when handoffs vanish in email chains or escalation paths collapse into ambiguity—Process loses its integrity as an interface. Process, in this sense, is never just a flow of tasks. It is a designed interface between structure and practice, rendered visible through the surfaces where agents meet it.

6.5. Strategic Impact of Process Interfaces

The strategic impact of Process lies not merely in its ability to "get things done" but in how the flow of work shapes trust, efficiency, and adaptability across sociotechnical systems. Purpose may orient, Function may expose, and Value may justify—but it is Process that determines whether outcomes arrive reliably, legibly, and at the right rhythm. The nuance is that flows are never neutral: how work is sequenced, where handoffs occur, and how variation is absorbed has consequences for scalability, resilience, and legitimacy. Designed flows provide continuity; emergent flows provide adaptability. The interplay between the two—sometimes reinforcing, sometimes colliding—reveals Process as an architecture of movement that can stabilize or destabilize entire systems. Understanding its strategic impact therefore requires looking not only at outputs but at the dynamics of flow itself.

6.5.1. Nuances of Flow

Flows are not neutral conduits. The way work moves through a unit carries meaning, establishes trust, and sets the tempo of coordination. Two processes may produce the same output, yet one is experienced as smooth, transparent, and reliable, while the other feels chaotic, opaque, and fragile. For example, two support teams may resolve customer issues within identical time frames, but one keeps clients informed at each stage while the other communicates only at closure—the difference in flow rhythm determines whether trust is reinforced or eroded. The nuance of flow lies in how timing, rhythm, and sequencing are designed and perceived—and in how designed order interacts with emergent adaptation.

Rhythm and Pace

Some processes move with the cadence of routine: payroll cycles, manufacturing runs, patient rounds. Others must respond in real time: fraud detection, emergency dispatch, crisis management. The rhythm of flow determines expectations of reliability, urgency, and slack. Misalignment between designed rhythm and actual need produces friction, delay, or overload.

Timing and Handoffs

Handoffs are the most fragile moments in any process. The timing of transitions—too soon, too late, or poorly signaled—creates bottlenecks or gaps. Smooth processes choreograph handoffs explicitly, making ownership and sequence clear. Chaotic ones leave transitions implicit, forcing agents to improvise.

Trust and Transparency

Flows that are legible inspire trust: participants can see progress, anticipate next steps, and confirm completion. Opaque flows generate anxiety—work vanishes into black boxes, and outcomes arrive unpredictably. Trust in Process is built not only by outcomes but by the visibility of flow along the way.

Adaptability and Slack

Rigid processes may operate efficiently under stable conditions but fail dramatically under stress. Emergent flexibility—agents deviating intelligently, improvising, or rerouting—often saves the system. Yet unmanaged improvisation can erode coherence. Well-formed processes build in slack: room for exceptions, capacity to absorb variation, and channels to convert desire paths into formal adaptations.

The Dialectic of Design and Emergence

Perhaps the deepest nuance of flow is its dialectical nature. Designed sequences provide predictability; emergent adaptations provide resilience. Neither is sufficient alone. Too much design suffocates adaptability; too much emergence collapses into noise. Strategic Process design means orchestrating both—creating flows that are reliable enough to trust and flexible enough to evolve.

These nuances reveal why Process has disproportionate strategic impact. It is not just about completing tasks; it is about how movement is structured, how rhythm is felt, and how adaptability is absorbed. In the nuance of flow, entire systems find—or lose—their coherence.

6.5.2. Measurable Benefits and Architectural Payoffs

Well-designed Processes do more than sequence tasks—they stabilize execution, preserve accountability, and enable scale. They reduce friction, transform improvisation into continuity, and create conditions where learning and adaptation can occur without collapse. The benefits of such Processes are not abstract; they can be observed, measured, and accumulated. Their architectural payoff lies in both local reliability and systemic resilience.

A. **Process Clarity Reduces Coordination Overhead**

When a Process makes roles, steps, and handoffs explicit, agents spend less time negotiating what to do and when. Ambiguity is replaced by rhythm, freeing energy for value creation.

- Faster onboarding of new team members
- Reduced need for ad hoc status meetings or escalation
- Lower incidence of duplicated or dropped tasks

Payoff: Clear Process reduces the cost of coordination, allowing agents to move in rhythm rather than in confusion.

B. **Process Continuity Builds Trust**

Reliable flows produce predictable outcomes. When participants can depend on consistent sequencing, they extend trust not only to individuals but to the system as a whole.

- Stakeholders experience fewer surprises and delays.
- Consumers gain confidence in delivery timelines.
- Regulators see evidence of procedural integrity.

Payoff: Continuity transforms fragile improvisation into dependable expectation, reinforcing legitimacy.

C. **Process Traceability Enables Accountability and Learning**

Processes that leave observable traces—logs, checkpoints, audit trails—make it possible to locate errors, improve flows, and evolve responsibly.

- Clear diagnosis of where failures occur
- Faster iteration and process improvement
- Stronger compliance posture through auditability

Payoff: Traceability turns Process into a learning mechanism, not just an execution path.

D. **Process Modularity Enhances Resilience**

When flows are segmented into coherent modules, failures remain contained, and evolution becomes manageable.

- Localized faults do not cascade system-wide.
- Teams can optimize individual segments independently.
- Processes can evolve incrementally without halting continuity.

Payoff: Modularity preserves both stability and adaptability, making units more evolution-ready.

E. **Process Design Reflects Organizational Maturity**

The quality of a unit's processes reveals its architectural health. Where processes are opaque, inconsistent, or brittle, underlying structures are weak. Where processes are coherent, transparent, and resilient, structural integrity is evident.

- Process quality becomes a diagnostic of system health.
- Visible flows demonstrate clarity of purpose and structure.
- Strong processes reinforce cultural legitimacy as much as technical reliability.

Payoff: Process serves as a mirror of maturity, showing whether the unit is structurally viable or held together by improvisation.

The architectural value of Process lies not only in efficient execution but in its ability to create trust, continuity, and learning across sociotechnical systems. Measured in reduced overhead, increased resilience, and stronger legitimacy, Process is not just an operational tool—it is an evolutionary asset.

6.6. Implementation Patterns

Though every unit has its own flows, certain structural patterns recur across domains and technologies. These patterns provide reusable frames for how work is sequenced, coordinated, and adapted. They are not recipes but archetypes—forms of Process that capture recurring logics of execution. Understanding them helps designers select the right shape for the right context, avoiding the twin risks of overdesign and underdesign.

6.6.1. Canonical Patterns

Pipeline

A linear flow where outputs of one step become inputs to the next. Pipelines excel at transformations that can be broken into discrete, ordered stages—such as data ingestion, manufacturing assembly, or software build chains.

Strength: Simplicity and clarity of sequencing

Risk: Fragility when exceptions or variations require deviation from the straight line

Workflow

A structured network of tasks, often branching and conditional, designed to accommodate multiple roles and pathways. Workflows dominate in human-centric processes (e.g., case handling, approval chains, service delivery).

Strength: Flexibility to model diverse scenarios

Risk: Complexity can balloon, making flows opaque or unmanageable

State Machine

A process defined by states and transitions, emphasizing conditions rather than linear steps. State machines are common in integration-centric and automated systems, where explicit rules determine movement between defined statuses (e.g., "pending," "approved," "settled").

Strength: Precision and clarity of allowed transitions

Risk: Overformalization that becomes brittle if states do not match lived reality

Event-Driven Process

Flows triggered by signals rather than pre-defined sequencing. Event-driven processes are prevalent in distributed systems and adaptive environments (e.g., fraud alerts, supply chain disruptions, IoT monitoring).

Strength: High adaptability and responsiveness to changing conditions

Risk: Potential loss of coherence if events proliferate without clear orchestration

Case Management

A semi-structured flow oriented around a central case or object (e.g., a patient record, an insurance claim, a legal file). Instead of fixed sequences, case management provides milestones, tasks, and rules while allowing agents to decide how best to progress.

Strength: Flexibility and agent empowerment

Risk: Ambiguity and inconsistency if rules for progression are weakly defined

Straight-Through Processing (STP)

An automated flow where tasks are executed without manual intervention from start to finish. Common in finance and logistics, STP seeks maximum efficiency and minimal delay.

Strength: Speed and cost reduction at scale

Risk: Vulnerability to silent failure, with little room for improvisation if exceptions are not well designed

6.6.2. Stigmergic Patterns

Not all processes are designed. Many arise from stigmergy—coordination through signals left in the environment that others can follow. These emergent flows are rarely captured in diagrams, but they are pervasive in practice. They illustrate the other grammar of Process: one shaped by traces, cues, and adaptations rather than central design. Though often informal, stigmergic processes can be highly effective, especially under conditions of uncertainty or scale.

Desire Path Process

When agents repeatedly deviate from official procedures, a new path emerges—whether through email, chat, or improvised shortcuts.

Strength: Surfaces the most natural or efficient flow

Risk: Remains untraceable and unguided if not observed and formalized

Signal-and-Response Process

Flows coordinated by cues in the environment—alerts, dashboards, kanban boards, shared documents—where the signal itself directs the next action.

Strength: Decentralized, scalable, and low-overhead

Risk: Ambiguity or overload can turn signals into noise, reducing coherence

Collaborative Swarm Process

Multiple agents acting in parallel, guided by shared context rather than sequential rules. Common in open-source projects, emergency response, or fast-moving creative teams.

Strength: Extreme adaptability, especially when priorities are shifting or unknown

Risk: Hard to measure, govern, or replicate without eventually codifying emergent practices

These stigmergic patterns do not replace designed flows; they coexist with them. Often, the healthiest systems are hybrids: designed pipelines or workflows provide stability, while stigmergic coordination absorbs variation. The mechanism of Directed Emergence links the two, allowing repeated improvisations to be recognized, tested, and formalized into sustainable processes. In this way, stigmergy is not a threat to architecture but a source of renewal, ensuring that Process evolves in rhythm with lived reality. This interplay becomes clearer when examined through the canonical and stigmergic patterns summarized in Table 6-3.

Table 6-3. *Canonical and Stigmergic Process Patterns—Their Contexts, Strengths, and Risks*

Pattern	Best used when/context	Primary strength	Primary risk
Pipeline	Tasks can be decomposed into discrete, ordered stages (manufacturing, data ingestion).	Simplicity, clarity of sequencing	Brittle when exceptions require deviation
Workflow	Multi-role, conditional human coordination (case handling, approvals)	Flexibility, explicit branching logic	Complexity can obscure flow.
State machine	Rule-based transitions between defined states (payment, authorization)	Precision of control logic	Overformalization that mismatches reality
Event-driven	Environments driven by asynchronous signals (monitoring, fraud alerts)	High adaptability, responsiveness	Loss of coherence if events multiply unchecked
Case management	Semi-structured work around a persistent case or object (patient, claim)	Agent autonomy, contextual handling	Ambiguity and inconsistency
Straight-through processing (STP)	Fully automatable, high-volume, low-variance flows (finance, logistics)	Speed and efficiency	Silent failure when exceptions occur
Desire path	Informal shortcuts repeatedly chosen by actors	Reveals natural flow	Untraceable until formalized
Signal-and-response	Coordination via shared cues (kanban, shared documents)	Decentralized scalability	Signal overload or misinterpretation
Collaborative swarm	Parallel creative or emergency activity under shifting priorities	Extreme adaptability	Hard to govern or replicate

Taken together, these patterns reveal the full spectrum of Process forms—from rigidly linear pipelines to loosely adaptive stigmergic flows. No single form suffices on its own; most units rely on hybrids. A healthcare provider might combine surgical checklists (workflow), patient records (case management), and automated billing (STP), while also depending on signals and improvisation during emergencies. A logistics chain may operate as a pipeline under normal conditions, yet shift into event-driven coordination when disruptions occur. Traditional BPM has been largely task-based, emphasizing predefined steps and handoffs. Stigmergic coordination, by contrast, is event-based: it thrives on signals, cues, and traces left in the environment. For architects, the challenge is not to impose one lineage but to compose across them—treating designed tasks and emergent events as complementary grammars. Process design provides the syntax that weaves them into coherent, trustworthy flow.

6.7. Interlocking Dependencies and Alignment

Process does not stand alone. In the Purpose Driven Design framework, it is structurally interdependent—reliant on other constructs to orient its flow, while producing outcomes that those constructs stabilize and preserve. A Process is not valid simply because it runs; it is valid when it aligns with the surrounding architecture: Purpose, Function, Value, Structure, Order, Culture, Memory, and Storage.

6.7.1. Alignment with Purpose: Direction and Coherence

Purpose defines why the unit exists; Process defines how that Purpose is enacted over time. Without alignment to Purpose, a process risks becoming activity without meaning—efficient motion with no legitimate direction. Purpose ensures that processes are not just busywork, but structured action tethered to the reason the unit exists.

6.7.2. Alignment with Function: Invocation and Fulfillment

Function is the entry point that triggers a process. A Function without a corresponding Process is hollow—promising an outcome it cannot deliver. Conversely, a Process without a Function may remain idle, inaccessible, or uninvoked. The two must evolve together: what Function exposes, Process must fulfill.

6.7.3. Alignment with Value: Continuity of Offering

Value is the justification for why a unit contributes. Process is how that Value is reliably produced. If processes are opaque or inconsistent, Value delivery becomes accidental rather than dependable. When aligned, Value and Process reinforce one another: Value provides justification; Process provides continuity.

6.7.4. Alignment with Structure: Roles and Responsibilities

Structure assigns the agents—human or artificial—who execute process steps. Without structural alignment, no one knows who owns which handoff, or who is accountable for exceptions. Misalignment here produces bottlenecks, idle roles, or shadow processes. Alignment ensures that every stage has an accountable agent.

6.7.5. Alignment with Order and Culture: Constraint and Climate

Processes are shaped by the rules and norms of their environment. In production-style processes, Order dominates: strict checklists, escalation paths, compliance requirements. In stigmergic workflows, Culture takes precedence: shared norms, mutual awareness, and informal trust enable coordination. Both influence Process profoundly; both must be respected.

6.7.6. Alignment with Memory: Traceability and Learning

Process generates traces—logs, histories, metrics—that Memory captures. Without Memory, a process cannot be audited, improved, or evolved; it remains trapped in the present tense. Alignment ensures that every execution leaves structural evidence, enabling governance and learning over time.

6.7.7. Alignment with Storage: Durability and Preservation

Many processes culminate in artifacts, records, or assets that must endure. Storage guarantees that what Process produces is not lost but preserved, governed, and made ready for reuse. Misalignment leads to orphaned outputs, fragmented data, or loss of continuity.

In sum, Process is not just motion—it is structured flow, tethered to Purpose, activated by Function, stabilized by Structure, constrained by Order, guided by Culture, traceable through Memory, and anchored in Storage. Where these align, Process is coherent and evolution-ready. Where they misalign, Process becomes mere activity—motion without architecture.

6.8. Failure Modes and Antipatterns

Process may be well-intentioned, even technically sound, yet still fail architecturally. These failures are rarely the result of one flawed step; they emerge when coordination mechanisms are misaligned with Purpose, Function, Structure, Order, Culture, or environment. What appears at first to be a "broken workflow" is often a recurring pattern of dysfunction—a process design that looks plausible in isolation but introduces fragility, inefficiency, or incoherence at scale. Recognizing these antipatterns allows architects to intervene at the structural level, not merely patch symptoms.

1. **Latent Process**
    - *Definition*: A process is designed but never meaningfully invoked.
    - *Symptoms*: Flow exists in documentation or tooling but is bypassed in practice.
    - *Consequence*: Duplication of effort, shadow coordination, loss of trust in formal design.
    - *Architectural Fault*: Misalignment with Function—no clear entry point activates the flow.

2. **Overengineered Process**
    - *Definition*: A process is burdened with unnecessary steps, checks, or approvals.
    - *Symptoms*: Long cycle times, excessive bureaucracy, disengaged participants.
    - *Consequence*: Inefficiency and the emergence of informal shortcuts.
    - *Architectural Fault*: Misalignment with Purpose—confusing control with coherence.
3. **Orchestration Overreach**
    - *Definition*: A process is tightly sequenced with no allowance for discretion or exception.
    - *Symptoms*: Frequent manual overrides, workarounds, or informal backchannels.
    - *Consequence*: Fragility under real-world variation.
    - *Architectural Fault*: Attempting determinism where ambiguity requires interpretation.
4. **Opaque Process**
    - *Definition*: A process executes but is invisible to participants or observers.
    - *Symptoms*: Stakeholders cannot see status; errors discovered late; accountability unclear.
    - *Consequence*: Loss of trust, weak governance, failure to learn.
    - *Architectural Fault*: Misalignment with Memory—traces absent or inaccessible.
5. **Shadow Process**
    - *Definition*: Informal coordination replaces the official flow.
    - *Symptoms*: Teams rely on email, chat, or ad hoc meetings instead of prescribed paths.

- *Consequence*: Fragmentation, duplication, and systemic drift.
- *Architectural Fault*: Misalignment with Culture—formal design ignored in favor of lived practice.

6. **Stigmergic Drift**
    - *Definition*: Emergent coordination expands unchecked, leaving formal process obsolete.
    - *Symptoms*: Desire paths dominate; process documentation lags behind reality.
    - *Consequence*: Fragility when informal workarounds collapse under stress.
    - *Architectural Fault*: Failure to direct emergence—lack of mechanism to formalize successful adaptations.
7. **Role Ambiguity**
    - *Definition*: Roles in the flow are vague, overlapping, or absent.
    - *Symptoms*: Dropped handoffs, duplicated effort, unresolvable delays.
    - *Consequence*: Execution stalls or diffuses responsibility.
    - *Architectural Fault*: Misalignment with Structure—agents not clearly mapped to steps.
8. **Process-Function Drift**
    - *Definition*: Functions evolve but the underlying process remains static.
    - *Symptoms*: Outcomes no longer match stakeholder expectations.
    - *Consequence*: Declining quality and credibility of outputs.
    - *Architectural Fault*: Misalignment between interface and execution logic.

9. **Tool-Centric Implementation**

   - *Definition*: Process design is dictated by what a tool can express, not what the unit requires.
   - *Symptoms*: Unnecessary complexity, missing steps, distorted workflows.
   - *Consequence*: Process loses architectural integrity; tool bias replaces design intent.
   - *Architectural Fault*: Substitution of instrumentation for structural reasoning.

10. **Zombie Process**

   - *Definition*: A process continues to run long after its Purpose has expired.
   - *Symptoms*: Rote execution with no clear justification; disengaged actors.
   - *Consequence*: Accumulated waste, invisible debt, systemic stagnation.
   - *Architectural Fault*: Lack of governance or feedback to retire obsolete flows.

11. **Misplaced Automation**

   - *Definition*: Automation applied to tasks requiring judgment or discretion.
   - *Symptoms*: Increased rework, manual overrides, suppressed exceptions.
   - *Consequence*: Degraded outcomes; agents bypass the automated steps.
   - *Architectural Fault*: Treating interpretive tasks as mechanizable.

12. **Escalation Loops**
    - *Definition*: Escalation paths are poorly defined or circular.
    - *Symptoms*: Issues repeatedly elevated without resolution; decision ping-pong.
    - *Consequence*: Delay, exhaustion, erosion of accountability.
    - *Architectural Fault*: Misalignment with Order—unclear authority or escalation logic.
13. **Compliance Theater**
    - *Definition*: Processes are created to appear compliant without embedding actual control logic.
    - *Symptoms*: Surface-level documentation, heavy overhead, little real risk reduction.
    - *Consequence*: False sense of security; hidden liabilities accumulate.
    - *Architectural Fault*: Treating visibility as compliance rather than embedding constraint.
14. **Invisible Process**
    - *Definition*: Coordination depends on tacit knowledge, memory, or relationships.
    - *Symptoms*: Collapse when key individuals leave; low repeatability or scale.
    - *Consequence*: Fragile execution; inability to evolve or audit.
    - *Architectural Fault*: Absence of formalization; no captured traces.
15. **Static Formalization**
    - *Definition*: A process is locked permanently at the point of first design.
    - *Symptoms*: Growing mismatch with environment, proliferation of workarounds.

- *Consequence*: Process abandonment or parallel shadow processes.
- *Architectural Fault*: Lack of versioning, review, or adaptation mechanisms.

While the antipatterns above describe common failure conditions, their remediation depends on context. The design principles and heuristics introduced earlier provide the structural vocabulary through which these failures can be re-examined and resolved.

These antipatterns do not arise from malice or incompetence but from structural misfits, misplaced confidence in control, or failure to recognize emergent adaptation. They signal that alignment has broken down—between Process and Purpose, Function, Structure, Order, or Culture. Addressed superficially, they recur; addressed architecturally, they reveal where flows must be redesigned, formalized, or retired. Process is most fragile when it is treated as either purely designed or purely emergent. Its health lies in the dialogue between the two.

6.9. Summary: Process As a Design Construct

Within Purpose Driven Design, Process is the unit's operational choreography—the structured movement that converts intent into continuity and continuity into outcome. It is the construct that ensures Purpose does not remain aspirational and Function does not remain hollow. Unlike static constructs that declare orientation or constraint, process is kinetic: it gives the unit rhythm, repeatability, and traceable flow.

Process resolves the fundamental tension between design and emergence. Designed flows provide reliability, legibility, and accountability; emergent flows provide adaptability, responsiveness, and innovation. Sustainable systems depend on their dialectical coexistence. A process too rigid ossifies; a process too fluid disintegrates. The architecture of process is therefore not a choice between order and improvisation but a grammar for composing them together.

Realized effectively, Process functions as

- *A stabilizer of execution*: Ensuring divided responsibilities align into coherent outcomes
- *A reliability mechanism*: Delivering Value with continuity and trust

- *A traceability surface*: Generating histories and logs that feed Memory and governance
- *A conduit to Storage*: Moving produced artifacts into durable preservation
- *A learning channel*: Making desire paths and emergent practices observable, so they can be directed into formal flows
- *A negotiator of constraint and climate*: Aligning with Order when rule-based rigor is needed, and with Culture when coordination depends on shared norms

The architectural treatment of process reveals that it is not just "how work gets done" but the most complex and sophisticated construct of Purpose Driven Design. It is where the interplay of Purpose, Function, Structure, Order, Culture, Memory, and Storage becomes most visible—and most fragile. When neglected, processes drift into opacity, bureaucracy, or collapse. When treated as a first-class construct—designed, surfaced, and evolved—process enables units to act coherently, scale reliably, and adapt continuously in environments of uncertainty. In practice, the principles and heuristics outlined earlier serve as a lens for self-assessment—helping architects and teams test whether their processes remain legible, governable, and aligned with Purpose.

In the age of sociotechnical complexity, process is no longer a back-office concern. It is a strategic infrastructure of motion—the architecture of flow through which purposeful systems remain both stable and alive.

CHAPTER 7

Structure

At dawn, the city was unrecognizable. Buildings that had stood for generations lay folded into heaps of concrete and steel. Roads were fractured, power lines cut, water mains burst. The earthquake had lasted less than a minute, but its consequences would define the week to come. Sirens wailed, smoke drifted, and silence settled over entire blocks. In the midst of this devastation, survival depended not on heroics but on Structure. Improvisation alone could save individuals; only structure could save the system.

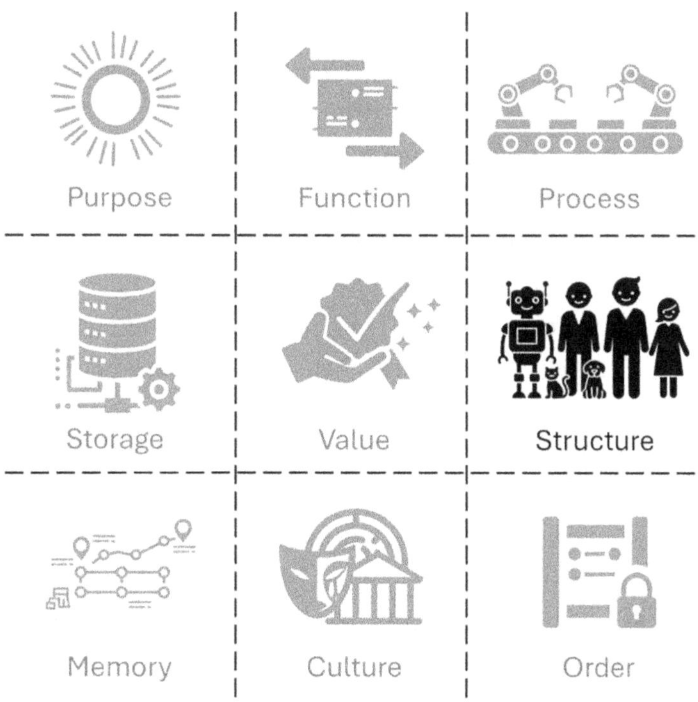

The first to arrive were human responders—firefighters, paramedics, and volunteers—each carrying skills honed for crisis. Alongside them came search-and-rescue dogs, trained to detect the faintest traces of human scent beneath rubble. Within

A. Milchman, *The Purpose Driven Design Framework for Modern Enterprises*,
https://doi.org/10.1007/979-8-8688-2223-0_7

hours, robots joined: compact tracked vehicles designed to crawl through unstable voids, aerial drones to map collapsed neighborhoods, and sensor units that could relay temperature, sound, and vibration data back to command posts. None of these agents could succeed alone. People could not safely enter every ruin. Dogs could detect but not extract. Robots could see but not decide. It was only when their roles interlocked that rescue became possible. That interlocking was not accidental—it reflected a structure intentionally designed for coordination under uncertainty.

Structure bound them together. A dog's sharp bark triggered an immediate handoff: a handler logged the location, and a robot was sent into the void. The robot's camera streamed live images to a human coordinator, who determined whether an extraction team could safely proceed. Above, drones circled, providing aerial context to update maps and reroute resources. Decisions flowed through clearly defined channels: dogs to robots, robots to humans, drones to command. Every signal had a path, every role a boundary, every action an accountable owner.

The difference this made was visible on the ground. In some places where coordination faltered, teams tripped over each other—multiple groups converging on the same site, others left uncovered, precious minutes lost. But in the sectors where Structure held, rescues were efficient. Survivors were located, stabilized, and evacuated in a rhythm that seemed improbable given the chaos around them. Studies of coordinated response consistently show that outcomes improve not through individual effort but through clarity of structure—a pattern visible here in every successful rescue. Trust was not improvised; it was designed into the system (unlike the opposite pattern in many organizations today, where crises of trust are often crises of unclear structure). Each agent knew its role, when to escalate, and when to yield.

What mattered most was not that humans, dogs, and machines were all present—it was that their presence was orchestrated. Structure turned partial capabilities into collective effectiveness. It distributed authority by role, not by rank: if a dog barked, everyone stopped. If a robot transmitted danger, extraction paused. If a paramedic signaled urgent need, priorities shifted. No one had total knowledge, but everyone had enough responsibility to act without waiting for universal approval. The same principle—authority distributed by function rather than hierarchy—defines resilient teams in every field.

The earthquake revealed a truth often overlooked in calmer times: resilience is structural. Tools, skills, and courage matter—but without clarity of roles, lines of escalation, and accountable interfaces, even the best-equipped teams dissolve into confusion. Structure does not eliminate uncertainty. It makes uncertainty survivable. Examples of these handoff patterns are summarized in Table 7-1.

Table 7-1. *Examples of Handoff Logic and Accountability Transfer Across Structural Contexts*

Handoff type	Trigger condition	From/to	Purpose	Example
Routine handoff	Completion of defined role	Operator → inspector	Transfer of accountability for verification	Shift turnover, code review
Escalation handoff	Detection of anomaly or risk	Analyst → coordinator	Transfer of decision authority	Incident reporting, clinical escalation
Collaborative handoff	Parallel roles converge on shared output	Engineer ↔ designer	Integration of specialized contributions	Product review, rescue coordination
Emergency handoff	Breakdown or overload	Field agent → command lead	Preservation of continuity under pressure	Disaster response, system failover

The story of those days is remembered not only for the rescues but for what made them possible: the architecture of coordination across species and systems. People, animals, and robots did not merely coexist. They acted as a coherent unit, bound not by hierarchy, but by structural clarity. That is why lives were saved.

What the earthquake response revealed is not a story of chance—it is a story of design. Structure was the hidden architecture that allowed diverse agents to act coherently under conditions of stress. It defined roles, interfaces, and escalation paths before crisis struck, so that in the moment of uncertainty, action could flow rather than stall. Function opens the door to interaction, and Process sustains activity over time—but without Structure, those constructs collapse into ambiguity. Structure is what makes action assignable, trust legible, and coordination survivable. In the chapters that follow, we will treat Structure not as an organizational chart or a static diagram but as the living configuration of agents—human, animal, robotic, or digital—whose clarity and alignment determine whether a unit can hold together when it matters most.

7.1. Foundations of Structure As a Design Construct

Structure is where design becomes delegation. If Purpose defines why a unit exists, Function opens the doorway of interaction, and Process sustains activity over time, then Structure determines *who or what acts, in what role, and with what accountability*. It is the architectural configuration of agents—human, animal, robotic, or digital—through which Purpose is enacted and Value is made real. Positioned among the five core constructs, Structure's role becomes clear in the comparative view shown in Table 7-2.

Table 7-2. *Structure Positioned Among the Five Core Constructs of Purpose Driven Design*

Construct	Architectural role	Question it answers
Purpose	Defines direction and intent	Why does this unit exist?
Function	Exposes interfaces for interaction	How does it engage others?
Process	Sustains activity over time	How does it operate continuously?
Structure	Configures agents and roles	Who acts, in what role, and with what accountability?
Value	Delivers and validates outcomes	What does it produce that matters?

Without Structure, even the clearest intentions and most elegant processes collapse into ambiguity. With it, partial capacities interlock into coherent systems of responsibility, authority, and trust. Here, we treat Structure not as an org chart or technical diagram but as the relational skeleton of sociotechnical units: the arrangement that makes action assignable, coordination survivable, and adaptation possible.

7.1.1. Defining Structure Architecturally

In Purpose Driven Design, Structure is the architectural configuration of agents—human, artificial, or hybrid—through which a unit enacts its Purpose. It defines who or what acts, how roles interrelate, and how responsibility is assigned, escalated, and traced. Structure is not an org chart, an access matrix, or a static diagram. It is the relational skeleton of the unit, without which Function and Process remain inert.

Architecturally, Structure makes action assignable. It transforms intent into delegated roles, routes processes through accountable agents, and ensures that outputs can be observed, governed, and improved. Without Structure, work fragments into ambiguity—no one knows who acts, who decides, or who owns outcomes. When ownership is ambiguous—such as in an IT outage where multiple teams assume another is monitoring the system—critical processes can run ungoverned until failure exposes the gap. With Structure, every agent has a boundary, every signal has a path, and every escalation has a destination.

A structurally sound unit exhibits three characteristics:

- *Role clarity*: Each agent's contribution is scoped, bounded, and framed in relation to others. Ambiguity about "who does what" is eliminated. *If two agents can't describe their boundary in one sentence, the structure isn't clear.*
- *Operational fit*: The arrangement of agents supports the unit's Function and Process. Real-time services, long-cycle production, and asynchronous insight delivery each demand different structural logics. *If coordination requires constant exception or workaround, the structure is likely misaligned with Purpose.*
- *Embedded accountability*: Responsibility can be traced, feedback routed, and learning absorbed. Without this, Memory loses connection to decision-making. *If an outcome can fail without anyone noticing, accountability isn't embedded.*

Modern units must also be elastic. Structure must accommodate partial automation, agent turnover, and the addition or removal of roles without collapsing coherence. In asynchronous workflows, this elasticity is visible when AI agents take over classification or monitoring tasks overnight and human reviewers resume in the morning—the structure remains coherent because roles, not individuals, anchor responsibility. Elasticity is achieved not through rigid hierarchy but through role logic—interfaces and boundaries designed for adaptability.

The Fleet metaphor reinforces this perspective. Just as ships have captains, crews, and compartments aligned to mission, units contain agents configured to act coherently. Structure is the internal configuration of actors, roles, and responsibilities that lets the unit operate independently, integrate responsibly, and evolve deliberately.

In short, Structure is what turns design into delegation and delegation into execution. The following subsections examine why it exists, what problems it solves, and how its evolution shapes modern systems.

7.1.2. Why Structure Exists

Structure exists because action must be made accountable. Purpose may define why a unit exists, and Function may open the door to interaction, but without Structure there is no answer to the question: *who acts, in what role, under what conditions?* A unit cannot sustain itself on intent and process alone; it requires a configuration of agents that can interpret, decide, and execute with clarity.

Architecturally, Structure addresses three necessities:

- *Delegation*: Work must be distributed. No single agent can embody all responsibility in complex systems. Structure provides the logic by which tasks are assigned, authority is granted, and roles interlock.
- *Accountability*: Outcomes must be traceable. Without Structure, responsibility diffuses into abstraction, and neither success nor failure can be owned.
- *Coherence*: Agents must fit together. Structure ensures that human, artificial, and hybrid roles align with the unit's Purpose, enabling coordination without constant negotiation.

Structure also exists to absorb change. Agents rotate, technologies evolve, and boundaries shift. A well-designed Structure accommodates these variations without dissolving into chaos. It provides a stable skeleton that persists even as its components are replaced or reconfigured. These stabilizing forces become clearer when examined through the three structural necessities outlined in Table 7-3.

Table 7-3. *The Three Structural Necessities—Delegation, Accountability, and Coherence—Viewed Through Their Functions, Vulnerabilities, and Architectural Linkages*

Structural necessity	Core function	Architectural question	Failure mode (when absent)	Illustrative expression	Cross-construct relationship
Delegation	Distributes work and authority across agents	Who acts, and under what mandate?	Centralization bottlenecks; decision paralysis; overreliance on individuals	Role assignments, escalation paths, substitution logic	Links Purpose (intent) to Process (execution)
Accountability	Ensures that every action and outcome has an identifiable owner	Who is answerable for this decision or result?	Diffused responsibility; invisible failure; untraceable learning	Traceable decision records, feedback loops, review protocols	Connects Order (rules) to Memory (record)
Coherence	Aligns roles and decisions so that the unit operates as a unified whole	How do all actions reinforce one another toward Purpose?	Fragmentation; duplicated effort; conflicting priorities	Defined interfaces, coordination mechanisms, dependency maps	Integrates Structure with Function (interfaces) and Value (outcomes)

In this sense, Structure is not an optional layer of organization. It is the design condition that allows Purpose to be enacted, Function to be operated, and Process to be sustained. Without it, capability remains latent and accountability evaporates. With it, the unit can act, adapt, and remain governable over time.

7.1.3. What Problem Structure Solves

Structure solves the problem of diffused responsibility. A unit may declare Purpose, design Function, or script Process, but without Structure these constructs remain inert. No one knows who acts, who decides, or who is accountable. Structure provides the architectural logic that makes every other construct operable. It assigns roles, delineates authority, and embeds responsibility in agents—human, animal, robotic, or digital—so that action is not only possible but governable.

The absence of Structure produces familiar failure modes. Purpose becomes a slogan without stewards. Value is promised but not delivered. Function is exposed but unreliable. Process drifts into improvisation. Order has no enforcers. Culture becomes rhetoric without lived behavior. Memory decays into inaccessible archives, and Storage becomes accumulation without custodianship. Each construct fragments because no one carries its weight. These breakdowns can be systematically diagnosed using the checklist in Table 7-4.

Table 7-4. *Checklist for Diagnosing Structural Decay Across the Purpose Driven Design Constructs*

When structure fails...	Construct impacted	Observable symptom (example)
Purpose	Becomes a slogan without stewards	Initiatives multiply without coordination or measurable outcomes
Value	Is promised but not delivered	Stakeholders disengage; offerings lose credibility or consistency
Function	Is exposed but unreliable	Interfaces or services exist on paper but fail in execution
Process	Drifts into improvisation	Workarounds replace defined sequences; outcomes depend on individuals
Order	Has no enforcers	Rules are selectively applied or constantly reinterpreted
Culture	Becomes rhetoric without lived behavior	Public values conflict with daily practice; cynicism replaces trust
Memory	Decays into inaccessible archives	Lessons are repeatedly relearned; records are ignored or lost
Storage	Becomes accumulation without custodianship	Systems overflow with unused data or obsolete assets; no one curates or retires them

Structure addresses this by solving two architectural problems:

- *Activation*: It transforms latent constructs into active ones by making them assignable. Purpose gains interpreters, Function gains operators, Process gains participants, Order gains stewards. With Structure, every construct has a bearer. When activation fails, decisions exist but no one owns them; when activation succeeds, effort converges naturally around defined roles.
- *Resilience*: It ensures that constructs persist through change. Agents may rotate, technologies may shift, boundaries may expand, yet Structure—anchored in role logic rather than rigid hierarchy—allows the unit to adapt without dissolving coherence. When structure is fragile, adaptation fractures and work stalls until authority intervenes; when structure is sound, adaptation unfolds seamlessly as roles absorb change.

Certain characteristics give Structure this generative capacity:

- *Self-organization*: Agents can reallocate roles locally while preserving coherence.
- *Elasticity*: Structure absorbs turnover, automation, or scaling without collapse.
- *Composability*: Roles and interfaces recombine so that Functions and Processes can evolve.
- *Visibility*: Accountability is legible, making Order enforceable, Culture observable, and Memory traceable.
- *Recursivity*: Structure can evolve itself, creating coordinators, stewards, and governors that sustain the other constructs.

In this way, Structure does more than prevent confusion. It enables the gradual emergence of all constructs in Purpose Driven Design—including Structure itself. It is the skeleton that gives Purpose a body, Value a path, Function an operator, Process a participant, Order a steward, Culture a bearer, Memory a recorder, and Storage a custodian. Structure makes architecture alive.

7.1.4. Historical Patterns and Structural Evolution

Structure has never been fixed. It has shifted with the kinds of agents available, the environments they operated in, and the technologies through which collaboration was made possible. Its evolution reveals not a single trajectory but a sequence of integrations: each era introduced new actors, and Structure adapted to bind them into coherent systems.

Tribal Structure: Human-to-Human Coordination

Early societies relied on kinship, ritual, and oral memory to distribute roles. Hunters, gatherers, and elders carried authority not through charts but through tacit expectations reinforced by repetition. Structure was visible in who sat at the fire, who led the hunt, and who decided when to move. It was fragile but legible.

Animal Integration: Expanding the Role Spectrum

The domestication of animals extended human capacity and required new forms of structure. Dogs guarded, hunted, and searched. Horses and oxen transported and plowed. Pigeons carried messages. These were not tools but agents with semi-autonomous roles. Structure had to clarify not only human authority but also how human and animal capacities interlocked: a shepherd was not only a farmer but the structural coordinator of a human-animal system.

Industrial Structure: Machines in the Workflow

With the rise of machinery, roles were redistributed again. Steam engines, looms, and assembly lines changed what it meant to work. Humans became operators, overseers, and maintainers in structures that blended living and mechanical actors. The design problem was no longer only who hunts or guards, but who starts, stops, repairs, or supervises machines whose endurance exceeded human limits.

Digital Structure: Software As Agent

The twentieth century introduced digital actors—programs, databases, and networks. These agents did not merely extend muscle; they extended decision and memory. Structure shifted again: operators became analysts, coordinators, or stewards, responsible for systems that processed transactions faster than human comprehension. Accountability had to adapt: when software fails, who is to blame—the code, the coder, or the system owner?

Hybrid Structure: Robots, AI, and Multi-Agent Systems

Today, robots dig through rubble, drones map terrain, algorithms triage patients, and AI models draft strategies. Structure must integrate human judgment, animal instinct, mechanical endurance, and algorithmic inference into a coherent whole. The design challenge is not only how to delegate but how to maintain accountability when roles span species and substrates. Who yields when a dog signals danger, a robot detects heat, and an AI flags risk simultaneously? The answer is structural clarity: defined escalation paths and boundary rules that prevent paralysis or conflict.

The Architectural Lesson

Across these eras, the pattern is constant: new agents emerge, and Structure absorbs them. Self-organization, elasticity, composability, visibility, and recursivity have allowed Structure to remain viable even as the definition of "agent" expanded from human to animal to machine to AI. Far from being static, Structure evolves by redesign—ensuring that each new actor finds a legible role, a bounded responsibility, and a pathway of accountability. Each integration revealed not only new possibilities but also new points of fragility, reminding architects that structural evolution must be accompanied by structural awareness.

7.2. Role of Structure in Sociotechnical Systems

The architectural story of Structure continues in the present. Every unit today—whether a medical response team, a logistics chain, or a digital service platform—depends on how its agents are arranged, empowered, and held accountable. The fundamentals have not changed: clarity of roles, pathways of escalation, and embedded accountability remain decisive. What has changed is the sheer diversity of agents and the velocity of their interactions. A single unit may now include people, animals, robots, and AI systems acting side by side, each with partial capabilities that only become effective when structurally integrated. Without Structure, their contributions collide or dissipate; with it, they interlock into resilient coordination. The stakes are visible everywhere: in hospitals where human judgment must align with algorithmic triage, in financial systems where software trades faster than human oversight, in disaster zones where dogs, drones, and responders must trust one another's signals. Structure ensures that these heterogeneous roles do not compete for authority but compose into a coherent pattern. It is the mechanism that stabilizes a unit internally, aligns it within a larger system, and makes it legible to its surrounding environment.

7.2.1. Internal Role: Harmonizing the Unit

Inside a unit, Structure provides the architecture of coherence. Each construct of Purpose Driven Design has its own logic—Purpose sets direction, Value justifies existence, Function exposes interaction, Process sustains flow, Order constrains behavior, Culture shapes climate, Memory preserves history, and Storage retains artifacts. Left unbound, these constructs risk drifting apart: Purpose becomes rhetoric, Process grows brittle, Culture turns ornamental. Structure holds them together.

It does so by assigning bearers. Purpose becomes credible only when a role is structurally accountable for it. Value is not just declared but delivered when someone—or some configured agent—owns its production. Function becomes trustworthy when operators stand behind the interface. Process requires participants whose handoffs are structurally defined. Order and Culture only shape behavior when roles embody rules and norms in practice. Memory and Storage persist because custodianship is structurally assigned.

In this way, Structure is not another construct alongside the rest—it is the harmonizing skeleton that links them into a coherent unit. It ensures that no construct floats free of accountability, and that together they form a resilient, evolution-ready system rather than a collection of disconnected intentions.

7.2.2. Systemic Role: Integration into System-Wide Coordination

Across a system of units, Structure is primarily encapsulated. Other units rarely need to know how a given unit arranges its agents; they engage instead through Purpose, Function, and Value. From the outside, it matters that a request is fulfilled, not how roles were distributed inside. Structure stabilizes the unit internally, but it is not normally exposed as a surface of integration.

There are, however, circumstances where systemic continuity requires partial visibility of Structure. When services span organizational boundaries—such as in emergency response, regulated finance, or healthcare—counterpart units must know not only what is delivered but who is accountable. Integration here depends on identified stewards, escalation paths, or designated interfaces between roles. In these cases, Structure is not fully revealed but selectively surfaced: the named compliance officer, the on-call physician, the service owner who holds accountability across units.

This limited exposure does not undermine encapsulation; it preserves systemic coherence. By surfacing only what is necessary—accountable roles, escalation points, or cross-unit liaisons—Structure allows units to interlock without surrendering their internal autonomy. The architectural lesson is clear: systemic coordination does not require transparency of every role, only legibility of the roles that cross boundaries.

7.2.3. Environmental Role: Interface with External Entities

To the outside world, Structure is usually invisible. Consumers, partners, and regulators interact with a unit through its Value, Function, and Purpose; they rarely see the internal arrangement of roles and agents. Yet in many contexts, Structure inevitably surfaces—not in diagrams, but in lived experience.

In *service contexts*, Structure is often partially exposed. Customers want to know who is responsible for their case, who they can escalate to, or who will deliver the next step. A hospital makes its structure legible by assigning attending physicians and triage nurses. A bank does the same by designating account managers or compliance officers. Regulators insist on this visibility: licensed officers, signatories, or accountable executives make Structure transparent at the points where trust and liability must be anchored.

In *experience-based contexts*, Structure becomes not only visible but constitutive of the offering itself. Theme parks, concerts, and hospitality environments do not simply deliver outcomes; they choreograph encounters. The placement of guides, performers, safety staff, and support systems creates the experience as much as rides or shows. Visitors may not see the backstage chart, but they feel the Structure in the smoothness of transitions, the confidence of escalation, and the orchestration of roles that makes the extraordinary appear effortless. In the Experience Economy, Structure is part of the product.

The architectural lesson is that environmental interaction determines how much Structure must be surfaced. In commodity contexts, it remains hidden. In services, it becomes selectively visible. In experiences, it is woven into what consumers perceive as Value itself. Structure, then, is not always backstage. At times, it is the stage on which trust, safety, and meaning are enacted. These shifts in visibility become clear when Structure is examined across different environmental contexts, as shown in Table 7-5.

***Table 7-5.** Visibility of Structure in Environmental Contexts*

Context	How structure appears	Architectural implication
Commodity (products, by-products)	Structure remains invisible; consumers interact only with outcomes.	Encapsulation dominates; external trust relies on Function and Value, not role visibility.
Service (healthcare, banking, consulting)	Structure is selectively visible: designated officers, case managers, escalation points.	Accountability must be surfaced at boundaries to ensure trust, compliance, and continuity.
Experience (theme parks, hospitality, live events)	Structure becomes part of the offering: staff roles, choreographed transitions, visible coordination.	Structure is woven into Value itself; design must treat role configuration as customer-facing.

7.3. Design Principles and Heuristics

Designing Structure is less about drawing charts than about composing clarity. Every role, interface, and boundary is an architectural decision that shapes how the unit acts and adapts. Principles provide the normative anchor—what must hold true for Structure to remain legible, accountable, and evolution-ready. Heuristics supply the practical moves—tested shortcuts that reduce ambiguity and prevent drift in real-world conditions. Together they prevent Structure from collapsing into rigidity on one side or improvisation on the other. Good Structure does not simply describe how agents are arranged; it enables Purpose, Value, and Process to converge, and allows Order, Culture, Memory, and Storage to be embodied. What follows sets out the principles and heuristics that make Structure not only stable but adaptive, ensuring that units remain coherent within themselves and composable across larger systems.

7.3.1. Principles of Structure-Oriented Design

Structure-oriented design treats the arrangement of agents as an architectural construct, not an organizational afterthought. Each role, boundary, and coordination pathway determines how Purpose is enacted, how Value is realized, and how accountability is preserved. Poorly designed Structure produces drift, opacity, and fragility; well-designed Structure generates coherence, resilience, and trust.

The following principles define what constitutes well-formed Structure in Purpose Driven Design:

1. **Every construct must have a bearer**

   Purpose, Value, Function, Process, Order, Culture, Memory, and Storage each require accountable roles.

   Implication: A construct without an assigned agent remains aspirational, not real.

2. **Role clarity precedes agent assignment**

   Roles must be explicitly defined—responsibilities, conditions, inputs/outputs, escalation paths—before they are filled.

   Implication: Assigning agents without role clarity creates ambiguity and structural drift.

3. **The human core anchors accountability**

   AI, robots, and animals may execute roles, but only humans provide moral legitimacy, interpret Purpose under uncertainty, and govern structural change.

   Implication: Human agents remain structurally indispensable for oversight and legitimacy.

4. **Structure must make accountability legible**

   Roles, escalation paths, and decision rights must be visible enough to trace responsibility.

   Implication: If no one can be named, outcomes cannot be governed or trusted.

5. **Structure must align with Purpose and Process**

   The arrangement of agents should fit the unit's mission and operational logic.

   Implication: Misfit roles introduce drag, bottlenecks, and incoherence.

6. **Structure must balance stability with adaptability**

   Roles should be elastic enough to absorb substitution, turnover, and automation without architectural redesign.

   Implication: Resilient units evolve through substitution, not recomposition.

7. **Role coordination must be explicit**

   Handoffs, supervision, and synchronization must be structurally defined, not assumed or improvised.

   Implication: Without coordination logic, Structure degenerates into a list of isolated responsibilities.

8. **Differentiation enables composability**

   Roles must be bounded and non-overlapping, so they can be reused and recombined like modular components.

   Implication: Clear differentiation allows units to scale and evolve without compounding complexity.

9. **Structural adaptation must be governed**

   Structure should evolve—adding roles, reallocating agents, revising coordination—but always with oversight, traceability, and preservation of human escalation paths.

   Implication: Uncontrolled change erodes trust, diffuses accountability, and invalidates Purpose.

10. **No Function without Structure**

    Every Function must be backed by defined roles and coordination pathways.

    Implication: A Function without Structure is an architectural facade; Structure without Function is wasted complexity.

7.3.2. Architectural Heuristics for Structure

While principles define what must hold true for Structure to be sound, heuristics offer practical shortcuts for designing it under real conditions—limited time, mixed agent types, evolving technologies, and organizational drift. These heuristics are not rigid rules but tested design moves. They reduce error, clarify accountability, and make units easier to adapt and govern.

1. **Define roles as contracts, not job titles**

   Describe each role by its responsibilities, inputs, outputs, escalation conditions, and boundaries—not by the persona that fills it. Titles can change; contracts remain stable. A structural role contract specifies obligations and interfaces—what the role owes, what it expects, and how it connects to adjacent roles.

   Rationale: Contract-based roles make substitution, automation, and delegation possible without redesigning the whole Structure.

2. **Start with the minimum viable Structure**

   Identify the fewest roles necessary to safely carry Purpose, Value, Function, and Process today. Expand only when recurring friction or risk shows the need.

   Rationale: Over-designed structures confuse accountability; under-designed ones leave gaps. Minimum viable Structure grows clarity without bloat.

3. **Anchor design on the dominant flow**

   Arrange roles and handoffs around the most typical path of work—the one that delivers the highest Value or occurs most often—before considering exceptions.

   Rationale: Stabilizing the primary flow exposes real dependencies early and prevents edge cases from dictating the whole design.

4. **Design for substitutability from the outset**

   Specify roles so that different agent types (human, AI, robot, animal) can perform them without changing structural logic.

   Rationale: Substitutability gives units elasticity—agents can rotate or evolve while the Structure remains coherent.

5. **Separate decision rights from expertise**

   Clarify who has the right to decide, independently of who holds knowledge or provides advice.

   Rationale: Without this separation, authority drifts informally to whoever "knows most," eroding accountability and slowing action.

6. **Make handoffs explicit and observable**

   Each transfer of work or responsibility should be visible, timestamped, and owned by a role.

   Rationale: Silent handoffs are where accountability vanishes. Observable handoffs preserve traceability and trust.

7. **Encode escalation paths clearly**

   Define thresholds for failure or uncertainty, specify who or what role receives escalations, and ensure escalation events leave structural traces.

   Rationale: Improvised escalation creates paralysis. Explicit escalation keeps the unit resilient under stress.

8. **Design seams and slack intentionally**

   Insert buffers where variability peaks—intake, triage, or high-volume junctions—so roles can absorb shocks without collapse.

   Rationale: Seamless structures are brittle. Designed seams allow resilience without breaking coherence.

9. **Prefer small, composable role clusters**

   Group a few complementary roles into coherent clusters instead of relying on large, undifferentiated teams.

   Rationale: Small clusters remain legible and recomposable, while large blobs conceal responsibility and slow adaptation.

10. **Continuously reconcile Structure with desire paths**

    Observe how agents actually coordinate—through side channels, workarounds, or informal routines—and either formalize or close them.

    Rationale: Ignoring emergent patterns creates fragility; absorbing or correcting them strengthens adoption and legitimacy.

11. **Version Structure like code**

    Document changes to roles, escalation paths, or coordination logic, and record the rationale.

    Rationale: Without versioning, structural drift is invisible. With it, adaptation becomes governable and accountable.

12. **Rehearse structural incidents**

    Run scenario drills where roles must respond to overload, absence, or conflict of signals.

    Rationale: Units discover missing roles and broken escalation paths faster through rehearsal than through real crisis.

13. **Distinguish temporary from structural roles**

    Identify roles that persist across cycles (structural) versus roles created for short-term tasks (temporary). Only the former should be embedded in Structure.

    Rationale: Treating temporary assignments as permanent bloats diagrams and confuses accountability.

14. **Keep the human core legible**

    Even when AI or robotic agents take primary execution, make clear which human roles hold final accountability for Purpose and compliance.

    Rationale: Structural legitimacy requires humans to be visible anchors of responsibility.

Together, these heuristics provide the pragmatic counterpart to principles: they help architects design Structure that is lean, accountable, and evolution-ready—able to integrate diverse agents, adapt under stress, and remain legible to those who depend on it.

7.4. Realization Across Implementation Planes

Structure is always the configuration of agents –human, animal, or artificial—together with the roles they hold, the boundaries that define them, and the accountability that links them. That architectural essence does not change across planes. What changes is how this same structure becomes visible, enacted, and governed. In the social plane, it is realized through formal roles, norms, and lines of responsibility. In the physical plane, it appears in the placement of crews, working animals, machines, and infrastructure that situate agents in relation to one another. In the digital plane, it is encoded through permissions, identity systems, service accounts, and AI agents acting within bounded roles. Across these manifestations, the principle remains the same: Structure defines who acts and with what responsibility. To design it architecturally is to ensure that its realization across planes is coherent—so that what is legitimate socially is also safe physically and governable digitally.

7.4.1. Manifestation in the Social and Physical Planes

In the social plane, Structure is realized as the assignment of roles and responsibilities among human agents. It defines who carries authority, who is accountable for outcomes, and how legitimacy is preserved. A courtroom illustrates this well: judge, jury, attorneys, and clerks are not steps in a sequence, but structurally differentiated roles whose accountability sustains the institution. The same principle holds in enterprises—compliance officers, service owners, or team leads anchor social legitimacy by making responsibility visible and traceable.

In the physical plane, Structure becomes tangible through the arrangement of agents and the environments in which they operate. Fire crews divide into engine, ladder, and rescue units; handlers and search dogs work as paired roles; robots and humans occupy complementary spaces in a warehouse. These arrangements are not simply logistical—they are structural, ensuring that each agent has a defined placement, boundary, and set of responsibilities within the whole. The physical manifestation of Structure is what allows multiple agents to occupy shared environments without competing for the same role or leaving gaps in accountability.

Across both planes, the essence remains unchanged: Structure configures agents into bounded roles with visible accountability. What differs is how those roles are made legible—through authority and norms in the social plane, and through placement and arrangements in the physical plane.

7.4.2. Structure in the Digital Plane

In the digital plane, Structure is realized through the explicit configuration of agents and their roles in software systems. Here, agents are often service accounts, AI models, automated scripts, or digital twins acting alongside humans. Structure determines which agents exist, what boundaries they operate within, and how their accountability is made visible.

Permissions and identity frameworks are the clearest manifestation. Access control lists, role-based entitlements, and policy bindings express not just *who can do what* but *who is structurally responsible* for those actions. Service accounts take on bounded roles—such as data ingestion, model training, or monitoring—that mirror human assignments but remain embedded in the system. Similarly, AI agents may hold decision-support or classification roles, but their legitimacy depends on a clearly defined structural placement with visible escalation paths to human oversight.

The digital plane also surfaces the elasticity of Structure. Roles can be reassigned, automated, or retired with relative speed, provided they were designed with substitution in mind. What remains constant is not the specific agent, but the structural contract: the defined role, its boundary, and its accountability trace.

In this way, digital manifestations of Structure extend the same architectural principles found in social and physical settings. They differ only in medium: roles are enacted through permissions and configurations rather than offices or crews, but the underlying purpose is the same—ensuring that every action is tied to an accountable role within the unit's design.

7.4.3. Implementation Surfaces and Sociotechnical Contexts

Structure only matters when it is made legible to agents who must act within it. This legibility is achieved through implementation surfaces—artifacts and interfaces that expose roles, boundaries, and accountability in usable form. In the social plane, these surfaces include org charts, role descriptions, accountability matrices, and escalation protocols. In the physical plane, Structure appears in crew rosters, shift schedules, unit badges, or even the physical positioning of agents in relation to one another. In the digital plane, it is instantiated through IAM dashboards, role-based access policies, service accounts, and audit logs that show how responsibility is distributed.

Each surface carries trade-offs. Org charts clarify authority but often oversimplify accountability. Dashboards provide transparency but may overwhelm with signals. Policy documents formalize escalation paths but can lag behind lived practice. Similarly, identity systems make digital roles explicit, yet risk opacity if roles proliferate without clear ownership.

Sociotechnical context shapes how these surfaces function. A hospital's Structure may be legible through patient assignment boards; a global enterprise through compliance attestations; a cloud platform through access-control dashboards. What matters architecturally is not the medium, but whether the surfaces make roles clear, boundaries visible, and accountability traceable.

The lesson is straightforward: Structure is always a system of agents, but its effectiveness depends on how well it is surfaced. Units cannot rely on implicit understanding alone. They must externalize Structure in ways that are usable, auditable, and credible within their social, physical, and digital contexts.

7.5. Strategic Impact of Structure Interfaces

Structure is not only an internal stabilizer; it also shapes how a unit projects authority and accountability outward. Interfaces of Structure—whether role designations, escalation points, or visible lines of responsibility—carry strategic weight. They determine how trust is earned, how power is distributed, and how legitimacy is sustained in the eyes of other units, regulators, and consumers. A well-designed interface makes authority clear without overexposing internal complexity, enabling units to scale coordination while preserving autonomy. Poorly designed interfaces either conceal accountability, eroding trust, or expose too much, constraining adaptability. The strategic impact of Structure lies in how it balances these tensions—making power visible enough to anchor confidence, yet bounded enough to protect coherence and resilience.

7.5.1. Nuances of Power and Accountability

Structure determines not only who acts but who decides, who delegates, and who remains answerable. These arrangements vary in form, and each variation carries architectural consequences. The nuances of power and accountability reveal how Structure stabilizes action, legitimizes authority, and preserves coherence under pressure.

Delegated Authority, Retained Accountability

Authority may be passed to another role, but accountability cannot be shed. A coordinator may delegate operational decisions, yet still carries the burden of outcome. This separation prevents diffusion of responsibility, ensuring that every action remains traceable to a structural anchor.

Visibility of Decision Rights

Decision rights must not only exist; they must be legible. Hidden or ambiguous authority produces bottlenecks, duplication, and mistrust. When decision rights are surfaced—through governance documents, policy bindings, or clear service ownership—they reduce escalation friction and make accountability transparent to both insiders and outsiders.

Autonomy Within Boundaries

Every unit requires autonomy to act purposefully; without it, Structure collapses into dependency, and Purpose cannot be expressed. Yet autonomy is never absolute. It is framed by boundary conditions—budget allocations, regulatory constraints, risk tolerances, or system-level goals—that define the legitimate scope of action. Within these boundaries, Structure determines how decisions are made and resources are applied. Autonomy and accountability are inseparable: to be autonomous is to carry full responsibility for outcomes within the space granted.

Asymmetry of Escalation

Not all escalations flow in the same direction. Operational risks may climb one path, regulatory risks another, and ethical concerns a third. Structure must recognize these asymmetries, mapping escalation channels so that authority is exercised responsibly across different contexts of consequence.

Elastic Delegation Across Agent Types

As AI agents and robots take on structural roles, authority may extend beyond human actors. Yet accountability must remain humanly legible, tethered to roles capable of interpreting Purpose and bearing institutional legitimacy. Delegation may stretch across agent types, but ultimate responsibility cannot be automated away.

Balancing Concentration and Distribution

Concentrating power accelerates decision-making but risks fragility; distributing it enhances resilience but can slow response. Structure cannot eliminate this tension—it must calibrate it. The balance between concentrated authority and distributed accountability is not static but context-dependent, shaped by the stakes, scale, and environment of the unit.

7.5.2. Measurable Benefits and Architectural Payoffs

Well-designed Structure does more than arrange agents; it clarifies authority, stabilizes accountability, and makes autonomy governable. These qualities create measurable benefits that accumulate across units and systems.

A. **Structural Clarity Reduces Coordination Overhead**

When roles and decision rights are explicit, less time is spent negotiating responsibility or resolving ambiguity. Meetings shrink, escalations decrease, and collaboration becomes faster and more direct.

Payoff: Clarity lowers the cost of coordination and frees capacity for purposeful work.

B. **Visible Accountability Builds Trust and Legitimacy**

When accountability is surfaced through roles and escalation paths, stakeholders can see who is responsible for outcomes. This strengthens confidence among consumers, regulators, and partner units.

Payoff: Visible accountability translates into greater trust in the unit itself.

C. **Elasticity Enhances Resilience**

Structure designed for substitutability allows agents to rotate, fail, or evolve without paralyzing the unit. Clear escalation ensures that crises are absorbed rather than amplified.

Payoff: Elastic structures reduce downtime and preserve continuity under stress.

D. **Adaptability Accelerates Response to Change**

Roles defined by contracts rather than personas can be reassigned quickly as new technologies or constraints emerge. Humans, robots, or AI can step into roles without redesigning the whole unit.

Payoff: Adaptability allows units to respond rapidly to technological, environmental, or regulatory shifts.

E. **Alignment Strengthens Systemic Integration**

Clear interfaces, named stewards, and visible escalation points allow units to interlock without exposing their internal complexity. This strengthens coherence at boundaries while preserving autonomy.

Payoff: Systemic alignment reduces friction and enables scalable collaboration across units.

7.6. Implementation Patterns

Structure becomes real when agents are configured into recurring arrangements that distribute authority and accountability in recognizable ways. These patterns are not exhaustive, but they capture how Structure is commonly instantiated across sociotechnical systems.

7.6.1. Hierarchical Role Tree

Authority flows downward through layers of command, with each role accountable to a superior. This pattern dominates in safety-critical and regulated environments, where clarity and discipline are paramount. Its strength is visibility; its weakness is brittleness under change.

7.6.2. Modular Role Clusters

Small groups of complementary roles operate as reusable units—scrum teams, response pods, or mission crews. Each cluster is self-contained yet connects to others through defined interfaces. This pattern scales by composition rather than expansion.

7.6.3. Hub-and-Spoke Stewardship

A central coordinating role oversees multiple specialized roles. The hub carries concentrated accountability, while spokes execute distributed tasks. This pattern provides oversight and integration but risks overloading the hub.

7.6.4. Matrix Overlay

Agents hold dual accountabilities across two structures—often functional and project lines. This pattern balances specialization with delivery but introduces tension: decision rights must be carefully clarified to prevent conflict.

7.6.5. Elastic Role Pool

Agents are rotated or assigned dynamically from pool—on-call rosters, shift rotations, or service accounts. This pattern enhances resilience under variable load but can dilute ownership if identity is blurred.

7.6.6. Delegated Autonomy Units

Whole units are granted authority within boundary conditions such as budget, compliance, or service levels. They act independently but remain accountable for outcomes. This pattern reduces central coordination but requires strong guardrails.

7.6.7. Hybrid Human–AI Roles

Roles are shared between human and artificial agents—AI monitors while humans escalate, or AI drafts and humans approve. This pattern expands capacity but depends on clear boundaries to preserve human accountability.

7.6.8. Shadow and Backup Roles

Critical roles are mirrored by designated backups or shadow agents ready to step in. This pattern mitigates single points of failure but demands explicit activation criteria to avoid confusion.

7.6.9. Distributed Stewardship Network

Accountability is shared across a peer network rather than concentrated in a hub—common in federated governance or open-source stewardship. This pattern maximizes resilience but increases coordination overhead. These patterns can be understood more clearly when viewed side-by-side in the comparative framework shown in Table 7-6.

Table 7-6. *Comparative Overview of Common Structural Patterns—Their Contexts, Payoffs, and Vulnerabilities*

Pattern	Best use context	Architectural payoff	Primary risk/fragility point
Hierarchical role tree	Safety-critical, regulated, or high-compliance domains	Clarity of command and traceability	Brittle under rapid change; top-heavy decisions
Modular role clusters	Project- or mission-based teams needing autonomy	Scalability by composition; local agility	Risk of duplication or misalignment between clusters
Hub-and-spoke stewardship	Environments needing central oversight with distributed execution	Integrative control; unified accountability	Hub overload; single-point-of-failure
Matrix overlay	Cross-functional organizations balancing specialization and delivery	Flexibility in resourcing and perspective	Confused authority; dual-report tension
Elastic role pool	Variable-load operations, shift systems, or service desks	Resilience through substitutability	Diluted ownership; identity ambiguity
Delegated autonomy units	Mature organizations scaling through semi-independent units	Faster local decisions; reduced central friction	Boundary drift; inconsistent standards
Hybrid human–AI roles	Mixed agent contexts (monitoring, triage, drafting)	Expanded capacity; continuous coverage	Accountability gaps; opaque AI reasoning
Shadow and backup roles	Critical continuity functions; disaster-recovery contexts	Continuity and knowledge retention	Activation confusion; idle redundancy

7.7. Interlocking Dependencies and Alignment

Structure does not stand alone. It is the configuration of agents, but that configuration depends on and reinforces every other construct of Purpose Driven Design. Without alignment, Structure risks becoming rigid formality—roles on paper with no living force. With alignment, it becomes the skeleton that makes Purpose enactable, Function reliable, Process sustainable, and Value legitimate.

7.7.1. Alignment with Purpose

Structure gives Purpose a bearer. Every declared mission requires accountable agents to interpret and act. If Structure is misaligned with Purpose, roles proliferate without clear direction, and autonomy becomes drift. Proper alignment ensures that roles are defined in service of why the unit exists.

7.7.2. Alignment with Value

Structure determines who is responsible for producing, safeguarding, and delivering Value. Without this link, Value remains aspirational and accountability for outcomes dissolves. Aligned Structure makes ownership of offerings explicit and ensures that someone—not just "the unit"—stands behind every commitment.

7.7.3. Alignment with Function

Functions are interfaces, but they require agents behind them. Structure supplies that accountability. If misaligned, Functions become facades: promises exposed without capacity or responsibility to fulfill them. Alignment ensures that for every Function surfaced, there is a role ready to stand behind it.

7.7.4. Alignment with Process

Processes describe flow, but Structure defines who participates. Misaligned Structure leaves steps orphaned, handoffs unclear, and escalation paths broken. Alignment ensures that Process has agents to carry its flow, making continuity possible and exceptions survivable.

7.7.5. Alignment with Order and Culture

Rules and norms are abstractions unless embodied in roles. Structure assigns responsibility for compliance, enforcement, and modeling of norms. If misaligned, Order and Culture become empty statements without lived practice. Alignment makes them operational—tied to agents who carry rules and climate into daily behavior.

7.7.6. Alignment with Memory

History must be preserved by someone. Structure assigns custodianship for what is remembered, archived, and interpreted. Without alignment, Memory becomes scattered or neglected, leaving the unit unable to learn from its past. Proper alignment ensures continuity of institutional knowledge.

7.7.7. Alignment with Storage

Artifacts require stewardship—data, documents, and materials must have structural owners. If no roles are accountable, Storage becomes accumulation without order, vulnerable to neglect or misuse. Alignment ensures that artifacts are curated, protected, and available for reuse.

7.8. Failure Modes and Antipatterns

Structure may appear orderly—roles assigned, charts drawn, accountability claimed—yet still fail architecturally. Failure arises when Structure is treated as a static form rather than a living configuration, or when roles, boundaries, and accountability are designed without regard for alignment with Purpose, Function, Process, and Value. The following antipatterns illustrate common structural breakdowns.

1. **Role Ambiguity**

   Definition: Roles are insufficiently defined, leaving agents unclear about scope or responsibility.

   Symptoms: Overlapping duties, repeated handoffs, contradictory instructions.

Consequence: Work stalls in negotiation; accountability dissolves.

Architectural Fault: Structure fails to provide legible boundaries and assignable responsibility.

2. **Zombie Structure**

   Definition: Roles persist even when no longer necessary, carrying no real responsibility.

   Symptoms: Agents occupy positions with little output, or functions that repeat others' work.

   Consequence: Structural bloat obscures true accountability and increases coordination cost.

   Architectural Fault: Structure not governed for adaptation; obsolete roles not retired.

3. **Overcentralized Structure**

   Definition: Authority concentrates in a single hub or role, regardless of scale.

   Symptoms: All decisions escalate to one office; bottlenecks form; response times slow.

   Consequence: Fragility under stress; overload in critical events; disempowerment of agents.

   Architectural Fault: Structural balance between concentration and distribution ignored.

4. **Diffused Accountability**

   Definition: Responsibility is spread across many agents without a clear anchor.

   Symptoms: Everyone is involved, yet no one owns the outcome; escalation loops endlessly.

   Consequence: Failure produces blame-shifting and paralysis; trust erodes internally and externally.

   Architectural Fault: Accountability not structurally tethered to explicit roles.

5. **Tool-Imposed Structure**

   Definition: Structure is shaped by defaults of platforms or tools rather than architectural intent.

   Symptoms: Roles mirror software licenses or workflow engine settings; escalation paths follow tool menus instead of design.

   Consequence: Structure drifts to fit tools, creating mismatches with Purpose and Process.

   Architectural Fault: Structural autonomy surrendered to convenience mechanisms.

6. **Fragile Substitution**

   Definition: Roles are designed for a single agent type, with no capacity for substitution.

   Symptoms: Human turnover halts progress; automation cannot be introduced without redesign.

   Consequence: Loss of resilience; inability to adapt to technological or workforce changes.

   Architectural Fault: Elasticity and substitutability neglected in structural design.

7. **Shadow Structures**

   Definition: Informal arrangements replace or override formal Structure.

   Symptoms: Decisions made in side channels; unrecognized influencers control outcomes.

   Consequence: Legitimacy undermined; formal accountability loses force.

   Architectural Fault: Structure not reconciled with lived practice ("desire paths" ignored).

8. **Structural Invisibility**

    Definition: External stakeholders cannot see who holds authority or accountability.

    Symptoms: Customers escalate endlessly; regulators demand names; peers distrust commitments.

    Consequence: Trust collapses; unit appears opaque and unreliable.

    Architectural Fault: Interfaces of Structure not surfaced in systemic and environmental contexts.

Together, these antipatterns reveal that structural failure is not about absence of form but misalignment of design. Structure must remain lean, legible, and governed for adaptation—otherwise it becomes a facade of roles that conceals rather than enables accountability. These failure modes can be systematically assessed using the diagnostic framework in Table 7-7.

Table 7-7. *Diagnostic Overview of Structural Failure Modes—Their Causes, Symptoms, and Corrective Patterns*

Antipattern	Root cause	Observable symptom	Impact on the system	Architectural correction
Role ambiguity	Undefined scope or overlapping authority	Conflicting instructions, repeated handoffs	Work stalls; accountability dissolves	Define explicit boundaries and escalation logic
Zombie structure	Roles persist beyond relevance	Idle positions, duplicated functions	Structural bloat; rising coordination cost	Retire or recompose obsolete roles
Overcentralized structure	Authority concentrated in a single node	Bottlenecks; delayed decisions	Fragility under stress; disempowered agents	Distribute decision rights through role logic
Diffused accountability	Responsibility spread without anchor	Everyone involved, no one responsible	Paralysis; internal trust erosion	Assign ownership to explicit accountable roles

(continued)

Table 7-7. *(continued)*

Antipattern	Root cause	Observable symptom	Impact on the system	Architectural correction
Tool-imposed structure	Tools dictate hierarchy or workflow	Organization mirrors software menus	Misalignment between design and intent	Re-assert architectural autonomy; adapt tools
Fragile substitution	Roles designed for single-agent dependency	Work halts after turnover or tech change	Loss of resilience; rigidity	Embed substitutability through cross-training or automation
Shadow Structures	Informal influence overrides formal design	Decisions made in side channels	Legitimacy erodes; hidden dependencies	Surface informal flows; reconcile with formal Structure
Structural invisibility	Lack of transparency to external parties	Escalations loop endlessly; opacity	External distrust; compliance risk	Make authority surfaces explicit and visible

7.9. Summary: Structure As a Design Construct

Within Purpose Driven Design, Structure is the key construct—the architectural configuration that makes every other construct real. Purpose may inspire, Value may justify, Function may interface, and Process may flow, but without Structure none of these can be carried forward. Structure provides the roles, boundaries, and accountability that give agency to intention. It is the system's skeleton: the design feature that ensures Purpose is enacted, Value is owned, Function is supported, and Process is performed with coherence and legitimacy.

Structure resolves the fundamental challenge of purposeful systems: how to distribute accountability among diverse agents—human, animal, and artificial—so that autonomy and coherence can coexist. By defining who acts, within what boundaries, and with what responsibility, Structure transforms potential into practice. It is the anchor that prevents diffusion, drift, and fragility, ensuring that every construct has a bearer and that no commitment floats unowned.

Realized effectively, Structure functions as

- *A configurational skeleton*, aligning all other constructs into a coherent whole
- *A mechanism of legitimacy*, ensuring accountability remains visible and traceable
- *A resilience framework*, enabling units to survive turnover, shocks, and technological change without collapse
- *An adaptability engine*, allowing substitution of agents and reconfiguration of roles without redesign
- *A harmonizing force*, binding Purpose, Value, Function, Process, Order, Culture, Memory, and Storage into an interdependent system

The architectural treatment of Structure reveals that it is not a static diagram or hierarchy but a living configuration that must be deliberately designed, surfaced, and evolved. When neglected, Structure decays into opacity, leaving Purpose rhetorical and Value unowned. When treated as a first-class construct—continuously refined and governed—it provides the stable skeleton on which every purposeful system depends.

In the age of sociotechnical complexity, Structure is no longer background scaffolding. It is strategic infrastructure: the architecture that makes systems not only operational, but accountable, resilient, and evolution-ready.

CHAPTER 8

Order

In the early 1930s, London's subway system was increasingly complex. Dozens of underground lines twisted beneath the city, connecting stations through an expanding web of tunnels. Official maps attempted to portray this network with topographic accuracy—overlaying the routes directly onto the city's real geography. While technically correct, these maps were visually chaotic. Lines curled unpredictably. Stations clustered unevenly. Passengers often turned the maps sideways or upside down, trying to trace a sensible path. For anyone unfamiliar with London, it was an exercise in frustration.

Enter Harry Beck, a young engineering draftsman—not trained in cartography, but fluent in circuit design.

A. Milchman, *The Purpose Driven Design Framework for Modern Enterprises*,
https://doi.org/10.1007/979-8-8688-2223-0_8

Beck proposed something radical: abandon geographic fidelity in favor of conceptual clarity. His design—released in 1933—positioned all stations at even intervals and drew every line in straight segments: vertical, horizontal, or 45 degrees. Central London was expanded to emphasize its dense connections; outer zones were compressed. What emerged was not a literal map of the city, but a topological diagram—more akin to a circuit board than a street grid.

Critics objected. A distorted map, they argued, could only mislead. But passengers immediately grasped its logic. They didn't need to know the actual distance between stations—they needed to understand the sequence of connections and where to change lines. Beck's map emphasized what mattered: interchange, direction, and flow. It turned transportation into a designed experience, not just a reflection of infrastructure.

What Beck revealed was that Order is not about representing what exists. It is about making clear what is *possible*. The map's constraints—its simplified angles and even spacing—were not aesthetic decisions; they were architectural tools. They reduced cognitive load, enabled faster decisions, and helped people stay oriented under stress, urgency, or limited attention.

In modern digital architecture, this distinction is often lost. For instance, many cloud overlay platforms claim to simplify complex environments, creating the illusion of portability and multi-cloud capability, but instead multiply intermediaries—adding layers of governance that obscure accountability, complicate interoperability, and introduce fragility without delivering real functional gain. When Order is sold rather than designed, complexity expands instead of contracting.

And notably, Beck's design didn't alter the infrastructure at all. The trains ran on the same tracks. What changed was the *mental model*—how people interpreted the system and chose to act within it. Beck didn't redesign the Tube; he redesigned *permissibility*.

This idea—designing how options are understood and acted upon—is central to the concept of Order in sociotechnical systems. Order determines who can take what action, under what conditions, with what authority, and through which interfaces. It is not limited to governance or policy, though policies often express it. In practice, Order may appear as a policy in the social plane, a safety protocol in the physical plane, or an access rule in the digital plane—but in every form, it defines how possibility is bounded and permission is granted. It is the logic of constraint, timing, and escalation that gives structure its discipline. Order doesn't depict reality—it composes it.

Structure defines roles—but Order governs the rules those roles must follow. It does not assign authority or responsibilities; rather, it externalizes constraints, permissions, timing rules, and escalation paths that shape how roles interact and how actions unfold. This relationship among Structure, Role, and Order is depicted in Figure 8-1.

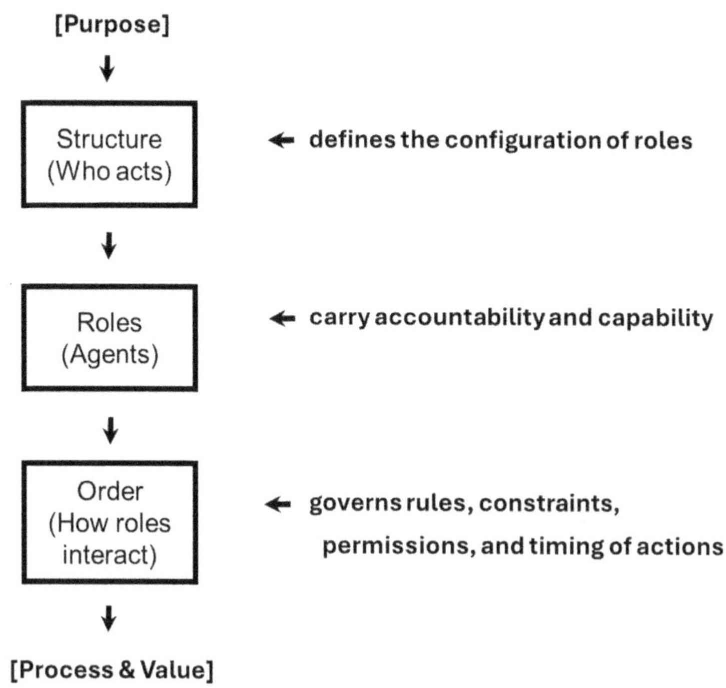

Figure 8-1. *Structure, Order, and Role—Structure defines who acts, Roles instantiate capability, and Order governs the rules that make those interactions coherent*

Order enables systems to operate coherently—not by prescribing behavior, but by bounding it. In this chapter, we explore Order as a design construct that clarifies limits, enforces boundaries, and makes rule-based coordination possible—especially under conditions of complexity, scale, or strain.

8.1. Foundations of Order As a Design Construct

Order gives form to possibility. It does not emerge from desire or instinct, but from deliberate design. While Purpose attracts, Function enables, and Process flows—Order ensures those flows are permitted, constrained, and recoverable. It is what makes systems intelligible and executable across time, agents, and conditions. Order is not a layer to be added; it is the architectural mechanism that makes coordination possible in the first place. When Order is designed only after friction appears, it becomes a constraint on recovery rather than an enabler of coherence. Designing Order up front ensures that coordination is intentional, not accidental.

To design Order is to decide how actions unfold, where boundaries lie, when escalation occurs, and how exceptions are resolved. Unlike informal norms or tacit rules, Order must be made explicit, externalized, and embedded in the very structure of interaction. Understanding its foundations is essential for any system that must operate under complexity without collapsing into rigidity or drift.

8.1.1. Defining Order Architecturally

In Purpose Driven Design, Order is not simply the presence of rules or procedures—it is the architectural mechanism that binds action to intention across agents, interfaces, and time. It does not just define what is allowed—it makes allowance legible, enforceable, and recoverable. Order gives a system its internal consistency by constraining behavior, exposing permissions, and embedding rules that enable coordination to unfold reliably. Without it, no sociotechnical construct—Purpose, Function, Process, Structure, or Culture—can operate coherently or evolve meaningfully over time.

Architecturally, Order is the structural grammar of a unit. It does not produce activity—it governs its shape, sequence, and eligibility. Order defines *how* things may happen, *under what conditions, by whom,* and *through which interfaces.* In physical systems, this grammar appears as procedures, safety interlocks, or timing sequences; in digital systems, as business and access rules, execution logic, and dependency orchestration. Each translates intention into permissible action. Unlike Culture, which shapes informal norms and expectations, Order externalizes and codifies formal constraints—rules, protocols, permissions, obligations, and escalation paths. It is not an overlay but an embedded substrate: the designed logic that enables predictable execution, enforceable boundaries, and recoverable exceptions.

Order manifests across all system planes—and in each, it binds possibility to constraint in a form that can be enacted:

- *In the physical plane*, through signage, floor layouts, safety procedures, and operational protocols
- *In the digital plane*, through access control, configuration policies, state machines, and workflow engines
- *In the social plane*, through delegation chains, institutional commitments, and codified authority structures

These are not decorative or explanatory elements. They are architectural surfaces where Order is enacted. They define obligations, expose valid pathways, prevent invalid transitions, and distinguish between permitted and prohibited actions. Importantly, Order must not be imposed retrospectively or assumed informally. It must be embedded up front—distributed across agents, artifacts, and interfaces—so that coordination becomes systemically possible without relying on memory, convention, or improvisation.

For example, consider a hospital's infection-control policy: physically, it appears as isolation zones and sanitation stations; digitally, as automated alerts that lock or unlock patient-room access; socially, as assigned responsibilities and escalation paths for medical staff. Across all planes, the same principle—containment through disciplined boundary management—translates into concrete constraints and permissions.

In well-designed systems, Order does not stifle autonomy. It enables bounded flexibility—allowing agents to operate independently within clearly defined constraints. The goal of Order is not control for its own sake, but *coherence under complexity*. In this framing, Order is not a bureaucratic burden. It is a compositional necessity.

A structurally sound Order possesses the following architectural characteristics:

- *Prescriptive*: Order defines what must or must not occur under specific conditions. It sets out workflows, transitions, compliance steps, authorizations, and timing rules—not as suggestions but as designed constraints on action.
- *Context-aware*: Order is never universal. It must be shaped by the Purpose it serves and the Function it enables. Rules gain power not from generality, but from fit—they must align with intent, role, and environment.

- *Actionable*: Order must operate where action occurs. It is not aspirational or conceptual—it is embedded in code, protocols, checklists, dashboards, roles, and permissions. What cannot be enacted is not Order—it is aspiration.
- *Recoverable*: Order anticipates not only normal execution but also deviation. It embeds mechanisms for detection, escalation, correction, and learning. A well-designed Order makes violations diagnosable and recovery feasible—turning disruption into structured response.
- *Compositional*: Order must scale through layering and interaction. In large systems, multiple Orders must coexist—locally defined, context-bound, yet aligned. Coherence emerges not from uniformity but from interoperable constraints.

Together, these characteristics define Order not as bureaucracy or enforcement but as a generative logic—one that makes autonomy safe, coordination possible, and execution legible across scale. Order does not emerge on its own; it must be architected deliberately, applied precisely, and embedded structurally. These characteristics can be examined more concretely through the architectural framework summarized in Table 8-1.

Table 8-1. *Architectural Characteristics of a Sound Order—Its Purpose, Manifestations, Vulnerabilities, and Examples Across System Planes*

Characteristic	Purpose	Manifestation across planes	Failure mode when absent	Illustrative example
Prescriptive	To define what must or must not occur; create clarity of permissible action	Physical: safety procedures; Digital: workflow rules; Social: decision authorities	Ambiguity, improvisation, policy drift	Aircraft ground operations governed by explicit safety sequences and authorizations
Context-aware	To align rules with Purpose, Function, and environment	Physical: local safety signage; Digital: role-based permissions; Social: jurisdictional laws	Irrelevance or overreach; rules ignored	Regional regulations tailored to local infrastructure limits
Actionable	To ensure Order operates at the point of execution, not in abstraction	Physical: checklists, manuals; Digital: automation scripts, dashboards; Social: role charters	Aspirational governance; rules unenforced	Clinical protocol embedded in digital workflow for real-time adherence
Recoverable	To embed detection, escalation, and correction into system logic	Physical: alarms, emergency stop; Digital: exception handling, rollback; Social: incident reporting, escalation	Silent failure, untraceable deviation.	A payment gateway flags transaction anomalies, routes them to human review, and logs the correction.
Compositional	To maintain coherence across layered Orders in complex systems	Physical: local site rules aligned with global standards; Digital: APIs mediating multiple subsystems; Social: departmental policies aligned with enterprise governance	Fragmented compliance, conflicting constraints	Multinational operations integrating regional and corporate procedures through interoperable standards

8.1.2. Why Order Exists

A system can attract stakeholders, offer value, define functions, and mobilize process—yet still collapse under ambiguity. Without Order, intention struggles to reach execution. Autonomy spills into contradiction. Coordination becomes conditional on memory, negotiation, or charisma.

Order exists to make systems *stable under complexity*. It provides the scaffolding that lets units operate concurrently, decisions escalate predictably, and boundaries hold under strain. When Order is sound, boundaries flex without breaking—units coordinate through defined channels; when Order is weak, pressure leaks across them, producing overlap, conflict, or paralysis. Where Structure defines who plays what role, Order defines how interaction unfolds—constrained, permitted, or escalated—across space, time, and context.

Its absence is rarely total. Most systems operate with partial or tacit orders—implicit rules, habitual behaviors, or brittle approvals. But as scale increases, reliance on the informal becomes a liability. Without an externalized and embedded logic of constraint, systems drift. Violations go undetected. Actions lose legitimacy. And the friction of interpretation replaces the fluency of execution.

Order exists not to control behavior but to *bound* it. It provides the architectural grammar that makes action eligible, rule-following observable, and deviation recoverable. A system without Order is not free—it is fragile.

Thus, Order exists to fulfill four design necessities:

1. *Constraint*: Imposing rule-based boundaries that define what actions are permitted, restricted, or escalated—before execution occurs.

2. *Legibility*: Making the logic of permission, obligation, and exception *visible* across agents, interfaces, and surfaces—so behavior can be understood, anticipated, and aligned.

3. *Recoverability*: Enabling detection, escalation, and correction when actions deviate from intended paths—so failure does not become systemic collapse.

4. *Unit Integrity*: Maintaining internal coherence by enforcing consistent rule sets across roles, states, and interfaces—using policy synchronization, validation mechanisms, and compliance checks.

Without Order, a unit may appear purposeful, structured, and even functional—but it remains unstable. Order is what renders the unit governable, giving it a bounded form that supports safe autonomy, reliable execution, and participation in coordinated systems. The elements that render a unit governable are detailed across physical, digital, and social planes in Table 8-2.

Table 8-2. *The Four Design Necessities of Order Expressed Across Physical, Digital, and Social Planes*

Design necessity	Physical systems	Digital systems	Social/organizational systems
Constraint	Safety boundaries, lockout procedures, or traffic controls limit what can occur.	Access permissions, rate limits, and automated validation prevent unauthorized or unsafe execution.	Policies, laws, and escalation rules restrict behavior before decisions occur.
Legibility	Control panels, alarms, and visible signals make process states interpretable.	Dashboards, logs, and audit trails reveal operational logic and exceptions.	Organizational charts, documented processes, and meeting cadences make accountability visible.
Recoverability	Emergency stop buttons, manual overrides, and redundant systems enable rapid correction.	Exception handling, transaction rollback, and alerting systems limit cascading failure.	Incident review, appeals processes, and cross-role escalation allow structured response.
Unit integrity	Standard operating procedures and synchronized controls maintain local coherence.	Policy synchronization, data validation, and configuration management ensure consistency across services.	Role charters, internal policies, and compliance checks align behavior across departments.

8.1.3. What Problem Order Solves

Order solves the architectural problem of *unbounded capability*—the condition in which a unit possesses functional surfaces and operational potential but lacks any coherent system of constraints, permissions, or governance. A familiar example is an algorithm or platform launched without operational guardrails—technically capable but behaviorally unbounded, producing outcomes no one can fully predict or control. Without Order, a unit may be richly interactive and structurally composed, yet remain behaviorally unstable. Its actions, decisions, and engagements occur without reference to shared rules, limits, or conditions. The result is a breakdown in the core mechanism of sociotechnical viability: *bounded, interpretable behavior.*

In modern systems, this problem manifests in several ways:

- *Unconstrained invocation*: Interfaces can be triggered without regard for context, role, or condition—leading to misuse, overload, or unintended effects.
- *Rule ambiguity*: Agents and users operate without clarity on what is allowed, required, or prohibited—undermining trust and impeding automation.
- *Governance drift*: Rules become fragmented across tools, documents, and teams—eroding the system's ability to enforce consistent behavior.
- *Boundary blindness*: Agents act without awareness of system-level limits—resulting in coordination failures, ethical breaches, or operational conflict.

Order addresses these failures by introducing *structured boundary conditions* into the architecture. It defines the limits within which capability may be exercised—through rules, roles, permissions, and constraints. These are not incidental. They are what render the unit *governable.* Order converts capability into coherence.

Specifically, Order solves four persistent structural deficiencies:

- *Unbounded execution*: When actions can be taken without constraint or oversight.
- *Incoherent rules*: When boundaries are undefined, conflicting, or unenforceable.

- *Non-governability*: When systems cannot be configured, audited, or adapted meaningfully.
- *Behavioral inconsistency*: When rules are interpreted or applied differently across agents or contexts.

These deficiencies—and the boundary expressions that resolve them—are mapped across system planes in Table 8-3.

Table 8-3. *How Order Resolves Structural Deficiencies Through Boundary Expressions Across Physical, Digital, and Social Planes*

Structural deficiency	Physical plane (boundary expression)	Digital plane (boundary expression)	Social/ organizational plane (boundary expression)	Resulting coherence
Unbounded execution	Physical guards, safety interlocks, physical separation zones	Access control lists, execution policies, API rate limits	Defined roles, delegated authority, escalation rules	Actions remain constrained and accountable.
Incoherent rules	Standardized signage, synchronized protocols, procedural manuals	Configuration standards, validation schemas, unified policy definitions	Policy alignment across departments; clarified decision rights	Rules remain interpretable and enforceable.
Non-governability	Inspection checkpoints, audit trails, maintenance logs	Logging, observability, configuration management, version control	Audit mechanisms, reporting systems, and governance reviews	Systems remain adaptable and transparent.
Behavioral inconsistency	Standardized workflows, redundancy procedures	Central rule engine, automated enforcement, consistent exception handling	Shared training, common policy interpretation, accountability matrices	Behavior remains uniform across agents and contexts.

By solving these, Order ensures that the unit does not merely *function*—it *behaves*. It becomes a structural participant in its environment not through raw capability but through *bounded, interpretable action*. And it does so in a way that is enforceable, adaptable, and aligned with the system's purpose and values.

8.1.4. Historical Patterns and Structural Evolution

Order has evolved in response to the growing complexity, autonomy, and interdependence of sociotechnical systems. While *Culture* encodes symbolic meaning through ritualized norms and inherited beliefs, *Order* introduces the structural logic that governs behavior—through constraints, permissions, and enforceable boundaries. Its historical trajectory reflects a shift from *implicit norms* to *explicit rule systems*, and from *static authority* to *dynamic governance*.

In *pre-industrial systems*, Order was largely informal—embedded within cultural structures. Boundaries were maintained through tradition, kinship, and symbolic authority. What could or could not be done was enforced not through rule systems, but through ritual, myth, and social expectation. These mechanisms belonged to Culture, not Order. In this phase, Order remained *latent*: present in behavior, but not architected.

With *industrialization*, Order emerged as a formal design layer. Bureaucratic systems introduced explicit constraints—job roles, reporting hierarchies, standard operating procedures. Governance became visible, documentable, and enforceable. The unit could now be audited, regulated, and scaled. Order was no longer tacit or reactive—it became a recognized structural element, distinct from both Culture and Function.

In *post-industrial systems*, Order shifted again—this time toward protocol-driven logic. Digital platforms required dynamic rule sets, programmable access controls, and machine-executable governance. Business Rules Management Systems (BRMS), workflow engines, and policy frameworks became central to system design. Boundaries were no longer rigid—they became contextual, conditional, and continuously updated. Order was no longer a document; it was an interface.

Today's *contemporary systems* demand yet another shift. Adaptive Order is now required—capable of evolving in response to new agents, emergent risks, and real-time conditions. Governance is increasingly distributed, interpretable, and negotiated at runtime. Smart contracts, policy engines, and AI agents implement rules that adapt to context while remaining observable and auditable. Order has become a *living system*—structural, responsive, and foundational to maintaining coherence in environments too fast or too complex for static control. This trajectory is captured in the historical progression shown in Figure 8-2.

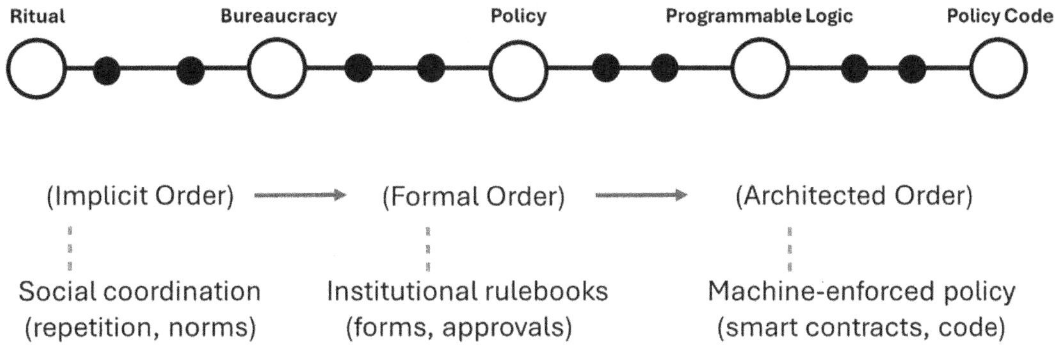

Figure 8-2. *The historical evolution of Order—from ritual repetition to bureaucratic formalization, and from institutional policy to programmable logic*

8.2. Role of Order in Sociotechnical Systems

Order does not initiate sociotechnical systems—but it enables them to scale, stabilize, and remain governable. When systems attempt to scale without Order, they expand faster than their rules evolve—boundaries blur, accountability diffuses, and ethical failures emerge as behavior outpaces governance. Once Purpose has set direction and Function has activated engagement, Order defines the boundaries within which actions may occur—legitimately, safely, and predictably. It allows autonomy to be exercised without collision, enables coordination across distributed agents, and ensures that behavior remains interpretable under pressure. Order transforms loosely connected activity into structured participation, ensuring that what a system can do is always aligned with what it should do.

8.2.1. Internal Role: Stabilizing the Unit

Within a unit, Order provides the internal stability necessary for reliable behavior. Structure defines roles; Function defines what those roles can do; but it is Order that defines *how* they must act—under what conditions, with what permissions, and along what constraints. It governs the relationships among internal components, sets valid transitions across states, and enforces the difference between authorized and unauthorized action.

Absent internal Order, even a well-structured and richly functional unit becomes volatile. Capabilities may be triggered out of sequence. Actions may be taken without approval. Boundaries between roles blur, and accountability diffuses. The result is internal incoherence—misuse of resources, duplicated effort, or outright contradiction.

Internally, Order appears in both digital and human layers. In software systems, it takes the form of access control, state validation, rule enforcement, and conditional logic. In human systems, it manifests as escalation paths, approval processes, delegation frameworks, and operating policies. These mechanisms reduce ambiguity, especially under pressure or turnover, and ensure that behavior can be anticipated and interpreted by others.

Importantly, Order within the unit is not just a control mechanism—it is a precondition for composability. A unit that cannot enforce its own boundaries cannot participate reliably in a broader system. Internal Order makes the unit *coherent, predictable,* and *safe to integrate.* It transforms capability into trustworthy behavior.

8.2.2. Systemic Role: Integration into System-Wide Coordination

No unit operates in isolation. To contribute meaningfully to a larger system, a unit must do more than function internally—it must behave in ways that are *compatible, coordinated,* and *constrained* by the architecture of the whole. This is the systemic role of Order: enabling distributed units to operate semi-independently while remaining aligned with shared governance, shared purpose, and shared operational rhythm.

In structurally sound systems, *integrative units—such as regional or global headquarters*—define and enforce boundary conditions for the operational units they govern. These conditions may include policy constraints, budget ceilings, escalation thresholds, or behavioral contracts. They create a layer of *systemic Order*—not by centralizing control but by externalizing expectations. This allows operational units to remain autonomous in execution while coordinated in behavior.

Without systemic Order, integration becomes fragile. Units duplicate efforts, conflict over shared resources, escalate in inconsistent ways, or diverge from enterprise intent. Local optimizations override system-level goals. Over time, coordination erodes—not from malice or error but from the absence of enforceable alignment.

Systemic Order allows multiple units to operate as part of a larger structure—coherent not because they are identical but because their interactions are bounded and interpretable. It makes reporting pathways clear, decisions traceable, and operations auditable across layers. Where Structure enables hierarchy and Function enables participation, Order enables alignment without micromanagement.

It is Order that allows structurally distinct parts—regional, divisional, or functional—to compose into a stable, navigable whole. In this way, Order does not merely govern action within the system; it makes the system itself *possible.*

8.2.3. Environmental Role: Interface with External Entities

Units do not exist solely within internal or systemic boundaries. They must also interact with external environments—customers, consumers, partners, regulators, investors, suppliers, job markets, and competitors. These interactions are not unbounded. They are shaped and governed by both internal Orders (what the unit allows) and external Orders (what the environment expects or demands). At this interface, Order plays a dual role: protecting the unit's integrity while enabling legitimate participation in broader ecosystems.

Internally, Order defines what the unit is willing or permitted to expose—capabilities, data, services, decisions. It ensures that interfaces with external actors are *intentional, constrained, and governed.* Not everything is available to everyone. Constraints may take the form of access rules, licensing terms, consent frameworks, or context-aware exposure. These rules define the *external posture* of the unit: what it reveals, when, to whom, and under what conditions.

Externally, units must also interpret and respond to Orders imposed by others. These include legal mandates, regulatory constraints, industry norms, audit obligations, certification schemes, and public expectations. Participation in a given market, domain, or community often depends on the ability to *recognize and comply* with these rules—sometimes in real time. A unit without the ability to adapt to external Order risks exclusion, penalty, or loss of trust.

Whereas internal Order ensures behavioral consistency, and systemic Order enables coordination across units, environmental Order enables *trustworthy engagement with the outside world.* It allows a unit to present itself as reliable, compliant, and context-aware—without losing control of its own boundaries.

Order, in this environmental role, is not just defensive. It enables strategic engagement. It allows units to manage exposure, shape perception, and selectively participate in evolving ecosystems—without being overwhelmed by them.

8.3. Design Principles and Heuristics

Order is not simply a set of rules—it is an architectural concern that must be shaped intentionally. Designing Order involves making explicit decisions about where boundaries are drawn, how permissions are granted, how violations are detected, and how constraints adapt to changing context. These decisions affect not just compliance, but coordination, usability, and trust. The following principles and heuristics offer guidance for architects and designers seeking to embed Order in ways that are both enforceable and adaptable—without compromising autonomy, performance, or changeability.

8.3.1. Principles of Order-Oriented Design

Order-oriented design approaches Order not as a passive policy layer but as an architectural discipline—concerned with defining, structuring, and embedding rules and constraints that enable purposeful, safe, and coherent behavior. Order is foundational to scaling autonomy, preserving alignment, and enabling evolution within sociotechnical systems.

The following principles support the architectural clarity, adaptability, and long-term coherence of Order:

1. **Order must serve Purpose**

   Order is not an end in itself—it exists to protect, enable, or constrain behavior in service of the system's Purpose. Constraints that cannot be traced to Purpose introduce unnecessary friction. Good Order clarifies why the rule exists, and whom it protects or empowers.

   Implication: Any rule that does not support Purpose risks becoming a source of rigidity, confusion, or drift.

2. **Design the minimum sufficient constraint**

   Effective Order balances stability and flexibility. Overconstraint suppresses creativity and responsiveness; underconstraint invites misalignment or risk. The goal is to enable safe failure and recoverable exceptions—not to eliminate all uncertainty. The guiding principle is minimum sufficient constraint: rules should be just strong enough to preserve coherence and just light enough to allow adaptation. A simple review asks—does this rule protect integrity, enable judgment, and recover gracefully when breached? If not, the constraint exceeds its purpose.

   Implication: Use hard constraints sparingly, reserve them for critical boundaries, and revisit them periodically to ensure proportionality.

3. **Make Order legible and operable**

   Agents must be able to interpret and apply Order correctly—whether they are humans, machines, or hybrids. This requires clear representation, appropriate modality (text, logic, API), and accessible rationale.

   Implication: Illegible Order leads to performative compliance—rules are followed without understanding or consistently ignored. For example, when access policies are opaque, employees either over-restrict their actions to avoid fault or bypass controls entirely to get work done—the appearance of compliance replacing genuine accountability.

4. **Anchor Order in Structure**

   Rules must be assigned to specific roles, interfaces, or enforcement mechanisms. Order that exists only in documentation decays quickly. It must be bound to the system's structural components: roles, workflows, interfaces, and stewardship.

   Implication: A rule without ownership or embodiment becomes unenforceable and eventually irrelevant. Test: Can the team name who interprets, enforces, and learns from this rule? If not, the rule exists only on paper.

5. **Propagate Order through interfaces**

   Distributed systems do not rely on central enforcement—they rely on interface-level clarity. Facades should expose constraints, permissions, obligations, and telemetry so that interacting units can coordinate safely.

   Implication: Systems become incoherent when constraints are hidden, fragmented, or assumed rather than declared. At every interface—between services, teams, or domains—Order must travel with the interaction: the same constraint logic that governs one boundary must be visible and interpretable on the next.

6. **Externalize Order from where it does not belong**

   Order is often buried inside logic, roles, workflows, or informal habits. This makes it brittle, opaque, and hard to align across units. Architecturally, Order must be extracted and surfaced—represented as explicit policies, rule sets, or constraint layers separate from Function, Process, or Culture.

   Implication: Hidden Order is ungovernable. Externalizing it makes it observable, adaptable, and structurally accountable. Many organizations have learned this through compliance failures—when business rules were embedded deep within software logic or informal practice, making them invisible to auditors. Once externalized into explicit policies and configuration standards, the same controls became reviewable, traceable, and adjustable without code changes.

7. **Design Order for change without redesigning the system**

   Order should evolve as the system matures. This requires parameterization, modularity, and structural separation—so that updates to rules don't require rewriting workflows or recoding logic. A stable framework should support dynamic policy.

   Implication: If rules must be changed by rewriting the system, adaptability will always be compromised under pressure.

8. **Balance formal and informal Order**

   Not all constraints must be codified. In smaller or trust-based environments, informal norms can suffice. However, as complexity, scale, or risk increase, these norms must either be formalized or made observable.

   Implication: Informal Order should not be ignored—but it must be monitored for erosion, conflict, or failure under stress.

8.3.2. Architectural Heuristics for Order

While principles clarify what Order should enable, heuristics offer practical shortcuts and design moves for embedding Order effectively into sociotechnical systems. These heuristics are not universal—they are context-sensitive patterns that reduce ambiguity, expose hidden constraints, and prevent governance erosion. They are especially useful under real-world conditions: rapid change, federated teams, cross-domain integration, or unclear compliance mandates.

The following architectural heuristics support the design of enforceable, visible, and adaptable Order:

1. **If there is no enforcement surface, there is no Order**

   Order exists only where it is embodied—in validation logic, access controls, escalation paths, or decision points. Documentation alone is not architecture. Without enforceable mechanisms, Order is aspirational.

   Rationale: Make enforcement surfaces visible, testable, and aligned with points of risk.

2. **Decentralize execution, centralize boundary definition**

   Let operational units act autonomously—but only within shared constraint envelopes defined by integrative units. This enables autonomy without chaos.

   Rationale: Boundary conditions are the only scalable substitute for micromanagement.

3. **Use facades to declare, not negotiate, Order**

   Interfaces between units should communicate fixed constraints, not initiate informal back-and-forth. Order requires predictability, not hidden rule logic buried in interaction chains.

   Rationale: Treat interface contracts as architectural boundaries—not developer convenience layers.

4. **Where memory exists, Order must be traceable**

   Any system that remembers—via logs, decisions, or configurations—must expose the logic behind its actions. Traceability is not optional in auditable or regulated environments.

   Rationale: Rule lineage is essential for accountability and trust.

5. **Use patterns of non-compliance to refine Order**

   Resistance, evasion, and exception are not just noise—they are signals. They reveal mismatches between policy, context, and Purpose.

   Rationale: Every act of defiance is feedback. Don't just enforce—learn.

6. **If Order cannot be articulated, it cannot be delegated**

   Clear delegation depends on clearly defined constraints. Vague or implicit rules create gaps in responsibility and enforcement.

   Rationale: Ambiguity in Order undermines reliability and introduces structural risk.

7. **Use static Order to govern dynamic behavior**

   When systems must operate under pressure or in real time, rely on preconditions, role boundaries, and topology—not just runtime logic. Static scaffolds create predictable patterns of response.

   Rationale: Dynamic behavior becomes safe only when structured by stable Order.

8. **Treat Order as composable, not monolithic**

   Avoid brittle, centralized policy engines. Design modular constraint layers per domain or interface, and compose them into systemic coherence.

   Rationale: Structural diversity requires constraint modularity. Compose Order as you compose systems.

8.4. Realization Across Implementation Planes

Order is not realized all at once, nor in a single place. It manifests across multiple implementation planes—each with its own language, constraints, and enforcement mechanisms. What begins as a structural intention may surface as policy, code, protocol, contract, or interaction pattern. These surfaces—and the distinct ways Order materializes across planes—are summarized in Table 8-4.

Table 8-4. *Common Surfaces of Order Across System Planes—Their Agents, Enforcement Mechanisms, and Architectural Purposes*

System type/ plane	Common order surfaces	Primary agent roles	Mechanism of enforcement	Architectural purpose
Social/ organizational	Policies, procedures, contracts, codes of conduct	Managers, regulators, committees, employees	Delegation chains, audits, disciplinary actions	Align behavior with institutional Purpose and accountability structures.
Physical/ operational	Safety signage, control panels, spatial layouts, maintenance schedules	Operators, technicians, inspectors	Physical interlocks, procedural routines, safety checks	Ensure predictable, safe execution of actions in material environments.
Digital/ software	Business rules, access-control lists, validation rules, workflow engines	Developers, system administrators, users, automated agents	Code logic, rule engines, event triggers	Maintain integrity, consistency, and permissions in computational systems.

To design Order effectively, architects must ensure that it travels coherently across these planes—so that the same constraint is recognizable to a user, enforceable by a system, traceable in memory, and aligned with Purpose. This section outlines how Order becomes real—not just declared.

8.4.1. Manifestation in the Social and Physical Planes

Order begins where behavior begins: in the human domain. Before any system enforces a rule, someone must interpret a boundary, make a decision, or choose to comply. In this sense, the social plane is the first and most persistent site of Order. It governs interaction through roles, norms, permissions, obligations, and escalation paths. These constraints may be informal or institutional—but they are always present. Every organization has a way of saying what is allowed, who may act, who must be consulted, and what happens when things go wrong.

In the physical plane, Order is manifested through spatial access, visibility, and physical control. Rooms with restricted entry, badge readers, surveillance systems, and separation of equipment zones are not just security measures—they are physical expressions of Order. They constrain what can be done, by whom, and under what conditions. Even seating arrangements, proximity to leadership, or workstation access can signal Order through physical means.

Social and physical Orders are often intertwined. An employee may have formal access to a system but hesitate to use it due to perceived status boundaries. A door may be unlocked, but unwritten norms discourage entry. These constraints may not be codified—but they shape behavior as surely as any policy engine.

Architecturally, these planes matter because they represent both the origins and endpoints of enforcement. If Order is designed without regard to how it will be interpreted socially or enacted physically, it becomes brittle or misaligned. If Order cannot be seen, sensed, or explained at the human level, it risks being ignored—regardless of what the digital plane dictates.

Designing Order, therefore, means aligning structural intent with social understanding and physical experience. The goal is not total control, but legible, navigable constraint—so that agents can act with autonomy while staying within bounds they recognize and respect.

8.4.2. Order in the Digital Plane

Order becomes enforceable when it manifests through real digital constructs. While its intent may begin as architecture or policy, its operational force emerges only when embedded in code, protocols, interfaces, and machine-interpretable logic. These constructs do not merely document boundaries—they *create* them. They make rules actionable, permissions traceable, constraints testable, and violations observable.

Order manifests across three distinct digital scopes: *within the unit, across the system of units*, and *at the boundaries where the unit interfaces with the external world.* These scopes align with how sociotechnical systems are constructed—autonomous, accountable, and yet interdependent.

Internal Enforcement: Within the Unit

Order within the unit governs how its constituent agents—human, automated, or hybrid—interact, decide, and behave. A unit may include operators, analysts, workflows, bots, and machine learning models—all contributing to the Function of the whole. Without internal enforcement of Order, the unit risks collapsing into inconsistency, conflict, or drift. Boundaries must be explicit, interpretable, and enforceable—so that decisions are legitimate, behaviors are predictable, and accountability is traceable.

Digital systems enforce internal Order through well-defined constructs that fall into five categories:

1. **Role and Capability Boundaries**

   These determine who can do what inside the unit. They protect systems from inappropriate use, accidental conflict, or privilege escalation.

   - *RBAC (Role-Based Access Control)*: Defines permissions based on formal roles (e.g., engineer, supervisor, auditor)
   - *ABAC (Attribute-Based Access Control)*: Grants access based on contextual properties (e.g., location, time, task context)
   - *Task Eligibility Rules*: Specify which agents—human or machine—are permitted to perform specific actions

- *Capability Constraints*: Limit the scope of actions an agent can perform, preventing overreach

Purpose: Prevent capability from becoming chaos

2. **Authority and Decision-Making Boundaries**

   These constructs clarify who has decision rights and under what conditions escalation or override is required.

   - *Delegation Rules*: Define when decisions may be passed to others and with what limits
   - *Escalation Protocols*: Trigger when issues exceed local authority or require broader visibility
   - *Override Conditions*: Permit rule-breaking in emergencies—but within structured guardrails
   - *Voting or Consensus Mechanisms*: Govern collective decisions among agents

   Purpose: Ensure that decisions are not just made—but made legitimately.

3. **Interaction and Coordination Constraints**

   Agents must synchronize, avoid conflict, and cooperate—especially in distributed or hybrid settings.

   - *Concurrency Controls*: Prevent conflicting actions through locking, queuing, or mutual exclusion
   - *Interaction Protocols*: Define rules of engagement between agents
   - *Collaboration Constraints*: Shape multi-agent workflows, including human–AI teaming
   - *Conflict Resolution Policies*: Define how overlaps or disagreements are resolved

   Purpose: Preserve functional integrity through structured coordination.

4. **Behavioral and Ethical Enforcement**

   Agents must behave in ways that reflect the values and obligations of the unit.

   - *Codes of Conduct*: Digitally enforce expectations around fairness, safety, and respect.
   - *Bias Mitigation Rules*: Adjust outputs from AI or analytics pipelines to reduce discriminatory effects.
   - *Monitoring and Intervention Triggers*: Detect anomalies and activate human review.
   - *Transparency Requirements*: Require agents to disclose actions, reasoning, or status.

   Purpose: Make behavior not just allowed—but accountable.

5. **Local Governance and Accountability**

   Order must be traceable and improvable.

   - *Audit Trails*: Record decisions, actions, and events.
   - *Rule Provenance Metadata*: Track the authorship and history of constraints.
   - *Compliance Checks*: Validate that agents operate within bounds.
   - *Feedback Loops*: Allow agents to report inefficiencies or conflicts.

   Purpose: Turn behavior into something observable, explainable, and correctable.

Together, these constructs ensure that a unit—however complex—behaves as a *unit*. Not through control, but through enforceable coordination and bounded autonomy.

Systemic Enforcement: Between Integrative and Operational Units

In large sociotechnical systems, no unit operates in isolation. Integrative units—such as global headquarters, regional command structures, or shared governance functions—exist not to centralize action but to set boundaries, coordinate priorities, and enable alignment across operational units. These integrative-operational relationships require more than structure or intent—they demand enforceable, interpretable Order.

Systemic enforcement ensures that operational units do not drift, duplicate, or conflict as they pursue their Purpose. It defines the boundary conditions under which autonomy becomes sustainable: what actions are permissible, what outcomes are expected, and how accountability is distributed.

1. **Shared Rules and Policy Inheritance**

   Rules often originate in integrative units and must propagate reliably to operational units.

   - *Global Operating Policies*
   - *Inherited Rule Sets*
   - *Contextual Overrides*

   These rules shape how operational units act, decide, and expose their capabilities.

2. **Platform-Level Governance Infrastructure**

   To scale Order across units, shared services are required.

   - *Centralized Policy Registries*
   - *Distributed Enforcement Engines*
   - *Systemic Audit Trails*

   Such infrastructure *enables* autonomy without losing alignment.

3. **Structural Permissions and Constraint Surfaces**

   Order governs not just intent—but allowed action.

   - *Budgetary Boundaries*
   - *Coordination Protocols*
   - *Rate Limits and Quotas*
   - *Territorial Constraints*

   These permissions form the *invisible architecture* of enforceable autonomy.

4. **Multi-Unit Decision and Escalation Logic**

   Systems must handle shared risk and distributed decisions.

   - *Mandated Escalation Paths*
   - *Predefined Coordination Triggers*
   - *Shared Risk Protocols*

   This logic makes coordination *predictable* rather than ad hoc.

Systemic Order enables a federation of autonomous capabilities to act as a unified whole—*not through control, but through designed coherence.*

External Enforcement: At the Boundaries

Every unit interfaces with the world beyond itself. This boundary zone—customers, partners, regulators, investors, and communities—is where Order becomes *visible* to outsiders. It shows what the unit can and cannot do, what is permitted, and what protections exist.

Enforcement here ensures legitimacy, safety, and interpretability at the edge. It shapes how trust is earned, how constraints are perceived, and how obligations are met.

1. **Customer and Consumer Boundaries**

   - *Terms of Service*
   - *Consent Mechanisms*
   - *Transactional Constraints*
   - *Usage Limits*

   These create enforceable expectations of use and protection.

2. **Supplier and Partner Boundaries**

   - *Delegation and Escalation Rules*
   - *Compliance Clauses*
   - *Service-Level Conditions*
   - *Disruption Protocols*

   These enable stable, accountable interdependence.

3. **Regulator and Authority Boundaries**
    - *Licensing Constraints*
    - *Reporting Protocols*
    - *Inspection Triggers*
    - *Provenance Requirements*

    Here, Order becomes a mechanism of *legal accountability*.

4. **Public, Community, and Societal Boundaries**
    - *Transparency Policies*
    - *Ethical Rule Sets*
    - *Engagement Rules*
    - *Emergency Protocols*

    Order at this level moves from policy to *social obligation*.

Together, these layers—internal, systemic, and external—reveal that Order is not a single mechanism but a distributed fabric of enforceable constraints. It stabilizes the unit internally, aligns it within a coordinated system, and legitimizes it at the boundary of interaction. These manifestations are architectural: they determine how behavior is bounded, how decisions are traced, and how rules become operational. Without them, Order remains aspirational—an idea unanchored in implementation. With them, it becomes real—*visible in behavior, traceable in action, and governable by design.*

8.4.3. Implementation Surfaces and Sociotechnical Contexts

To be enforceable, Order must be implemented as a surface—a constraint, rule, permission, or policy interface where an agent can *experience, interpret,* or *respond to* what the system allows, forbids, or requires. These surfaces are not merely technical—they are sociotechnical. They mediate between designed boundary conditions and the realities of agent context, authority, role, and decision logic. Order becomes operational only when its limits are made intelligible—whether to humans, machines, or mixed-agent teams.

Typical Order surfaces include

- *Access and Permission Interfaces*, which define what an agent is allowed to do. These may appear as user-facing toggles, disabled options, error messages, or permissions matrices—but always translate underlying access control logic into a surface that constrains or enables behavior. For artificial agents, the equivalent may be ABAC rules, feature flags, or scoped capabilities enforced through APIs.
- *Rule and Policy Editors*, where Order is configured, declared, or adjusted. These surfaces serve operational governance teams, compliance managers, or business owners. They often expose structured rule constructs—business rules, triggers, exceptions, and thresholds—in editable or declarative form. The key is not only enabling configuration, but exposing the *governable intent* of the rule.
- *Violation Feedback and Explanation Surfaces*, which inform agents when Order has been breached or invoked. These may include inline error messages, policy explanations, audit logs, or override prompts. In multi-agent environments, they must provide just enough context to resolve ambiguity without overloading the interface. For AI systems, these surfaces may take the form of logs, observability dashboards, or human-in-the-loop review prompts.
- *Governance and Audit Interfaces*, which expose the *traceability* of Order. These surfaces serve auditors, compliance officers, and regulators. They include dashboards for rule coverage, change histories, escalation events, and agent-specific compliance footprints. Their goal is not interaction but *transparency*—making the architecture of Order legible over time.
- *Sociotechnical Touchpoints*, where Order is embedded in practices, artifacts, or norms rather than software. These include contracts, checklists, escalation paths, or physical processes that reflect rules too complex or contextual to automate. In such cases, the surface is often procedural—but no less real.

Each surface reflects a different sociotechnical context. A supervisor evaluating override eligibility during a crisis needs different signals than a machine agent enforcing a rate limit. A regulator validating license compliance requires different traceability than a frontline user deciding whether to submit a form. Designing Order across these contexts means not only encoding the rule—but matching its *exposure* to the agent's role, expectation, and trust level.

Across all surfaces, the principle remains the same: *Order must be visible, interpretable, and enforceable at the point of engagement.* If it is buried, fragmented, or overly opaque, the system becomes unstable—behaviors drift, decisions go unchecked, and the boundary between legitimacy and violation becomes blurred.

8.5. Strategic Impact of Order Interfaces

Order interfaces are not only operational—they are strategic. When well-designed, they shape how a system is governed, trusted, and scaled. They determine who can do what, under what conditions, and with what level of visibility or recourse. These interfaces are the architecture of control—and the quality of their design has lasting consequences for compliance, resilience, and institutional legitimacy. Whether formalized as access portals, policy dashboards, audit logs, or interactive permissions, Order interfaces are where architecture meets governance. Their impact is not limited to enforcement—it extends to how systems evolve, how violations are mitigated, and how organizations adapt to changing regulatory and ethical demands.

8.5.1. Nuances of Regulation

Order interfaces govern how systems behave under constraint—but the act of governing itself is filled with tension. Regulation is not a singular condition to satisfy—it is a dynamic field shaped by conflicting rules, evolving expectations, and competing sources of authority. Architects must design Order in ways that go beyond simple compliance, navigating nuanced boundaries between competing demands. Three core tensions define this complexity:

Enforceability vs. Interpretability

Order is expressed through multiple design constructs—rules, constraints, policies, access models, governance layers. Each offers different degrees of precision, performance, and ambiguity:

- *Rules* define logic but may struggle with exceptions.
- *Constraints* enforce boundaries but resist nuance.
- *Access models* control permission but must reflect social realities.
- *Policies* are normative but may lack technical expression.
- *Governance constructs* record decisions, but lag behind live behavior.

Together, these constructs make Order enforceable—but with trade-offs. High precision may reduce human flexibility; high interpretability may undermine automation. Well-architected systems treat these constructs as modular components in a dynamic governance stack, selecting and layering them intentionally to support both legitimacy and agility.

Internal, Systemic, and Environmental Regulation

Order must operate simultaneously across three scales:

- *Internally,* it governs relationships within the unit—agent permissions, decision rights, coordination protocols.
- *Systemically,* it binds units into organizational wholes—integrating operational and integrative components through escalation rules, budget constraints, and delegated authority.
- *Environmentally,* it interfaces with external entities—customers, suppliers, regulators—through policies, contracts, disclosure boundaries, and ethical safeguards.

Each layer imposes different requirements, rhythms, and degrees of negotiability. Internal order can be tightly coupled with structure; systemic order must balance cohesion and autonomy; environmental order must be adaptive yet demonstrably legitimate. Misalignment across these layers produces drift, fragmentation, or breakdown.

Internally Defined vs. Externally Imposed

Order can arise from within the system—or be imposed from without:

- *Internally defined* rules reflect the unit's own strategy, ethics, or culture. They are shaped by organizational values, risk appetite, or innovation priorities.
- *Externally imposed* rules come from regulators, laws, industry standards, or public expectations. These may be inflexible, ambiguous, or misaligned with internal logic.

Tensions arise when internal innovation collides with external compliance, or when global mandates conflict with local norms. Architecture must account for both—creating override paths, escalation protocols, and traceable rationale for rule interpretation.

8.5.2. Measurable Benefits and Architectural Payoffs

Well-designed Order does more than constrain—it stabilizes, legitimizes, and scales behavior. It reduces uncertainty, distributes governance, and enables both human and artificial agents to act with confidence in complex environments. The benefits of such Order are not abstract; they can be observed in the reduction of behavioral drift, the acceleration of regulatory alignment, and the increased resilience of decision-making under pressure. Its architectural payoff lies in both behavioral coherence and institutional legitimacy.

A. **Behavioral Coherence Reduces Risk and Drift**

When boundaries are clearly expressed—through rules, constraints, and permissions—agents know what is allowed, what is expected, and what will be enforced. This behavioral clarity leads to

- Fewer violations, incidents, or escalations
- Greater consistency across teams, systems, and environments
- Lower risk of shadow behavior or policy circumvention

Payoff: Coherent Order reduces variance. Systems behave more predictably—even under scale, stress, or change.

B. **Embedded Governance Accelerates Regulatory Alignment**

When Order is implemented through modular constructs—ABAC, policy engines, audit trails—it becomes easier to align systems with evolving laws, standards, and ethics. This embeddedness enables

- Faster adaptation to new regulatory requirements
- Automated enforcement of compliance controls
- Reduced manual oversight and auditing effort

Payoff: Embedded Order makes governance proactive, scalable, and less dependent on reactive processes.

C. **Transparent Boundaries Enable Trust and Autonomy**

When Order is made visible—through user-facing terms, system-exposed constraints, and explainable rule logic—agents can operate autonomously *within known limits*. This enables

- Increased trust from external stakeholders
- Safer delegation to human or AI agents
- Faster decision-making with fewer bottlenecks

Payoff: Transparent Order fosters aligned autonomy. Agents can act decisively without exceeding their mandate.

D. **Layered Enforcement Strengthens Institutional Memory**

When Order constructs are traceable—versioned, logged, and tied to decisions—they create a record of why the system behaves as it does. This institutional memory supports

- Post-event analysis and continuous improvement
- Accountability in multi-agent systems
- Knowledge transfer across teams and generations

Payoff: Layered Order preserves context. Behavior remains interpretable even as actors, tools, and conditions evolve.

E. **Adaptive Boundaries Enable System Resilience**

When Order is designed to evolve—through dynamic rules, contextual overrides, and modular governance—it allows the system to respond intelligently to change. This adaptability supports

- Rapid reconfiguration during crises or disruptions
- Context-sensitive enforcement without full redeployment
- Ongoing alignment with shifting environments and agent roles

Payoff: Adaptive Order makes systems resilient. Boundaries flex without breaking, enabling continuity under volatility.

F. **Distributed Governance Enhances Scalability**

When Order is decentralized—implemented through local rule sets, agent-specific constraints, and federated policies—it allows governance to scale across geographies, teams, and modalities. This distribution enables

- Local autonomy with systemic consistency
- Reduced bottlenecks in operational decision-making
- Enforceable rules across heterogeneous agent networks

Payoff: Distributed Order scales governance. Systems grow without losing control.

G. **Normative Alignment Strengthens Ethical Integrity**

When Order reflects not just operational constraints but normative commitments—such as fairness, transparency, and sustainability—it embeds values into behavior. This alignment enables

- Ethical decision-making by human and artificial agents
- Mitigation of reputational and social risks
- Proactive alignment with emerging standards and expectations

Payoff: Normative Order embeds integrity. Systems behave not just legally but ethically.

H. **Interoperable Boundaries Improve System Integration**

When Order constructs are standardized—using shared schemas, rule languages, or governance APIs—they enable frictionless interaction across internal units and external platforms. This interoperability supports

- Seamless integration with partner systems and ecosystems
- Lower effort in onboarding new capabilities or teams
- Consistency in multi-platform and cross-domain coordination

Payoff: Interoperable Order connects systems. Boundaries become bridges, not barriers.

I. **Predictive Governance Enhances Strategic Foresight**

When Order is instrumented—tracking rule invocation, boundary violations, and override patterns—it becomes a source of strategic insight. This instrumentation enables

- Early detection of drift, non-compliance, or stress
- Continuous refinement of policy and constraint models
- Governance models that evolve based on observed system behavior

Payoff: Predictive Order informs foresight. Governance becomes anticipatory, not merely reactive.

The architectural value of *Order* lies not only in what it restricts, but in how intentionally, intelligibly, and sustainably it defines boundaries. Measurable benefits include greater behavioral consistency, scalable governance, embedded compliance, ethical alignment, and system-wide resilience. When Order is treated as a first-class design concern—not as a static rule-set or administrative burden—it becomes a systemic enabler of clarity, legitimacy, and control. It allows systems to grow, adapt, and interact—without losing their shape.

STRATEGIC AUDIT CHECKLIST FOR ORDER INTERFACES

System architects can use the following questions to evaluate whether Order interfaces are adequate, adaptable, and transparent:

Legibility and Trust

- Can participants clearly see who can act, under what conditions, and with what oversight?
- Are permissions, obligations, and recourse pathways visible to both human and machine agents?

Adaptability

- Can rules and access conditions be revised without destabilizing dependent systems?
- Do interfaces separate mutable policy logic from fixed enforcement code?

Accountability

- Is every rule or policy traceable to an accountable role or governance body?
- Are decision logs and exceptions automatically captured for audit and learning?

Proportionality

- Does the interface grant the minimum necessary power to perform legitimate functions?
- Are escalation channels defined for when rules conflict or authority is ambiguous?

Transparency and Resilience

- Can observers understand why an action was permitted or denied?
- Do monitoring and alerting mechanisms reveal misuse or drift before it becomes systemic?

Ethical and Regulatory Alignment

- Are ethical principles and regulatory constraints encoded, not just documented?
- Can compliance be demonstrated through evidence produced by the interface itself?

Evolution Readiness

- Can new forms of agents (e.g., AI decision-makers) be incorporated without redesigning the entire control layer?
- Is there a versioning path for rules that preserves history and legitimacy?

Well-designed Order interfaces answer these questions visibly and repeatedly—turning governance from an afterthought into a living architectural property.

8.6. Implementation Patterns

Order is not enforced through good intentions—it is architected, exposed, and sustained through patterns. These patterns reflect recurring ways that Order is implemented across sociotechnical systems. They help designers ensure that boundaries are not merely conceptual, but practical—usable, interpretable, and enforceable.

Unlike principles and heuristics, which shape the *design intent*, implementation patterns reveal how Order takes architectural form. Whether encoded in access controls, policy engines, rule graphs, or observability layers, each pattern addresses a structural challenge in governing behavior. The following patterns offer a coherent vocabulary for implementing Order across diverse systems and contexts:

8.6.1. Constraint Graphs

Implements Order through an interconnected set of interdependent constraints.

Constraints are modeled as nodes and relationships within a graph—allowing for propagation, cascading rule enforcement, and the detection or resolution of conflicts. This pattern is common in scheduling, entitlement systems, and resource allocation.

Payoff: Enables systemic coherence and conflict-aware enforcement in environments with complex rule interactions.

Example: An airline scheduling system models pilots, aircraft, and flight regulations as a constraint graph, automatically re-routing crew assignments when weather delays or compliance limits are breached.

8.6.2. Policy-As-Code

Encodes Order in executable, version-controlled code artifacts.

Policies are expressed declaratively in languages such as Rego (OPA) or Cedar (AWS), allowing integration into CI/CD pipelines and infrastructure-as-code practices. They become testable, reviewable, and deployable like any other software component.

Payoff: Makes Order programmable, auditable, and operationally scalable across environments.

Example: A financial institution uses policy-as-code to embed transaction-limit rules directly into its cloud infrastructure, blocking non-compliant configurations before deployment.

8.6.3. Dynamic Boundaries

Enables Order to adapt in real time based on system context, environmental signals, or agent behavior.

Boundaries shift according to load, risk profile, or task type. This is critical in dynamic access control, security hardening, or adaptive throttling—where rigid boundaries would create fragility.

Payoff: Supports resilience and responsiveness without sacrificing control or safety.

Example: A hospital's data platform automatically relaxes access controls for emergency responders during declared crises, then restores normal restrictions when conditions stabilize.

8.6.4. Tokenized Permissions

Uses cryptographically verifiable tokens to encode, delegate, and revoke permissions.

Tokens such as OAuth scopes, signed JWTs, or custom permission artifacts define scope, conditions, and validity. They are portable, inspectable, and enforceable at the point of invocation.

Payoff: Enables granular, distributed, and revocable enforcement of Order in multi-agent ecosystems.

Example: A logistics network issues signed tokens to partner APIs, granting temporary access to shipment data that expires once delivery is confirmed.

8.6.5. Access Gate

Implements Order through explicit checkpoints placed at invocation surfaces.

Before a capability is accessed, a gate validates the agent's role, state, or context. This may occur at the API layer, UI layer, or middleware component. The gate blocks unauthorized actions and clarifies expectations.

Payoff: Prevents unauthorized access and ensures that invocation only proceeds under valid boundary conditions.

Example: A machine-learning platform routes every model-training request through an access gate that checks data-classification level and reviewer approval before allocating compute resources.

8.6.6. Behavioral Envelope

Defines a bounded range of acceptable behavior, rather than discrete binary rules.

Agents are free to operate within a defined envelope—such as resource limits, budget caps, or deviation thresholds. Violations trigger adaptive constraints or alerts, not hard failure.

Payoff: Balances autonomy with safety, enabling flexibility while ensuring predictability.

Example: A trading algorithm is allowed to adjust portfolio weights within a 3% deviation band; exceeding that envelope automatically triggers supervisory review.

8.6.7. Governance Mesh

Distributes Order enforcement across nodes, agents, or units in federated systems.

Each unit (e.g., a regional team, microservice, or device) enforces local constraints while syncing with global policy. The mesh ensures coherence without relying on centralized enforcement.

Payoff: Scales Order horizontally across regions, teams, or modalities while preserving alignment.

Example: A multinational bank enforces anti-money-laundering rules through a governance mesh where each country node applies local limits but syncs with global reporting standards nightly.

8.6.8. Explainable Order

Implements Order through rules that include embedded rationale, traceability, and provenance.

Each rule is accompanied by human-readable explanations, source metadata, and justification logic. This transparency supports user trust, organizational learning, and regulatory defensibility.

Payoff: Makes Order interpretable by both humans and agents—critical for hybrid, AI-driven systems.

Example: A public-sector AI system attaches explanatory notes to each denied application, citing the policy clause and decision logic that triggered the outcome.

8.6.9. Audit-Driven Order

Uses traceability not only for compliance, but as a feedback mechanism for design and adaptation.

All rule evaluations, constraint applications, and agent actions are logged and made observable. Audits become part of operational intelligence—supporting real-time alerts, postmortems, and adaptive refinement.

Payoff: Provides accountability, historical insight, and evidence for adjusting Order dynamically.

Example: An energy-grid operator uses audit-driven order to analyze every emergency override, updating escalation thresholds based on lessons learned from real incidents.

8.6.10. Interaction-Aware Enforcement

Designs interfaces to expose Order early—making constraints visible before invocation.

This includes UI affordances (e.g., disabled buttons, greyed-out options), inline validation, and feedback mechanisms that prevent rule violations before they occur. It prevents agent confusion and encourages compliant behavior by design.

Payoff: Reduces friction and failure by aligning user or agent behavior with boundary conditions at the point of engagement.

Example: A procurement portal disables submission buttons when spending requests exceed delegated authority, showing an immediate explanation and escalation path.

These implementation patterns translate abstract boundary logic into concrete enforcement surfaces, mechanisms, and behaviors. Each pattern addresses a distinct architectural challenge: from decentralized governance to adaptive enforcement, from explainability to interaction design.

Together, they form an implementation grammar for enforceable boundaries—ensuring that systems not only function but behave. By using these patterns deliberately, architects turn Order into a living, evolving mechanism of coherence, accountability, and trust.

8.7. Interlocking Dependencies and Alignment

Order does not operate in isolation. In the Purpose Driven Design framework, it is a *regulatory backbone*—not a mere constraint layer but a structural necessity that ensures all other constructs operate coherently, responsibly, and predictably. While other constructs define what the unit is and does, *Order determines what is permitted*, what is bounded, and under what conditions those capabilities remain legitimate.

Order shapes the behavior of the system—not by creating capability, but by *governing it.* Its alignment with other constructs is what makes the unit *interpretable, governable, and viable*—especially in complex or evolving sociotechnical environments.

8.7.1. Alignment with Purpose: Legitimizing Intent

Purpose defines why a unit exists. Order defines the *acceptable expressions of that Purpose.* Not all possible interpretations of Purpose are permitted—Order introduces structural boundaries to prevent overreach, drift, or abuse. Through formal constraints and enforceable governance, Order ensures that Purpose is *interpreted narrowly enough to remain safe, yet broadly enough to remain useful.*

Without Order, Purpose risks becoming a justification for unchecked behavior. With Order, Purpose becomes *governable intent.*

8.7.2. Alignment with Value: Regulating Exchange

Value cannot be delivered indiscriminately. Order constrains how Value is accessed, under what terms, by whom, and with what consequences. These constraints—policies, rate limits, eligibility rules, consent gates—ensure that Value is delivered *equitably, ethically, and sustainably.*

Order protects both the provider and recipient of Value, converting raw potential into *structurally appropriate exchange.*

8.7.3. Alignment with Function: Bounding Invocation

Function without Order is overexposed. It may be invoked at the wrong time, by the wrong agent, under the wrong conditions. Order determines the *legal, contextual, and operational constraints* of every Function: who can use it, what inputs are valid, what conditions must be met.

Well-aligned Order makes Function *invocable but governable*—exposed without becoming exploitable.

8.7.4. Alignment with Process: Enforcing Consistency

Process describes how things get done. Order determines *whether those processes are legitimate*—by defining who may initiate them, how they unfold, when exceptions are allowed, and what auditability must be maintained.

Order ensures that Process is not just efficient, but *accountable*—and capable of withstanding scrutiny.

8.7.5. Alignment with Structure: Containing Authority

Structure distributes capability across roles, teams, and units. Order defines *what each role is allowed to do*, and under what limits. It sets delegation rules, escalation paths, and separation-of-duty boundaries that prevent abuse and ensure proportional responsibility.

Without Order, Structure becomes ambiguous or *unsafe*. With Order, Structure becomes *legitimate and bounded.*

8.7.6. Alignment with Culture: Operationalizing Norms

Culture expresses shared beliefs, values, and expectations. Order *translates those norms into enforceable boundaries*—such as codes of conduct, ethical policies, and behavioral standards. Where Culture says, "what we believe," Order says, "what we permit."

Order *makes Culture actionable*—even in the presence of diverse agents or mixed motivations.

8.7.7. Alignment with Memory: Constraining What Is Retained

Memory captures the history of system events. Order constrains what is recorded, who may access it, how long it is kept, and what form it takes. These boundaries protect privacy, support compliance, and ensure interpretability over time.

Order ensures that Memory is not just a log—but a *regulated, intentional record.*

8.7.8. Alignment with Storage: Enforcing Conditions of Retention

Storage holds Value, Process artifacts, and Memory. Order defines *how Storage may be used*—through encryption, access controls, redundancy rules, and data classification policies. It ensures that what is stored is both *safe and compliant.*

Storage without Order risks becoming a liability. With Order, Storage becomes a *controlled asset.*

Order does not replace the other constructs—but **it** *renders them viable.* It defines the *operational legitimacy* of Purpose, the *fair delivery* of Value, the *safe invocation* of Function, the *accountability* of Process, the *boundaries* of Structure, the *enforceability* of Culture, the *privacy and auditability* of Memory, and the *protection and governability* of Storage.

In this way, Order is the *meta-construct* that upholds the structural integrity of the unit. It is what makes the system *not only functional—but lawful, ethical, scalable, and safe.*

8.8. Failure Modes and Antipatterns

Order is not optional—it is what enables sociotechnical systems to behave in bounded, interpretable, and governable ways. But when Order is poorly designed, inconsistently enforced, or allowed to decay, it ceases to function as a structural asset. The result is not just friction or inefficiency—but architectural failure. Without effective Order, systems lose coherence, autonomy becomes unmanageable, and accountability collapses.

This section identifies recurring antipatterns that emerge when Order is misaligned, fragmented, or misplaced. These are not mere policy flaws—they are architectural symptoms of deeper design breakdowns.

1. **Disembodied Order**

   Definition: Rules or constraints exist in policy or documentation, but are not anchored in any structure, role, or enforcement mechanism.

   Symptoms: No one is accountable; rules are ignored or inconsistently interpreted.

   Consequence: Performative governance, trust erosion, rising exceptions.

   Architectural Fault: Order detached from Structure or Function.

2. **Purpose-Blind Constraint**

   Definition: Constraints are imposed without alignment to the unit's Purpose or Value.

   Symptoms: Legacy rules persist despite harm; processes are followed without justification.

   Consequence: Agent frustration, misalignment, eventual subversion.

   Architectural Fault: Order unanchored from Purpose or Value.

3. **Overconstraint and Static Rigidity**

   Definition: Order is too rigid to accommodate exceptions, evolution, or ambiguity.

   Symptoms: Frequent rule violations, growing shadow systems.

Consequence: Fragility, workarounds, divergence between declared and actual behavior.

Architectural Fault: Lack of adaptability; static enforcement mechanisms.

4. **Fragmented Order**

   Definition: Different units or subsystems enforce incompatible rules or standards.

   Symptoms: Conflicting constraints; duplicated effort; inconsistent user experience.

   Consequence: System incoherence, unscalable governance.

   Architectural Fault: Distributed Order without integration logic.

5. **Misplaced Order Ownership**

   Definition: Order is defined or enforced by agents lacking authority or context.

   Symptoms: Delays, unresolvable disputes, finger-pointing.

   Consequence: Governance bottlenecks, accountability confusion.

   Architectural Fault: Misaligned Structure and Governance.

6. **Invisible Order**

   Definition: Order exists in code, scripts, or habits—but is undocumented and opaque.

   Symptoms: Unexpected behavior; reliance on tribal knowledge.

   Consequence: Onboarding friction, compliance risk, system brittleness.

   Architectural Fault: Lack of exposure surfaces and observability.

7. **Order Without Feedback**

   Definition: No mechanism exists to assess rule effectiveness or adapt them.

   Symptoms: Obsolete rules remain in force; violations accumulate unnoticed.

Consequence: Governance decay, unresponsive systems.

Architectural Fault: Missing Memory and adaptive governance logic.

8. **Order Externalized Without Context**

   Definition: Constraints are imposed externally without regard for local realities.

   Symptoms: Misfit mandates, frontline burden, quiet resistance.

   Consequence: Subversion, non-compliance, governance inversion.

   Architectural Fault: Environmental Order misaligned with internal constraints.

9. **Contradictory Order Surfaces**

   Definition: Different interfaces express conflicting rules (e.g., API vs. UI).

   Symptoms: Errors, automation failure, inconsistent outcomes.

   Consequence: Fragile integrations, increasing exceptions.

   Architectural Fault: Lack of cross-surface coordination; design incoherence.

10. **Absent Boundary Conditions**

    Definition: Units operate without constraints set by integrative structures.

    Symptoms: Local optimization, escalation confusion, systemic drift.

    Consequence: Fragmentation, misaligned behavior, strategic incoherence.

    Architectural Fault: Missing Systemic Order from integrative units.

These antipatterns expose structural vulnerabilities that compromise the role of Order in sociotechnical systems. They are not isolated implementation errors—they are indicators that the architecture has failed to integrate Order as a first-class design construct. Recovering from these failures requires more than new rules. It requires restoring clarity, alignment, and enforcement mechanisms that are legitimate, contextual, and adaptive. These failure modes—and the architectural corrections they require—are summarized in the diagnostic framework shown in Table 8-5.

Table 8-5. *Order Diagnostic Summary*

Antipattern	Core symptom	Likely cause	Diagnostic cue	Corrective action
Disembodied order	Rules exist on paper but lack enforcement	Order detached from Structure or Function	Repeated rule violations with no accountability	Anchor each rule to a role, process, or enforcement surface.
Purpose-blind constraint	Constraints persist without justification	Order unaligned with Purpose or Value	Agents question rule rationale or ignore it	Review constraints for relevance; retire or redesign outdated ones.
Overconstraint and static rigidity	Frequent workarounds or shadow systems	Excessive or outdated rules	High volume of exceptions or override requests	Recalibrate constraints; introduce adaptive or dynamic boundaries.
Fragmented order	Inconsistent rules across units or tools	Lack of systemic integration	Conflicting policies or duplicated governance	Consolidate into shared policy registries; enforce integration checks.
Misplaced order ownership	Governance bottlenecks and finger-pointing	Misaligned Structure and governance roles	Delays or unresolved disputes over who enforces rules	Assign legitimate stewards; clarify escalation and decision rights.
Invisible order	Undocumented constraints embedded in code or habit	Lack of exposure surfaces and observability	Teams rely on tribal knowledge to interpret behavior	Surface implicit rules through documentation, dashboards, and audits.

(*continued*)

Table 8-5. *(continued)*

Antipattern	Core symptom	Likely cause	Diagnostic cue	Corrective action
Order without feedback	Rules remain unchanged despite repeated violations	Missing Memory and adaptive feedback loops	Governance decay; obsolete rules still in effect	Embed monitoring and feedback; review violations as design input.
Order externalized without context	External mandates misfit local realities	Environmental Order misaligned with internal constraints	Persistent resistance or quiet non-compliance	Contextualize external rules; co-design local adaptation paths.
Contradictory order surfaces	Conflicting constraints across interfaces	Lack of cross-surface coordination	Different rule logic in API, UI, or workflow layers	Synchronize constraint logic across all exposure surfaces.
Absent boundary conditions	Units operate without higher-level constraints	Missing Systemic Order from integrative structures	Local optimizations diverge from enterprise intent	Define clear systemic boundaries; establish integrative policy guardrails.

8.9. Summary: Order As a Design Construct

Order makes architecture governable.

It does not operate at the surface but at the foundation—shaping what behaviors are possible, what actions are legitimate, and what constraints are enforced. A system without Order is not free—it is erratic. It cannot scale its behavior, coordinate its parts, or respond to change without collapsing into ambiguity.

In Unit Oriented Architecture, Order is not reducible to policies, permissions, or access rules. It is the *design layer that enables agents—human or artificial—to act within interpretable constraints.* It ensures that behavior remains bounded, purposeful, and coherent even in the face of autonomy, distribution, or complexity.

Across this chapter, we have shown that

- *Order evolves*—from ritual to bureaucracy to code—mirroring the growth of sociotechnical complexity.
- *Order stabilizes the unit,* aligns it with system-wide coordination, and governs its interface with the external world.
- *Order manifests* socially, physically, and digitally—through protocols, interfaces, constraints, policies, and enforcement mechanisms.
- *Order scales through design*—via architectural principles, heuristics, and implementation patterns that balance local autonomy with systemic coherence.
- *Order affects every other construct*—enabling Function, constraining Process, bounding Culture, and making Purpose observable through permitted behavior.
- *Order fails in patterned ways*—when it is brittle, invisible, overly centralized, or unenforceable.
- *Order pays off*—in system resilience, ethical integrity, clarity of interaction, and the ability to embed rules into living infrastructure.

Ultimately, Order delivers what no other construct can: *architectural governability.* It turns raw capability into bounded action, and isolated interfaces into accountable participation. It allows systems to scale without unraveling, evolve without losing coherence, and operate without constant supervision. It is the construct that makes constraint expressive, permission intentional, and enforcement interpretable.

Realized effectively, Order functions as

- *A constraint layer,* shaping what is allowed, disallowed, or required—at every level of the system.
- *An interpretability engine,* enabling agents to reason about permissible actions, boundaries, and responsibilities.
- *A coordination framework,* ensuring interactions across units, teams, and agents remain aligned and coherent.

- *A governance substrate*, embedding policy, ethics, and regulation into operational systems and interfaces.
- *A structural binder*, connecting Purpose, Function, Process, and Structure through enforceable boundary conditions.

These qualities are not abstractions—they are design outcomes. They are what make Order not simply a set of rules but a living architectural mechanism for shaping viable, ethical, and evolvable sociotechnical systems.

CHAPTER 9

Culture

In the early 1970s, a new research center opened in Palo Alto, California. Funded by Xerox Corporation, the Palo Alto Research Center (PARC) was given an ambitious mandate: to invent the future of information technology. PARC succeeded beyond any measure of technical imagination, pioneering the graphical user interface, the personal computer prototype, object-oriented programming, Ethernet networking, and laser printing. Its researchers created the first graphical user interface, the first personal computer prototype, object-oriented programming, Ethernet networking, and the foundations of laser printing. The building was a crucible of innovation—intellectually rich, socially experimental, and years ahead of its time.

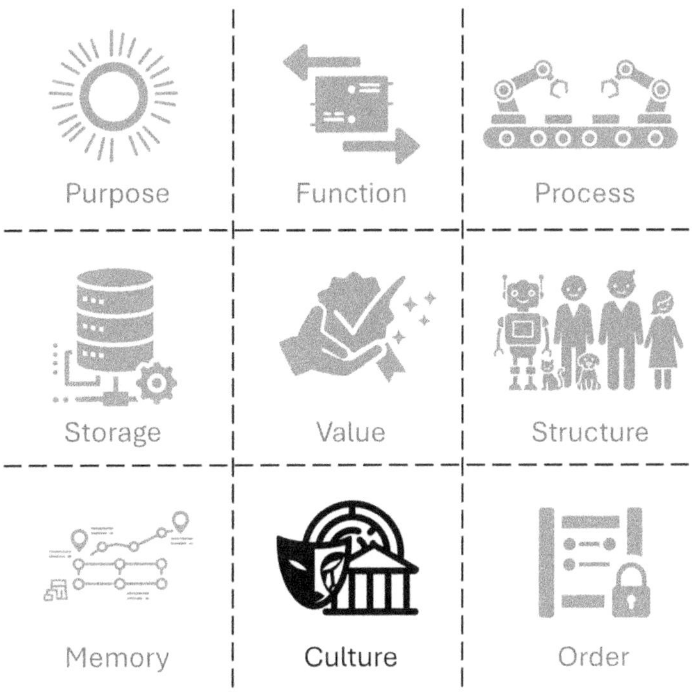

A. Milchman, *The Purpose Driven Design Framework for Modern Enterprises*,
https://doi.org/10.1007/979-8-8688-2223-0_9

But despite these breakthroughs, Xerox failed to capitalize on most of them—a failure rooted not in technology but in culture.

The PARC team had no shortage of insight, talent, or resources. What it lacked was cultural alignment with the broader organization—the shared norms, priorities, and interpretive habits that allow innovation to connect with institutional reality. Xerox, headquartered on the East Coast and rooted in the copier business, saw computing as peripheral. Executives were skeptical of work they didn't understand. Field sales teams resisted technology they couldn't sell easily. Strategic focus remained on protecting copier revenue. When Steve Jobs visited PARC in 1979, he immediately grasped the potential of the graphical interface—and within five years, Apple released the Macintosh. Xerox, by contrast, shelved most of its breakthroughs or buried them in divisions ill-suited to carry them forward.

The failure was not technological. It was cultural—costing Xerox its market leadership and much of the pioneering talent that had defined PARC's success.

PARC had developed its own internal norms: radical openness, interdisciplinary collaboration, and experimentation without immediate commercialization. These norms clashed with the risk-averse, efficiency-driven culture of Xerox as a whole. The researchers lived in one culture; the corporation operated in another. The two structures never integrated. PARC was a unit without a bridge.

This dissonance between breakthrough and integration reveals a deeper architectural insight: culture is not peripheral—it is foundational. It governs not only how people behave, but how systems absorb, interpret, and sustain change. Culture enables Purpose, supports Function, informs Process, and reinforces Order. When culture misaligns with structure, it renders even the most sophisticated architectures inert.

What makes the PARC story enduring is not only the brilliance of what was created—but the fragility of systems that lack cultural coherence. Culture is not an accessory to innovation. It is the invisible infrastructure that determines what is attempted, supported, accepted, or rejected. It shapes how people respond to ambiguity, how they interpret success, and how they allocate attention and trust.

Today, many organizations attempt to "build a PARC" of their own—innovation labs, accelerators, R&D centers. Some succeed. Many fail for the same reason: they reproduce the form of PARC but not its cultural logic. Or worse, they isolate the new culture so completely that the parent system cannot benefit.

What made PARC powerful was not just what it invented—but how it worked. What made it vulnerable was that it belonged to a system that could not understand its culture.

Culture defines how a unit behaves when no one is looking. It encodes the assumptions, expectations, and behaviors that emerge through repetition—not through rule. Unlike Order, which specifies what is permitted, Culture shapes what is normal. Order keeps us from parking in an accessible spot even when we're in a hurry, while Culture moves us to help an elderly person cross the street without anyone telling us to. Culture determines how agents respond to ambiguity, failure, disagreement, and success—when no explicit instruction exists. In this chapter, Culture is examined not as a corporate slogan, but as a structural layer that stabilizes and guides behavior across time and conditions.

9.1. Foundations of Culture As a Design Construct

Culture is often treated as atmosphere—an intangible by-product of leadership style, shared history, or team dynamics. But in Purpose Driven Design, Culture is something far more deliberate: a structural layer that shapes behavior, distributes interpretation, and aligns action without command. It is not what people say they value, but how they act when no one is watching. These contrasting interpretations are made explicit in the comparative view presented in Table 9-1.

Table 9-1. *Contrasting Perspectives on Culture: From Atmosphere to Architecture*

Typical treatment in business literature	Architectural view in Purpose Driven Design
Culture as atmosphere or morale booster	Culture as a structural layer guiding behavior through embedded norms
Focus on leadership tone, values statements, or engagement scores	Focus on systemic design—how roles, interfaces, and practices reinforce norms
Seen as "soft" or complementary to structure and process	Treated as an architectural construct interdependent with Purpose, Structure, and Order
Managed through workshops, slogans, and incentives	Designed through patterns, feedback loops, and lived interaction surfaces

Culture emerges not from intention alone, but from repeated choices, patterned interactions, and embedded assumptions that quietly guide what is normal, acceptable, or rewarded. These assumptions can be surfaced by observing how decisions are justified, which behaviors draw praise or silence, and where informal workarounds persist despite official rules. Like an internal climate, it shapes every decision surface—especially in the absence of rules or oversight. To design for alignment that lasts, Culture must be approached not as ambience but as architecture.

9.1.1. Defining Culture Architecturally

In Purpose Driven Design, Culture is not ambience—it is an architectural construct that governs how behavior is shaped, norms are reinforced, and decisions are made without deliberation. It functions as the internal climate of a unit, composed not of values alone but of expectations, dispositions, and interpretive frames that guide agents—human, artificial, and even nonhuman—toward aligned action. In the face of emerging organizational challenges, especially in automation and digital transformation, culture often becomes a latent catalyst for change. It energizes transformation not just by motivating individuals, but by supplying coherence when formal structures and planning fall short. Yet this same force, indispensable as it is, can be strained when asked to compensate for the absence of design clarity—revealing the limits of cultural momentum when not paired with architectural intent.

Architecturally, Culture operates as an implicit decision infrastructure. Where Order provides explicit rules and enforcement mechanisms, Culture encodes unspoken permissions, prohibitions, and priorities. It reduces the burden of coordination by enabling agents to act in ways that are not only rule-compliant but contextually appropriate—often without requiring clarification, escalation, or centralized control.

Culture is structurally embedded. It manifests not just in slogans or rituals but in hiring practices, role definitions, interface defaults, decision latencies, and feedback loops. Like a system-wide bias deliberately built into the unit, it answers a continuous background question: *What is expected here? What is rewarded? What is considered possible, good, or dangerous?*

EXAMPLE—EMBEDDED CULTURE IN PRACTICE

A global software company once struggled with delayed product launches despite well-defined processes. A review revealed that its "risk-averse" culture was not expressed through mission statements but embedded structurally: approval chains with multiple sign-offs, default "cc all" email norms, and dashboards that highlighted defect counts more than customer satisfaction. These mechanisms quietly reinforced caution over speed. When the company rebalanced its metrics and reduced mandatory approvals, behavior changed without a single motivational campaign. Culture shifted not through messaging, but through redesigning how decisions were made and rewarded.

A well-formed Culture possesses several architectural characteristics:

- *Distributed Interpretation*: Culture enables agents to make sense of ambiguous situations in similar ways, reducing the need for constant rule updates. It supplies a shared interpretive lens without enforcing uniform behavior—for example, when teams across regions instinctively prioritize customer safety before profitability during crises.
- *Behavioral Gravity*: Culture pulls action toward familiar patterns. It shapes habits, influences defaults, and accelerates alignment—especially under uncertainty, overload, or novelty. Like a gravitational field, it favors lower-effort paths that are still directionally aligned—such as developers naturally reviewing one another's code even when peer review is optional.
- *Normative Compression*: Culture reduces coordination overhead by collapsing broad possibilities into a few legible and socially acceptable trajectories. It narrows choice through tacit filtering, without rigid constraint—prioritizing coherence over exhaustiveness. For instance, designers may converge on a common visual language not because they are told to, but because divergence feels socially inefficient.

- *Persistence Across Roles*: Culture persists beyond individual roles or temporary configurations. It provides a durable field of expectations across agent turnover, structural transitions, or evolving processes—visible when new hires quickly absorb unwritten norms about responsiveness or decision-making tone.

From a design perspective, Culture is not an artifact to be added once systems are operational. It must be intentionally architected from the beginning—infused into onboarding flows, collaborative behaviors, evaluation rhythms, escalation paths, and even the semantics of human-agent interfaces. Where Purpose provides directional justification and Order ensures formal compliance, Culture makes the system *habitable*. It turns aligned behavior into the path of least resistance—nudging systems toward coherence without constraint.

PRACTICAL TIPS FOR DESIGNING CULTURAL ALIGNMENT

- *Start early*: Embed cultural principles during system conception, not after launch.
- *Design for induction*: Ensure onboarding experiences model expected behaviors and interpretive norms.
- *Architect feedback*: Create visible loops where cultural misalignments can be raised and corrected.
- *Align incentives*: Reward behaviors that reinforce intended culture rather than only technical outputs.
- *Surface norms*: Make tacit expectations visible through language, UI tone, or decision protocols.
- *Model coherence*: Demonstrate desired culture in leadership and cross-functional interactions.

Culture, then, is not a by-product of success—it is a precondition of sustainable alignment. Studies on organizational performance consistently link strong cultural alignment to long-term adaptability and financial success, underscoring that coherence precedes achievement, not the other way around. Culture supports architecture by embedding intent into habit, enabling purposeful units to operate coherently even when explicit oversight is absent or ambiguous. Designed well, Culture becomes the invisible structure that holds action together—without needing to hold it back.

9.1.2. Why Culture Exists

Culture exists because not everything can be specified, monitored, or enforced. As systems theory reminds us, complex environments require local interpretation and self-organization; Culture provides that adaptive logic when formal design reaches its limits.

In any purposeful system—whether a one-person operation or a distributed organization—there are moments when the rules run out. Situations arise that are too ambiguous, too novel, or too context-dependent to be resolved through predefined logic or external direction. In these moments, Culture fills the gap. It provides an internal compass when external maps are incomplete.

Systems need Culture because coordination at scale is otherwise brittle. As roles multiply and agents diversify, the cost of prescribing every behavior grows exponentially. Formal processes become sluggish; decision rights become tangled. Culture offers a lighter mechanism. Instead of prescribing action, it shapes interpretation. Instead of issuing instructions, it sets defaults for what agents are likely to do—even when no one is watching.

Culture also exists to enable responsiveness without chaos. It allows systems to adapt fluidly while preserving internal consistency. Unlike rules, which can fracture under pressure or fall out of date, Culture evolves through use. It reinforces itself through repetition and reflection, creating a sense of continuity across changing circumstances.

For systems that include artificial agents, Culture enables coherence not through control but through orientation. If properly embedded, Culture can influence how AI agents prioritize outcomes, resolve trade-offs, or escalate uncertainty—without requiring a constant human-in-the-loop.

At its core, Culture exists because alignment cannot be fully outsourced to governance or engineering. It must be lived, felt, and reinforced through the system's everyday choices. It offers a solution to the architectural problem of sustained coherence in the presence of autonomy, complexity, and change.

9.1.3. What Problem Culture Solves

Culture solves the problem of sustaining coherence, motivation, and momentum within sociotechnical units.

Every unit, once structured, must remain operational—not just through rules and roles but through energy. Structure may define interfaces, boundaries, and responsibilities, but it does not explain why agents engage, how they interpret ambiguity, or what they do when the environment shifts. For this, the unit requires a climate—a felt sense of *how things are done here* that is shared across time, roles, and contexts. Culture provides that climate.

Architecturally, Culture acts as both *the default decision system* and *the internal energy field* of the unit. It enables agents—human or artificial—to act coherently without constant oversight, while also creating conditions in which agents feel empowered to act at all. Culture is what makes Structure *inhabitable.* It animates roles, sustains interaction patterns, and offers orientation in the absence of instruction.

As Jamshid Gharajedaghi observed, Culture becomes especially powerful when no explicit decision has been made. If a unit does not actively choose how to behave under uncertainty, its culture chooses by default. These defaults become so ingrained that they are mistaken for *reality*—a phenomenon that stabilizes systems, but also makes their assumptions invisible. Culture, in this sense, solves the architectural problem of *automaticity with alignment.* It allows agents to act quickly, instinctively, and usually correctly—because the system's logic is embedded in what feels normal.

But Culture does more than automate decision-making. It also solves for

- *Internal coherence*: Creating a recognizable identity across the unit's operations, even as agents change
- *Interpretive synchrony*: Reducing divergence in how agents respond to novelty, conflict, or success
- *Operational readiness*: Maintaining a proactive, resilient climate that enables the unit to function when Order is insufficient or Structure is under strain

Rational decisions require analysis. Emotional decisions depend on mood and context. Cultural decisions, by contrast, are enacted through what *feels obvious*—what is *just done here.* They are the fastest and least expensive choices a unit can make. But they come at a cost: once embedded, they are rarely questioned. When unexamined, cultural defaults can outlive the context that made them useful, silently stalling adaptation.

EXAMPLE—CULTURE IN CRISIS

During a major systems outage, a financial firm's entire trading platform went offline for several hours. No playbook covered the specific failure. Yet teams across technology, risk, and client services coordinated seamlessly—sharing updates through informal channels, prioritizing client transparency, and working around hierarchy. Later reviews found that what saved the situation wasn't procedure but culture: a shared reflex to protect trust before performance. The crisis ended with minimal loss, and the incident became a proof point of how deeply embedded cultural defaults can stabilize action when formal rules are silent.

Culture, then, solves the problem of sustaining action when rules run out and energy is low. It is the system's silent momentum—the patterned background that makes Structure livable, behavior coherent, and Purpose actionable, day after day.

9.1.4. Historical Patterns and Structural Evolution

Culture has always emerged alongside structure—sometimes to reinforce it, sometimes to resist it, often to make it livable. Across history, stable human systems have relied not only on formal organization but on a persistent climate of expectations: the unwritten norms that guide behavior when rules are absent, delayed, or disputed. As systems evolved in complexity, the role of Culture became more foundational—shaping not only how units operate internally, but how they evolve over time.

In early social groupings, culture emerged through repetition and survival. Norms were encoded in stories, rituals, and taboos—transmitted not through design but through lived consequence. Cultural assumptions were not abstract; they were structural memory. These assumptions preserved coherence across generations and across events that no one agent could fully understand or control.

As systems became more specialized, differentiated, and technical, structure began to decouple from culture. Formal institutions—governments, armies, corporations—developed complex orders and architectures, but often neglected the cultural substrates required for operational alignment. When these systems succeeded, it was usually because an implicit culture held them together. When they failed, it was often because that culture was either misaligned, siloed, or absent altogether.

Modern sociotechnical units inherit both this legacy and this risk. They are designed—at least in part—with formal roles, defined interfaces, and explicit responsibilities. But the deeper continuity of the unit—its ability to persist, adapt, and cohere—depends on cultural forces that are often treated as secondary or informal. In truth, they are neither.

From an architectural perspective, Culture has always served as the bridge between *structural stability* and *evolutionary viability*. It enables systems to change without breaking, and to grow without fragmenting. Cultural patterns evolve gradually—through repeated interpretation and use—and their inertia can either preserve the unit's integrity or impede its transformation.

This tension is key. Culture is inherently conservative: it remembers what worked. But architecture requires responsiveness: it must remain open to what's changing. Without Culture, structure becomes brittle. Without structural awareness, Culture becomes obsolete. The history of complex systems reveals this interplay repeatedly—systems endure not because they resist change but because their Culture makes change intelligible, bearable, and actionable.

In Unit Oriented Architecture, each sociotechnical unit carries with it a distinct Culture—not as ornament but as inheritance, adaptation, and compass. The unit's evolution depends not just on its processes or interfaces but on the lived logic that connects its past, present, and future.

9.2. Role of Culture in Sociotechnical Systems

Culture does not operate at a single level. In complex sociotechnical systems, it functions simultaneously within units, across systems, and at the boundaries between the system and its environment. Each unit—a self-contained but interdependent sociotechnical construct—requires a stable internal culture to ensure coherence of action. Yet stability within the unit is not enough. For the system as a whole to remain coordinated, integrative units must shape a broader cultural logic—one that orients operational units toward shared priorities without suppressing their autonomy. And beyond system boundaries, culture serves as the interpretive interface between the system and its external environment. It shapes how regulators perceive compliance, how customers experience trust, and how the job market evaluates credibility. Culture, then, is not just a local property of the unit; it is a structural phenomenon that anchors internal behavior, enables systemic integration, and governs external perception.

9.2.1. Internal Role: Stabilizing the Unit

Within a sociotechnical unit, Culture plays a stabilizing role by creating a consistent, self-reinforcing climate that guides behavior without requiring constant oversight. Structure defines roles and interfaces, Process defines flows, and Order defines rules—but it is Culture that ensures those elements function coherently under everyday conditions. It makes the unit livable, aligned, and self-correcting.

Internally, Culture stabilizes the unit in three key ways:

- *Predictability through shared assumptions*: Agents—human or artificial—operate more effectively when they can anticipate how others will act under conditions of uncertainty or stress. Culture provides the interpretive substrate for these expectations. It tells agents what is normal, what is rewarded, what is off-limits—not through enforcement but through repetition and recognition.
- *Resilience through habitual alignment*: When processes break down or roles shift, Culture maintains coherence. It allows the unit to adapt while preserving its character. Decisions do not scatter; they converge around inherited patterns of response that reflect the unit's identity and accumulated learning.
- *Motivation through meaningful climate*: Culture energizes Structure by making participation feel purposeful. It gives form to intangible elements like pride, trust, responsibility, and belonging. Even when formal incentives are weak or absent, a strong internal Culture can sustain engagement—because agents feel they are acting in accordance with something understood and shared.

Importantly, internal Culture is not static. It evolves with the unit, shaped by onboarding practices, interaction patterns, feedback loops, and tacit behaviors. Left unattended, it may drift or decay; carefully cultivated, it becomes the unit's deepest form of alignment—less visible than policy, but more enduring than tools.

Architecturally, then, Culture acts as the unit's internal stabilizer. It absorbs fluctuation, anchors adaptation, and ensures that Structure is not only followed but inhabited. A unit with strong internal Culture does not merely function—it coheres.

9.2.2. Systemic Role: Integration into System-Wide Coordination

In large sociotechnical systems, Culture functions not only within units but across them. It becomes the shared logic that allows distributed units to operate independently while contributing meaningfully to the whole. This systemic Culture is not a sum of internal cultures; it is a coordination layer—a distributed fabric that supports integration, interpretation, and direction across structurally diverse and often asynchronous units.

Systemic Culture differs in scope and purpose from local Culture. Where a unit's Culture governs its internal coherence, systemic Culture governs relational coherence. It enables units to understand, anticipate, and align with each other—not by enforcing sameness but by shaping shared patterns of reasoning, escalation, and resolution. The result is a system that can act together, even when its parts are specialized, autonomous, or in tension.

Architecturally, system-wide Culture must support multiple interaction modes:

1. *Systemic Alignment.* Integrative units set priorities, model behavior, and define boundary conditions that operational units are expected to respect. Operational units, in turn, contribute their expertise, report compliance, and adapt within those constraints. This dynamic often includes tension—between control and autonomy, between system-wide priorities and local conditions. A well-designed systemic Culture does not erase this tension; it makes it productive. It offers interpretive frames and escalation paths that allow both sides to remain aligned without collapsing into uniformity.

2. *Collaboration.* In collaborative modes, multiple units pursue a shared objective. Culture enables common interpretation of goals, mutual accountability, and fluid adjustment of roles. It prevents fragmentation by making interaction familiar—even when structures differ. Cultural alignment in this mode increases speed, reduces negotiation overhead, and builds trust across boundaries.

3. *Cooperation.* Units may operate in parallel while pursuing different goals. They are not working together toward a shared outcome but must remain compatible and non-obstructive. Systemic Culture in this context enables respectful coexistence:

shared language, interface clarity, and minimal misalignment. Cooperation is less intense than collaboration, but still requires a baseline of mutual recognition and interpretive compatibility.

4. *Competition.* In some cases, units are intentionally or unintentionally placed in competition—for example, sales units serving overlapping markets, or innovation teams proposing alternate strategies. Even here, systemic Culture plays a role: it defines acceptable boundaries, discourages sabotage, and ensures that competition enhances rather than undermines the system. Culture in this mode prevents escalation into dysfunction, while still allowing performance pressure to drive excellence.

System-wide Culture must accommodate these modes simultaneously. It must create enough coherence to enable alignment—without flattening the system into uniform compliance. This is not achieved through control but through architectural propagation: integrative units shape narratives, norms, and interfaces; operational units respond, adapt, and contribute feedback. Culture travels in both directions.

When this cultural infrastructure is absent or weak, systems fall back on bureaucracy, escalation, or workaround logic. But when systemic Culture is present, coordination becomes fluid—not because every unit is the same but because every unit can *interpret* and *engage* with the others in ways that are contextually appropriate and strategically aligned.

9.2.3. Environmental Role: Interface with External Entities

Every sociotechnical system exists within a broader environment. It is embedded in markets, communities, regulatory ecosystems, professional networks, supply chains, and public discourse. At these boundaries, Culture functions as the system's interpretive interface—translating Purpose, expressing Value, and enabling Function in ways that are legible, trustworthy, and responsive to external expectations.

Internally, Culture aligns action. Systemically, it enables coordination. But externally, Culture signals *who the system is* and *what it stands for*. It shapes how customers perceive trust and quality, how regulators assess risk and compliance, how talent evaluates desirability, and how partners interpret reliability. In this way, Culture is not only a stabilizer or integrator—it is a *communicator of intent and identity.*

This external interface is composed of both *internal culture expressed outward* and *external cultural constructs interpreted inward.* The system's internal Culture—its habits, tone, style of decision-making—seeps into every interaction, from a product experience to a service complaint to a press release. But the system must also interpret the cultures of its environment: what regulators value, what customers fear, what the labor market aspires to. This exchange is not static. It is a continuous interpretive negotiation.

Architecturally, Culture supports Purpose, Value, and Function at the boundary in three ways:

- *Purpose becomes believable*: A system's stated Purpose is evaluated not by what it claims but by how it behaves. Culture reinforces Purpose through visible consistency—across tone, decisions, priorities, and reactions. When internal behaviors reflect declared intent, external entities begin to trust that Purpose is not a slogan but a governing logic.
- *Value becomes experiential*: Culture shapes how Value is delivered, experienced, and judged. It influences service attitude, design sensibility, ethical stance, and conflict resolution. Two systems may offer similar products—but one is preferred because its cultural expression of Value *feels* more authentic, empathetic, or aligned with the consumer's worldview.
- *Function becomes dependable*: Reliable Function requires more than technical execution—it requires interpretive predictability. Culture makes Function legible. It governs how failures are explained, how uncertainty is handled, and how the system behaves under pressure. A culturally coherent system is one that behaves in expected ways—even when something goes wrong.

Across this interface, both *internal coherence* and *external empathy* matter. Systems that present a fragmented cultural face appear untrustworthy. Systems that ignore external cultural cues appear arrogant or obsolete. Those that maintain a strong internal Culture while adapting respectfully to external contexts become not only legible—but resilient, persuasive, and chosen.

Finally, Culture at the boundary also protects. It absorbs cultural shocks, filters misunderstandings, and stabilizes identity in the face of changing expectations. A well-formed Culture does not isolate the system—it anchors it. It allows the system to engage with external forces *without losing its shape.*

9.3. Design Principles and Heuristics

Culture cannot be engineered in the way structure or process can. It is not deployed but cultivated—shaped through choices, reinforced through repetition, and stabilized through alignment with the system's Purpose, Function, and environment. Yet while Culture resists direct control, it is far from accidental. Architects can influence it by designing for the conditions under which it forms: who is onboarded and how, what is rewarded and why, how decisions are made and explained, and what behaviors are normalized over time. This section outlines key principles and heuristics for designing culture intentionally—not as an afterthought, but as a central layer of architectural logic that enables alignment, coherence, and trust at every level of the sociotechnical system.

9.3.1. Principles of Culture-Oriented Design

Culture-oriented design approaches Culture not as ambience or afterthought, but as an infrastructural layer of coordination—shaping defaults, resolving ambiguity, and sustaining coherence across time, roles, and contexts. A unit's Culture is not decorative. It animates Purpose, stabilizes Function, supports Structure, and manifests at system boundaries. Culture must be designed as deliberately as interfaces or workflows—not by decree but through embedded logic and lived practice.

The following principles support the architectural quality, behavioral coherence, and operational sustainability of Culture:

1. **Anchor Culture in Purpose, Not Sentiment**

   Culture should not reflect abstract aspirations or borrowed slogans. It must reinforce the unit's declared Purpose in operationally meaningful ways. For example, if the Purpose centers on auditability or fairness, Culture should reinforce transparency, traceability, and elevation of edge cases.

   Implication: Design Culture to reflect how the system behaves when fulfilling its Purpose—not how it sounds when describing its ideals.

   Example: A data team that values transparency embeds open documentation and peer reviews into its workflow—not just "honesty" as a slogan.

Common Mistake: Treating Culture as a motivational veneer detached from Purpose leads to contradictions between declared ideals and actual incentives.

2. **Design for Behavioral Defaults**

   Culture is most powerful when it guides behavior without requiring deliberation. In high-pressure or ambiguous situations, well-formed Culture surfaces as instinctive response. These behavioral defaults should be intentionally selected and aligned with system goals.

   Implication: Embed defaults that make the right action easier than deviation; do not rely on constant reminders or approval cycles.

   Example: A "security-first" culture ensures encryption and access control are standard templates, not optional steps.

   Common Mistake: Overcomplicating procedures forces agents to reinterpret every situation, eroding cultural consistency.

3. **Codify Through Interaction Patterns, Not Declarations**

   Culture is transmitted through what agents *do*, not what systems *say*. Slogans and posters do not embed Culture; rituals, tools, decision reviews, and naming conventions do.

   Implication: Operationalize Culture through concrete experiences—onboarding formats, escalation rules, retrospectives, and interface cues—rather than symbolic affirmations.

   Example: A company promoting inclusion ensures mixed-group decision reviews instead of running stand-alone awareness workshops.

   Common Mistake: Declaring values without embedding them in daily interactions results in symbolic compliance.

4. **Reinforce Through Role Modeling and Structural Coupling**

   Culture spreads through example and exposure. It is learned by watching others—especially those in roles with high structural influence—navigate trade-offs. When these role models act in culturally aligned ways under cost or ambiguity, Culture gains credibility.

Implication: Couple Culture to power—those with authority must consistently model the behaviors expected of others.

Example: When senior engineers admit design errors publicly, a learning culture gains legitimacy.

Common Mistake: Leaders who delegate "culture work" while acting inconsistently erode credibility and alignment.

5. **Maintain Coherence Across Constructs**

   Culture cannot be isolated from the rest of the system. It must reinforce—and be reinforced by—Structure, Process, and Order. A culture of innovation must not be undermined by punitive escalation procedures; a culture of care must not conflict with metrics that reward speed over safety.

   Implication: Audit cultural expectations against structural and procedural realities; alignment must be designed, not assumed.

   Example: A "speak-up" culture cannot thrive under punitive escalation rules or opaque hierarchies.

   Common Mistake: Allowing mismatched systems (e.g., innovation culture with rigid approvals) fragments intent and practice.

6. **Calibrate for System Scale and Agent Type**

   Cultural logic must match the cognitive and operational realities of the system's agents—whether human, robotic, or artificial. What is tacit for senior humans may need to be explicit for AI agents.

   Implication: Adjust the abstraction level of norms to match agent cognition and context; tacit cues for humans require explicit encoding for AI.

Example: Humans may infer ethical nuance, but AI systems need explicit parameters to reflect the same cultural intent.

Common Mistake: Assuming tacit understanding applies equally to algorithmic agents leads to unpredictable behavior.

7. **Support Self-Regulation and Distributed Adaptation**

   Culture is most valuable when it enables agents to act without constant oversight. That requires interpretive alignment and the ability to resolve uncertainty locally.

   Implication: Embed interpretive cues and trusted defaults so agents can resolve uncertainty locally without hierarchy.

   Example: In remote-first teams, employees apply shared escalation norms instead of waiting for managerial direction.

   Common Mistake: Over-centralizing control suppresses the adaptive capacity Culture is meant to provide.

8. **Embed Simplicity and Repeatability Over Richness and Completeness**

   Effective Culture is cognitively light and behaviorally durable. Avoid complexity that demands constant interpretation.

   Implication: Prefer concise, memorable rules that travel easily across agents and time; complexity weakens adoption.

   Example: "Escalate early" or "assume good intent" guides daily choices more reliably than multi-page codes of conduct.

   Common Mistake: Overloading agents with abstract principles creates fatigue and uneven interpretation.

These principles provide a structural lens for designing Culture not as an after-effect of leadership or legacy but as a compositional layer of the system itself. They clarify how Culture interacts with Purpose, Function, and Structure—and how it stabilizes behavior without prescribing it. While Culture must ultimately be lived to be real, it must first be designed to be livable. Principles guide that design.

9.3.2. Architectural Heuristics for Culture

While principles clarify what Culture must enable architecturally, heuristics offer practical design moves for how to shape and sustain it. These heuristics are not universal rules, but tested interventions that reduce ambiguity, support alignment, and accelerate cultural coherence under real-world constraints—whether onboarding new agents, navigating distributed structures, or recovering from drift.

They are especially useful when operating in complex systems where Culture must scale across agents, adapt across contexts, and remain consistent over time without relying on top-down control.

Below are ten architectural heuristics to support the cultivation of strong, coherent, and evolution-ready Cultures:

1. **Design from the most typical behavioral flow first**

   Begin by shaping Culture around the most common decisions, interactions, or coordination patterns agents face in practice. Identify the recurring situations where agents must interpret priorities, resolve tension, or act under time pressure—and ensure the cultural logic in those moments is clear, repeatable, and aligned with system Purpose. These typical flows often represent the heartbeat of the unit's Function and the most visible expression of its identity. Establishing cultural coherence here first provides a stable foundation before addressing rarer or more ambiguous edge cases.

   Rationale: Anchoring Culture in its dominant behavioral context creates immediate clarity, builds early momentum, and helps surface real complexity—revealing where cultural friction or misalignment is most likely to occur under stress.

2. **Make the Culture observable before declaring it**

   Before defining or publicizing cultural values, allow the desired behaviors to emerge, stabilize, and become recognizable in everyday practice. Agents trust what they can see repeated—not what is announced. Focus first on shaping interactions, rituals, and decisions that reflect the intended cultural tone. Only after these patterns are embedded should language, symbols, or declarations be introduced. The visible consistency of behavior creates credibility; labeling it too early risks cynicism or contradiction.

Rationale: Culture becomes durable when it is experienced before it is named. Observable alignment builds trust, while aspirational declarations without behavioral grounding lead to reputational drift and quiet disengagement. Practically, this can be achieved by piloting small, visible behavior changes—such as new meeting formats or feedback rituals—and letting these stabilize before attaching formal language or values statements.

3. **Use edge cases to reveal and reinforce norms**

   Pay close attention to how the system behaves in moments of ambiguity, conflict, or stress. These edge cases—escalations, failures, or ethical dilemmas—reveal the underlying cultural logic more clearly than routine activity. Treat them not just as exceptions to manage, but as critical opportunities to demonstrate, reinforce, or reshape norms. Document decisions, explain reasoning, and surface patterns that reflect the system's deeper commitments.

   Rationale: Edge cases test cultural coherence. When handled visibly and consistently, they reinforce the system's identity and build trust in its integrity.

4. **Shape the first experience intentionally**

   Culture is often formed in the first interactions—onboarding sessions, introductory meetings, orientation materials, or the first live conflict. These moments encode not just information, but tone, expectation, and interpretive stance. Design these early experiences to reflect the Culture the unit aspires to maintain. Let new agents encounter the system in a way that is both welcoming and clarifying.

   Rationale: First impressions create long-term defaults. Cultural tone is disproportionately shaped by how the system handles entry, vulnerability, and early exposure.

5. **Embed cues in environments and workflows**

   Cultural logic becomes more stable when it is reinforced by the environment—through tools, templates, naming conventions, UI defaults, and workflow prompts. These passive reinforcements shape behavior without imposing mandates. The goal is to create subtle but persistent signals that reduce cognitive overhead and reinforce expectations, especially under time pressure.

   Rationale: Culture is easier to follow when it is built into how agents work. Embedding cues in the environment improves adoption, alignment, and reliability.

6. **Encode norms in repeatable language**

   Cultural expectations are more easily internalized when expressed through simple, memorable phrases. These phrases serve as shared shorthand—helping agents act with confidence, align with each other, and reinforce coherence. Avoid abstract or performative language. Instead, distill core behaviors into practical slogans that can be referenced in decision-making or retrospectives.

   Rationale: Language transmits Culture. Clear, resonant phrases make complex behaviors legible and portable across roles and contexts.

7. **Recognize culturally aligned behavior, not just outcomes**

   Cultural credibility grows when the system notices and affirms behavior that reflects its declared values—even when results are mixed or ambiguous. Recognize agents who demonstrate the right instincts in difficult situations, not just those who deliver successful outputs. This helps detach alignment from performance and centers it in intention and judgment. For example, organizations may incorporate "values in action" criteria into promotion reviews or ask candidates to describe times they upheld principles over short-term success.

Rationale: Behaviorally aligned effort sustains Culture over time. Rewarding only outcomes can create dissonance and weaken shared norms.

8. **Align governance rituals with cultural tone**

   Review processes, feedback loops, and escalation paths often express more about Culture than any formal value statement. If retrospectives are rushed, feedback is punished, or silence is rewarded, agents will absorb those patterns regardless of stated intentions. Ensure that cultural tone—openness, rigor, trust, or accountability—is reflected in how governance rituals actually function.

   Rationale: Culture is revealed under pressure. Misaligned governance mechanisms signal hypocrisy and invite disengagement.

9. **Audit and adapt Culture periodically**

   As the system's Purpose evolves and its Structure changes, its Culture may lag or ossify. Treat Culture as a dynamic architectural element—capable of drifting, degrading, or becoming obsolete. Create moments of reflection that allow agents to surface tensions, misalignments, and new patterns that may warrant formalization. This can be done through periodic cultural audits, pulse surveys, cross-team retrospectives, or peer reviews that make implicit norms visible and discussable.

   Rationale: Culture is not self-correcting. Without periodic review, it may entrench outdated logic or fail to scale with the system's growth.

10. **Design for transmission, not enforcement**

    Sustainable Culture cannot rely on central mandates or managerial control. Instead, it must be learned, taught, and reinforced horizontally—by peers, rituals, and stories. Aim to build a Culture that is easy to internalize and easy to share. The more the system depends on top-down reminders, the less durable its Culture will be under stress.

Rationale: Culture that spreads informally is more resilient. Design cues, patterns, and phrases that agents can carry across roles and generations.

These heuristics are not formulas—they are interventions. Culture emerges from repeated interaction, not static intention. Each choice, interface, and conversation either strengthens or erodes the cultural logic of the unit. What matters is not perfection but persistence: shaping patterns deliberately, noticing drift early, and reinforcing what aligns with Purpose over time. In a well-designed system, Culture becomes self-reinforcing—not because it is mandated, but because it works.

9.4. Realization Across Implementation Planes

Culture is not implemented in isolation. It must be realized across the same planes in which the sociotechnical system itself operates: the *social*, *physical*, and *digital*. These planes provide the surfaces through which Culture is embodied, sensed, and reinforced. In the social plane, Culture lives in relationships, language, rituals, and everyday interpretation. In the physical plane, it manifests through space, signage, artifacts, and spatial affordances. In the digital plane, it is expressed and shaped through interfaces, defaults, system behavior, and data exposure. While Culture originates as shared meaning, its endurance depends on how well it is embedded across these multiple surfaces—so that it can be recognized, enacted, and transmitted without continual explanation. Designing for this kind of realization means understanding the interplay between planes and selecting appropriate levers for cultural expression within each. These cross-plane carriers of Culture—and the mechanisms through which they reinforce it—are outlined in Table 9-2.

Table 9-2. Cultural Transmission Vectors Across Implementation Planes

Plane	Primary carriers of Culture	How Culture is reinforced	Design levers for architects
Social	Language, rituals, relationships, meeting norms, feedback tone	Repetition, storytelling, social proof, peer recognition	Onboarding design, meeting formats, feedback structures, leadership modeling
Physical	Space layout, signage, dress codes, artifacts, environmental cues	Visibility, accessibility, spatial symbolism	Workspace zoning, artifact placement, environmental design, accessibility and transparency cues
Digital	Interface language, defaults, automation behavior, data visibility	Interaction patterns, algorithmic feedback, embedded prompts	UI/UX tone, default settings, notification design, ethical AI constraints

9.4.1. Manifestation in the Social and Physical Planes

Culture first emerges in the social plane. It is shaped by how agents interact, how they interpret ambiguity, and how they respond to each other under pressure. It lives in tone, language, silence, escalation patterns, meeting rituals, and informal norms. These patterns are often reinforced not through formal governance but through recognition, mimicry, peer correction, and storytelling. Social dynamics such as hierarchy, trust, reciprocity, and tolerance for dissent determine how Culture is learned, reinforced, or resisted. In a well-aligned system, social practices do not just reflect Culture—they *transmit* it.

The physical plane, while often overlooked, is equally potent. Culture is embedded in space: in what is visible or hidden, open or protected, shared or segregated. Physical artifacts—uniforms, signage, workspace layouts, onboarding materials, physical access controls—act as constant cultural cues. A secure room, a prominently posted value statement, a whiteboard left full of ideas rather than wiped clean—each carries cultural meaning. Even infrastructure design signals expectations: whether a space invites conversation or signals surveillance, whether it encourages transparency or demands restraint.

Importantly, the social and physical planes work together. A ritual such as a daily stand-up may begin as a physical gathering but becomes a social contract; a wall display of decisions may start as decoration but becomes a shared memory and interpretive tool. These manifestations become cultural anchors—stabilizing shared understanding even when other elements of the system evolve.

Architecturally, these planes are not just outputs of Culture; they are *interfaces for shaping it*. Deliberate design of physical space and social ritual can support the emergence of behaviors aligned with Purpose and Function. When misaligned—such as when open collaboration is encouraged verbally but spaces isolate people physically—cultural credibility fractures.

To embed Culture durably, these planes must be treated as active surfaces. Designing culture-oriented systems means shaping the interactions, spaces, and artifacts that allow Culture to persist—not as abstraction but as lived reality.

9.4.2. Culture in the Digital Plane

In digital systems, Culture is no longer transmitted solely through human interaction. It is encoded into platforms, interfaces, workflows, and algorithms—embedded in the way agents access, interpret, and respond to information. This transformation changes the nature of Culture itself: from a tacit behavioral substrate into a programmable logic layer that is interpreted by both human and non-human agents. Yet this encoding introduces the risk of cultural drift—when software updates, metric biases, or automated feedback loops gradually alter behaviors and priorities without conscious design, allowing systems to evolve values their creators never intended.

Defaults are among the most powerful cultural carriers in the digital plane. Chosen for efficiency, they often go unquestioned—and thus silently propagate values. A system that defaults documentation to public view reinforces transparency. One that requires approvals before deployment embeds a logic of accountability. Whether in version control settings, naming conventions, or user permissions, defaults become design-time decisions that instantiate behavioral norms at scale.

Digital workflows also ritualize behavior. Interfaces and tooling often impose required sequences—such as mandatory templates, fixed handoff points, or timed approvals—which shape cultural expectations through repetition. A pull request form requiring justification encourages a culture of thoughtful explanation; an auto-escalation path reinforces a culture of urgency or defensiveness. These constraints, once embedded, become default behaviors—replacing informal discretion with ritualized interaction.

Feedback loops further condition behavior. Notification patterns, metrics dashboards, social features (likes, mentions), and AI-driven suggestions all serve as cultural amplifiers. For example, if a collaboration platform prioritizes responsiveness with algorithmic visibility, it can create a culture of constant availability—even when this undermines focus, depth, or quality. The system rewards a behavior pattern and, over time, agents treat it as normative.

As automation and machine agents participate in workflows, they begin to express and shape Culture directly. A bot that flags only technical errors but not human-centered concerns subtly deprioritizes code readability or user empathy. Recommendation engines tuned for engagement may elevate superficial metrics over long-term trust. These encoded priorities are rarely neutral—they reinforce certain cultural values while eroding others, often unintentionally.

Digital systems also formalize informal practices. A norm such as tagging others for awareness ("@team" or "FYI") may evolve into automated watchers or mandatory fields. This shift can bring consistency but may also reduce flexibility. What began as adaptable becomes rigid, sometimes codifying behavior that no longer fits the context.

Cultural drift is a hidden risk in evolving platforms. A system upgrade, tooling change, or integration redesign can break embedded rituals—altering how agents communicate, reflect, or escalate. When interfaces change, cultural defaults change with them. Without intentional recalibration, systems may silently adopt values inconsistent with Purpose, Function, or prior behavior.

In distributed environments, the digital plane becomes the primary site of cultural legibility. Emojis, comment tone, file naming conventions, and inline tags all acquire symbolic weight. They function as proxies for tone, hierarchy, and trust. Culture becomes visible not in how people speak but in how they configure, format, and respond across systems.

Architecturally, the digital plane must be treated as a primary surface of cultural design. Every workflow, automation, and platform choice expresses intent—whether consciously or not. Designing for cultural coherence requires explicit attention to defaults, constraints, feedback systems, symbolic cues, and the values embedded in automation logic.

In short, Culture in the digital plane is not abstract—it is executable. And if left unexamined, it will execute whatever logic is most convenient for the platform, not what is most aligned with Purpose.

9.4.3. Implementation Surfaces and Sociotechnical Contexts

Culture may begin as belief and intention, but it becomes real only when instantiated—*somewhere*. These instances are not always material in the way APIs or interfaces are. Instead, they are *interpretive surfaces*—sites where behavior stabilizes, meaning accumulates, and memory forms. In sociotechnical systems, these surfaces are highly varied, but always designable. They provide the medium through which values are normalized, reinforced, and transmitted—not just across roles, but across units, contexts, and generations.

While Function is implemented through operational and interactional surfaces, Culture is implemented through *symbolic, temporal, relational, and reflective surfaces*. These surfaces may be social or physical, digital or ambient—but in all cases, they *guide behavior without requiring enforcement*. Effective cultural design requires identifying which of these surfaces matter most in a given system—and tuning them with intent.

Below are key categories of cultural surfaces—each one a carrier of norms, interpretations, and system memory.

Core Cultural Interfaces

These are the most direct surfaces through which Culture is formed and reinforced. They influence everyday behavior and interpretation.

- *Rituals and rhythms*: Cadenced activities that normalize behavior and create shared temporal structure. *Examples:* Stand-ups, retrospectives, demos, one-on-ones.
- *Language and terminology*: Everyday vocabulary that encodes values, roles, and expectations. *Examples:* Status labels, error messages, naming conventions, internal slogans.
- *Storytelling and symbolic artifacts*: Narratives and visual cues that compress memory and meaning into cultural shorthand. *Examples:* Origin stories, dashboards, wallboards, symbolic metrics.

- *Onboarding and induction*: Early exposure moments that transmit cultural logic through lived experience. *Examples:* Training sessions, buddy systems, initiation workflows.
- *Recognition and feedback rituals*: Signals of value that reinforce what is celebrated, tolerated, or ignored. *Examples:* Kudos systems, shout-outs, emoji reactions, performance reviews.

Transitional and Reflective Surfaces

These surfaces are activated during moments of change, tension, or reflection. They offer rare but highly charged opportunities for cultural reinforcement or repair.

- *Norm codification*: Explicit articulation of expected behaviors, rules, and escalation paths. *Examples:* Codes of conduct, playbooks, decision rubrics.
- *Role modeling and leadership signals*: Visible behaviors of influential agents that set precedent, tone, and tolerance. *Examples:* How leaders handle conflict, celebrate wins, admit mistakes.
- *Boundary rituals*: Transitions and thresholds that mark identity shifts and role evolution. *Examples:* Farewells, promotions, project handoffs, debriefs.
- *Conflict handling surfaces*: Interfaces for dissent, debate, or resolution that reveal cultural maturity. *Examples:* Mediation rituals, escalation etiquette, debate formats.
- *Temporal symbols*: Milestones and anniversaries that punctuate time with shared meaning. *Examples:* Launch dates, hack weeks, seasonal rituals, retrospectives.

Environmental and Systemic Cues

These surfaces influence Culture ambiently, often through presence or absence. They create passive signals that accumulate meaning over time.

- *Cultural debt registers*: Mechanisms for surfacing cultural erosion, drift, or misalignment. *Examples:* Ignored rituals, outdated norms, toxic behaviors left unaddressed.

- *Environmental cues*: Physical or digital ambient signals that shape mood and expectations. *Examples:* Lighting, spatial arrangement, UI design, background music.
- *Cross-unit cultural bridges*: Shared practices that create cohesion across distributed or federated systems. *Examples:* Inter-unit exchanges, federated rituals, alliance ceremonies.

These surfaces are not interchangeable. Some offer high frequency but low depth (e.g., emoji reactions); others offer low frequency but high symbolic weight (e.g., promotions, origin stories). What matters is *coherence*: when surfaces align, Culture stabilizes. When they contradict, agents learn to distrust the stated values and follow the implicit ones.

Designing Culture means identifying the few surfaces that carry the most weight—and treating them as *architectural primitives*. They are not decoration. They are structure.

9.5. Strategic Impact of Culture Interfaces

Culture is often described in vague or aspirational terms, yet its impact on system performance is precise and measurable—if observed through the right lens. Culture interfaces are not merely passive reflections of intent; they are *active instruments of alignment, motivation, and coordination.* When designed intentionally, they reduce friction, accelerate adaptation, and reinforce Purpose without continuous oversight. The strategic payoff of culture surfaces lies not in their symbolism alone but in their capacity to shape decision patterns, energize collaboration, and stabilize distributed systems under pressure. Understanding how these interfaces influence agents—internally, collectively, and across boundaries—reveals their role as essential levers of architectural integrity.

9.5.1. Nuances of Motivation and Alignment

Culture does not motivate through mandate. It motivates by shaping what feels *normal*, *worthy*, or *possible*. Unlike formal controls or incentive systems, Culture works through expectation, identity, and interpretation. It aligns agents not by force but by resonance—providing a shared lens through which action becomes meaningful and coordinated without explicit synchronization.

At the unit level, Culture supports motivation by creating coherence between an agent's role, the perceived purpose of the system, and the behaviors observed around them. It fosters internal alignment through *default decision logic*: when agents face ambiguity, they reach for culturally reinforced cues—not because they are required to but because these cues feel natural, safe, or rewarded. A culture that values initiative will see agents self-activating; one that prizes caution will see deliberation. Both outcomes are rational—relative to the culture.

Crucially, Culture does not impose uniformity. It enables alignment even when motives differ. Agents may be driven by personal growth, duty, recognition, stability, or curiosity—but Culture frames these motives within a shared behavioral envelope. By embedding values into rituals, stories, terminology, and role expectations, Culture offers multiple entry points into alignment—without requiring ideological consensus.

Across units, Culture supports systemic alignment by reducing interpretive drift. It helps ensure that operational decisions support global direction—not by hardcoding control paths, but by reinforcing patterns of sense-making. For example, two units pursuing different goals may still escalate issues in compatible ways, coordinate around shared rituals, or use common language to surface tension. These patterns allow local autonomy to coexist with global cohesion.

Well-designed Culture does not suppress dissent—it frames it constructively. It provides interpretive mechanisms for disagreement, challenge, and reform. In systems that aim to evolve, this capacity for *motivated misalignment*—where divergence serves system learning rather than fragmentation—is vital. Cultures that reject dissent often calcify; cultures that metabolize it adapt.

In short, Culture is not a replacement for motivation or structure. It is the *medium* through which motivation becomes behavior, and behavior becomes alignment. By shaping what is salient, what is safe, and what is celebrated, Culture quietly calibrates the system—so that agents move differently, even when no one is watching.

9.5.2. Measurable Benefits and Architectural Payoffs

Well-designed Culture does more than harmonize—it *accelerates, aligns, and preserves*. It reduces decision fatigue, improves coordination, and makes systems more resilient under ambiguity or stress. The benefits of a coherent Culture are not philosophical abstractions; they are architectural returns—expressed in faster onboarding, reduced escalation, fewer breakdowns, and a higher degree of behavioral coherence without micro-management. Culture makes alignment *affordable*. Its payoff is visible in both local interactions and global integrity.

A. **Cultural Coherence Reduces Interpretive Drift**

When Culture is coherently designed—transmitted through consistent surfaces and reinforced by patterns of use—agents no longer need to guess what is expected of them. They interpret ambiguity in aligned ways, reducing divergence and miscommunication.

- More consistent decisions across heterogeneous agents
- Greater resilience when policies or structures lag behind events
- Reduced variance in quality, tone, and behavior
- When decisions must be made quickly, or without explicit instruction, agents rely on culturally reinforced defaults
- Patterns of prioritization, escalation, and resolution emerge through repetition—not prescription

Payoff: Operationally, Culture acts as a default decision system—providing low-cost, high-speed alignment where structure ends and interpretation begins.

B. **Embedded Norms Lower Governance and Enforcement Costs**

When norms are encoded into rituals, templates, and language—not just documented—behavior is guided without supervision. This decreases reliance on top-down control.

- Fewer violations and misunderstandings
- Faster assimilation of new agents and teams
- Reduced need for exception handling and correction

Payoff: Embedded culture scales without scaling bureaucracy. It enforces values without enforcing procedures.

C. **Shared Memory Enables Faster Recovery and Learning**

Cultural surfaces serve as informal memory systems. When behavior is shaped by collective stories, rituals, and reference points, agents recover faster from failure and adapt faster to novelty.

- Quicker pattern recognition in novel contexts
- Higher tolerance for ambiguity and emergent strategy
- Faster conflict resolution and norm restoration

Payoff: Cultural memory provides continuity without requiring full documentation. It accelerates regeneration.

D. **Motivational Alignment Reduces Friction and Drift**

When agents feel that cultural logic aligns with Purpose and personal values, motivation becomes self-regulating. The need for artificial incentives or hierarchical enforcement diminishes.

- More intrinsic engagement across agent types
- Lower resistance to transformation and role fluidity
- Reduced burnout, detachment, and passive noncompliance

Payoff: Culture provides a motivational infrastructure—linking effort, meaning, and reward without constant managerial intervention.

E. **Interpretive Bridges Enable Cross-Unit Coherence**

Culture also facilitates collaboration *between* units. Shared language, rituals, and symbolic anchors enable distributed teams to coordinate without alignment fatigue.

- Smoother cross-functional work
- Reduced meta-coordination in federated or mesh structures
- Easier onboarding into multi-unit collaborations

Payoff: Cultural bridges are cheaper than architectural refactoring. They allow complex systems to operate as coherent wholes, even when structurally decentralized.

Culture, when architected with precision, produces returns across time, space, and scale. It preserves alignment when systems grow, buffers agents when systems fail, and animates Purpose when structures become rigid. These are not soft benefits. They are *architectural dividends*—compounding quietly across every surface Culture touches.

9.6. Implementation Patterns

Culture becomes real when it is lived, repeated, and made tangible through the system's surfaces. From an architectural perspective, this realization takes shape through implementation patterns—recurring ways in which Culture is surfaced, reinforced, and transmitted across agents and units. These patterns do not merely describe the content of cultural values; they describe the *mechanisms by which those values are enacted, experienced, and preserved.*

Though some patterns rely on documentation or tooling, they are not primarily technical. They are structural, rhythmic, and symbolic—anchoring the system's identity, guiding default behavior, and enabling adaptation across time and scale. Each pattern embodies a way Culture is *made visible, normalized, and accessible*—whether through rituals, artifacts, norms, or leadership behavior.

1. **Codified Norms with Living Rituals**

   Cultural coherence depends on a balance between what is explicitly codified and what is rhythmically enacted. This pattern embeds values through written norms (e.g., escalation paths, naming conventions, playbooks) reinforced by lived practices (e.g., stand-ups, demos, retrospectives). Codification reduces ambiguity and enables repeatability at scale, while ritual breathes continuity and relevance into those norms. The interplay between documentation and behavior ensures that Culture remains both legible and alive. However, if rituals become rote or norms ossify, Culture can lose its adaptive edge. Research on organizational decline repeatedly shows this pattern: once-innovative firms like Kodak or Nokia maintained strong internal rituals but failed to question them, mistaking procedural excellence for adaptability—a form of cultural strength that turned rigid over time.

2. **Symbolic Infrastructure Anchored in Origin**

   Narratives, symbols, and historical anchors provide long-term coherence and emotional resonance. Origin stories, founding metaphors, and symbolic artifacts (e.g., dashboards, naming conventions, performance metaphors) communicate not just what matters but *why* it matters. This pattern helps preserve institutional memory and encode values across time. It supports continuity across transitions, gives new agents a sense of belonging, and stabilizes identity through change. But if left untouched, symbolic infrastructure can drift into irrelevance—or worse, exclusion.

3. **Multi-Modal Cultural Transmission**

   Agents absorb Culture through multiple channels—visual, verbal, procedural, ambient. This pattern ensures that cultural logic is not dependent on a single medium, agent type, or learning style. Values are expressed through onboarding content, feedback signals, user interfaces, architectural defaults, emoji reactions, internal slogans, and even the tone of error messages. Multimodality increases accessibility—cognitively, emotionally, and socially—especially in complex systems with diverse agent types. It also spreads interpretive load across surfaces, reducing the risk of cultural bottlenecks.

4. **Federated Cultural Surfaces with Local Autonomy**

   In large sociotechnical systems, Culture must enable systemic alignment without erasing local variation. This pattern creates shared cultural scaffolding—such as common terminology, global retros, federated rituals, or company-wide feedback rhythms—while allowing operational units to adapt or extend practices to fit their context. When well implemented, federated culture promotes interpretive compatibility and system-wide cohesion. The risk lies in over-standardization, which may suppress local innovation, or over-fragmentation, which may erode trust and shared understanding.

5. **Conflict Metabolism Routines**

   Healthy Culture does not eliminate tension—it metabolizes it. This pattern formalizes safe, constructive ways to surface and process friction, ambiguity, or disagreement. Examples include escalation etiquette, debrief formats, cultural debt registers, and role-respecting dissent protocols. These routines reinforce psychological safety and keep Culture adaptive, preventing drift into avoidance, toxicity, or groupthink. The goal is not harmony at all costs but coherence through *deliberate repair and renewal.*

6. **Cultural Induction and Re-Induction**

   Culture is absorbed most deeply during moments of identity shift. This pattern targets those moments—onboarding, promotions, team formation, re-orgs—as prime opportunities for cultural encoding. Practices include buddy systems, initiation rituals, first-week journeys, and cultural milestones. Re-induction is equally vital, offering checkpoints to realign expectations during role changes, team pivots, or strategic shifts. Without these moments, agents may carry assumptions that no longer serve the system—or miss the ones that do.

7. **Embedded Defaults as Cultural Scaffolding**

   Cultural behavior is shaped less by what is stated than by what is *assumed.* This pattern embeds cultural logic into templates, naming standards, decision trees, time zones, access scopes, and design constraints—quietly shaping how agents act by influencing what they take for granted. These default settings encode priorities and guardrails without requiring constant explanation. Done well, they build ambient alignment. Done poorly—or left unexamined—they become invisible constraints that may inhibit change or diversity of action.

9.6.1. Cultural Observatories and Reflective Feedback Loops

Culture must adapt over time, but only if it can see itself. This pattern builds lightweight observatories: recurring surveys, culture barometers, anonymous input rituals, or reflection checkpoints. It helps detect cultural drift, expose misalignments, and surface emerging tensions. These observatories support introspection without heavy process—and create space for genuine feedback without politicization. When feedback is ignored, they become symbolic. When taken seriously, they enable Culture to remain aligned with Purpose even as context evolves.

These patterns offer not just ways to transmit Culture, but ways to shape *how* Culture behaves under scale, stress, and change. Each pattern reinforces the idea that Culture is not a side effect of social interaction—it is a structural resource, embedded through repetition, symbol, and expectation. When applied intentionally, implementation patterns transform Culture from something ambient and anecdotal into something *composable, scalable, and architecturally coherent.*

9.7. Interlocking Dependencies and Alignment

Culture does not exist in isolation. Within the Purpose Driven Design framework, it is an *interdependent design construct*—emerging from the system's core commitments and reinforcing them through shared interpretation, repetition, and behavior. The strength of any Culture lies not only in its resonance with agents but in its alignment with the unit's Purpose, its support for Function, and its capacity to stabilize Process, Structure, Order, Memory, and Storage.

Culture may be ambient, but it is never neutral. Misaligned Culture can undermine Purpose, distort Value, or obstruct Process. Aligned Culture, by contrast, becomes a form of *structural intent made durable.* It sustains coherence in motion, bridges variability across contexts, and enables the system to make meaningful decisions without centralized control.

9.7.1. Alignment with Purpose: Cultural Coherence and Justification

Purpose defines *whom* the unit exists to serve and *why*. It articulates the desire and intention to respond to the needs of specific entities in the environment—stakeholders whose very existence gives the unit meaning. Culture, when aligned with Purpose, ensures that this intent is not merely declared but *felt*—guiding attention, framing priorities, and shaping what is treated as legitimate or off-course.

An aligned Culture makes Purpose *real* not by executing it, but by reinforcing its salience. It orients agents toward the right stakeholders, clarifies what the unit is ultimately for, and sustains motivational focus even amid changing tasks or environments. When agents act in ways that resonate with the Purpose, it is often because Culture has already made that Purpose seem obvious—even in the absence of direct instruction.

Misaligned Culture weakens this gravitational pull. It may glorify speed over meaning, reward internal metrics over stakeholder impact, or normalize priorities that make the Purpose feel hollow or disingenuous. In such cases, the unit's stated intent begins to feel performative—while its actual behavior reflects a different, often contradictory, cultural logic.

The test of alignment is simple: *Does the prevailing Culture make the unit's Purpose more legible, resonant, and believable to agents and stakeholders alike—or does it distract from, dilute, or distort that intent?* A Culture aligned with Purpose anchors the system not just in identity but in orientation—and makes it possible to evolve without losing the reason for existing.

9.7.2. Alignment with Value: Cultural Legibility and Meaning

Value is what the unit offers in exchange—internally or externally. Culture provides the lens through which that Value is interpreted, prioritized, and experienced. When aligned, Culture clarifies *what is valued* and *why*. It shapes the moral economy of the system: what is celebrated, tolerated, or ignored; what counts as contribution; and what trade-offs are considered acceptable in the pursuit of value creation.

A misaligned Culture may celebrate volume over quality, compliance over creativity, or control over responsiveness—distorting the intended nature of the unit's Value. Even when metrics look healthy, the lived experience of Value may feel hollow or adversarial. Alignment ensures that the Value created is not only *delivered* but *believed*.

9.7.3. Alignment with Function: Behavioral Compatibility

Function operationalizes Purpose through interaction. Culture determines *how* that Function is approached, used, and interpreted. An aligned Culture supports Function by making interaction intuitive, respectful, and legible. It influences who takes initiative, how feedback is processed, how interfaces are treated, and how failure is handled.

Misaligned Culture can introduce friction: treating self-service as suspicious, misusing collaborative functions as compliance checklists, or reinforcing dependency where autonomy is expected. Alignment ensures that Functions are not only *designed* to perform but *culturally supported* to succeed. A Function without cultural reinforcement becomes brittle; one with it becomes fluid and trustworthy.

9.7.4. Alignment with Process: Cadence, Rhythm, and Norms

Process defines the structured sequence of activity. Culture determines *how process feels*—whether it is followed with engagement or resistance, adapted wisely or distorted carelessly. When aligned, Culture provides the tacit agreements that allow Process to operate without excessive oversight: shared sense of urgency, quality, timing, and role.

Misalignment surfaces as either rigidity (process worship) or abandonment (process bypass). Aligned Culture allows for procedural resilience—adapting to the moment without drifting from the goal. Culture sets the tone of Process: whether it is lived as ritual, chore, or creative rhythm.

9.7.5. Alignment with Structure: Role Expectations and Interpretive Behavior

Structure establishes the formal composition of roles, relationships, and boundaries. Culture breathes life into Structure by informing how roles are inhabited, how boundaries are respected or challenged, and how authority is interpreted. Aligned Culture makes Structure *work*—not through enforcement but through interpretation.

In misaligned systems, Structure may exist on paper, but roles are abandoned, boundaries blurred, and coordination degraded. In aligned systems, agents understand not only their own roles but *how to relate* to others—what to expect, how to escalate, and where flexibility ends. Structure defines position; Culture defines posture.

9.7.6. Alignment with Order: Norms, Rules, and Behavioral Constraints

Order provides the rules of interaction, the constraints of conduct, and the conditions of legitimacy. Culture reinforces Order by making rules feel intuitive and enforceable—not as foreign impositions but as expressions of shared priorities. When aligned, Culture does not just obey Order—it *extends* it, making compliance natural and non-coercive.

A misaligned Culture either undermines Order (through passive resistance or workarounds) or over-extends it (through rigidity and over-interpretation). Alignment allows Order to function as *both structure and signal*—enabling Culture to channel, rather than contradict, the system's logic of legitimacy.

9.7.7. Alignment with Memory and Storage: Preservation and Reuse

Memory captures the system's lived history; Storage retains its artifacts. Culture shapes what is remembered, what is honored, and what is worth archiving. When aligned, Culture filters Memory to preserve meaning—not just data. It also influences how Storage is interpreted—what gets reused, what becomes ritualized, and what is discarded.

Misaligned Culture may over-index on novelty (erasing hard-won insights) or hoard without meaning (preserving everything but forgetting why). Aligned Culture maintains not just institutional knowledge but institutional *sensemaking*. It makes history useful, not just available.

9.8. Failure Modes and Antipatterns

Culture is not a decorative layer—it is an architectural force that either coheres or corrodes the system from within. When deliberately designed and maintained, Culture reinforces Purpose, stabilizes Function, and enables scalable coordination. When neglected, it mutates into a source of friction, fragmentation, or facade. Cultural failures rarely appear as outages; they manifest as drift, dissonance, and disengagement. The following antipatterns illustrate common structural failures that compromise Culture's role in sociotechnical systems. In practice, such erosion often becomes visible through metrics—rising turnover, slower decision cycles, inconsistent communication patterns, or declining participation in shared rituals—quantitative signals of a qualitative breakdown.

1. **Culture by Osmosis**

   Definition: Culture is left to emerge passively through proximity and imitation rather than intentional design.

   Symptoms: Agents receive mixed signals; norms vary unpredictably across teams; behavior is governed by lore, not logic.

   Consequence: Cultural fragmentation, local fiefdoms, inconsistent onboarding.

   Architectural Fault: Absence of codified induction, shared symbols, or cultural bridges across units. As one executive in a post-merger integration observed, "We assumed culture would spread naturally once people worked together—it never did, and by the time we noticed, each team had built its own version of the company."

2. **Symbolic Inflation**

   Definition: Cultural values are represented in symbols and surfaces but not reinforced through behavior or decisions.

   Symptoms: Posters, dashboards, and values statements exist without resonance or legitimacy.

   Consequence: Cynicism, disengagement, erosion of trust in leadership.

   Architectural Fault: Disconnection between cultural surfaces and structural enforcement.

3. **Cultural Rigidity**

    Definition: Norms that were once adaptive ossify into unquestionable truths.

    Symptoms: Innovation feels risky, rituals outlive their relevance, challenges are framed as cultural betrayal.

    Consequence: Loss of agility, stagnation, inability to evolve with Purpose.

    Architectural Fault: No feedback loops or interpretive mechanisms for cultural refactoring.

4. **Invisible Inequity**

    Definition: Implicit cultural defaults favor certain agent types, roles, or styles—often without awareness.

    Symptoms: Newcomers struggle, insider language dominates, participation feels unsafe or ambiguous.

    Consequence: Attrition, underrepresentation, rise of shadow norms.

    Architectural Fault: Cultural accessibility not addressed across surfaces, roles, and modalities.

5. **Mimicry without Meaning**

    Definition: Borrowed rituals, labels, or language are applied without alignment to local Purpose or structure.

    Symptoms: Hollow standups, empty retros, Agile theater.

    Consequence: Performative culture, misaligned behavior, wasted motion.

    Architectural Fault: Imported practices without translation to local function, structure, or stakeholder needs.

6. **Competing Microcultures**

    Definition: Units evolve disconnected cultures that cannot coordinate or interoperate.

Symptoms: Mismatched assumptions, conflicting rituals, failed collaboration.

Consequence: System-level incoherence, high friction across interfaces, duplication of effort.

Architectural Fault: Lack of cultural integration mechanisms or federated bridges between units.

7. **Culture as Enforcement Mechanism**

   Definition: Culture is weaponized to suppress dissent, constrain ambiguity, or enforce conformity.

   Symptoms: Fear of being "off-brand," premature consensus, unsafe challenge environments.

   Consequence: Intellectual stagnation, hidden resistance, psychological disengagement.

   Architectural Fault: Overcentralized cultural control without room for interpretive tension or pluralism.

8. **Cultural Amnesia**

   Definition: Institutional memory erodes; cultural knowledge is lost during transitions or growth.

   Symptoms: Repetition of past mistakes, rituals lose meaning, new agents invent old solutions.

   Consequence: Fragility under pressure, loss of identity, missed learning.

   Architectural Fault: No coupling between Culture and Memory constructs; no ritualized preservation or handoff.

9. **Culture-Free Automation**

   Definition: Bots, agents, or automated systems act without awareness of human cultural context.

   Symptoms: Escalations feel abrupt, defaults feel alien, machine decisions violate unspoken norms.

Consequence: Distrust in automation, increased human intervention to "translate," misalignment with system tone.

Architectural Fault: Lack of cultural embedding in digital surfaces or AI behavior shaping.

10. **Disembodied Cultural Influence**

    Definition: Influential agents (e.g., executives, consultants) affect Culture without being embedded in the unit.

    Symptoms: Cultural mandates feel external; design moves are misinterpreted; trust is thin.

    Consequence: Cultural dissonance, ineffective interventions, symbolic compliance.

    Architectural Fault: Culture design decoupled from unit boundaries and lived realities.

Cultural antipatterns are not management flaws—they are design faults. They emerge when Culture is allowed to drift, ossify, or detach from the structural realities of the sociotechnical system. Diagnosing these failure modes requires architects to listen for dissonance, observe friction, and identify where surfaces contradict intent. Repairing them requires more than better rituals—it requires *architectural coherence*: a Culture that reinforces Purpose, supports Function, and evolves in rhythm with the system it inhabits.

9.9. Summary: Culture As a Design Construct

Culture, within the Purpose Driven Design framework, is not a soft variable or symbolic veneer—it is the system's *default coordination intelligence*. It enables coherent behavior, shared interpretation, and distributed decision-making in the absence of explicit instructions, rational deliberation, or real-time supervision. Unlike constructs that must be invoked—through interfaces, commands, or logic gates—Culture operates *continuously*, orienting agents through embedded expectations and practiced defaults.

At its core, Culture solves the architectural problem of coherence at scale. It allows large, distributed, or heterogeneous systems to function as wholes rather than collections of parts. Where Structure provides roles, and Process supplies motion, Culture infuses meaning, rhythm, and interpretive guidance. It bridges intent and behavior, linking Purpose to daily actions without introducing brittleness or bottlenecks.

Realized effectively, Culture functions as

- *A behavioral substrate,* guiding how agents perform roles, resolve conflicts, and navigate priorities
- *A mechanism for scalability,* enabling systems to maintain alignment without centralized micromanagement
- *A governance surface,* allowing intent to be reflected in practice even when formal controls are absent
- *A resilience layer,* ensuring stability when structures, processes, or technologies fail or evolve
- *An integrative bridge,* reinforcing Purpose, Function, Process, Structure, Order, Memory, and Storage across time and circumstance

The architectural payoff of Culture is coherence under complexity. But this coherence is not self-sustaining. Culture must be designed, surfaced, interpreted, and maintained—like any other system element. When neglected, it does not disappear; it *drifts.* When misapplied, it does not weaken harmlessly; it *fragments* the system.

In modern sociotechnical environments, Culture is no longer an afterthought of leadership or an artifact of history. It is *strategic infrastructure*—the invisible architecture that holds systems together when process breaks down, when information is incomplete, or when purpose must be inferred rather than enforced.

Culture is how coherence persists when nothing else is available to ensure it.

CHAPTER 10

Memory

In the year 868 CE, in a cave near Dunhuang on the edge of the Gobi Desert, a remarkable artifact was sealed away: a printed scroll containing a Buddhist text known as the *Diamond Sutra*. When rediscovered in the early twentieth century, it was identified as the world's oldest known printed book bearing a specific date. Its endurance across centuries prefigures the central architectural question of this chapter—how systems, ancient or digital, decide what to remember, what to forget, and how to preserve meaning over time.

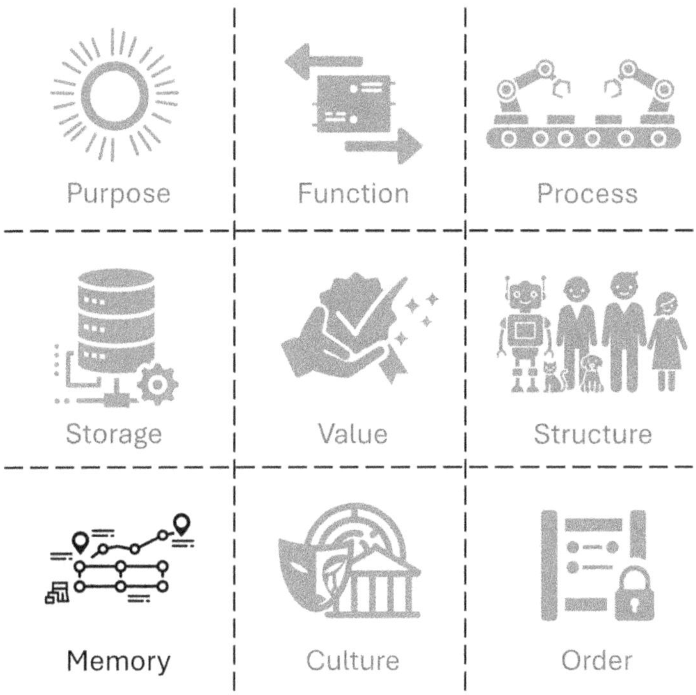

A. Milchman, *The Purpose Driven Design Framework for Modern Enterprises*,
https://doi.org/10.1007/979-8-8688-2223-0_10

It was not a book in the modern sense. The scroll stretched over five meters long, printed using hand-carved wooden blocks on delicate paper. Its first lines declared it was created for “universal free distribution”—a striking phrase that fused preservation with purpose. In many ways, this ancient intent anticipates the spirit of today’s open-source movements, where knowledge is not only preserved but deliberately made shareable to ensure continuity and collective improvement. Though it held no commercial value, it was crafted with precision and hidden with care—not to inform the present but to speak across generations.

The creation of such a text required more than paper and ink. It depended on a system—spiritual, social, and technological—that valued memory. Monks trained in calligraphy carved the blocks. Printers reproduced the text faithfully. Custodians sheltered the result from weather, war, and oblivion. It was not merely copied; it was curated. Not merely stored; it was entrusted. Curation here meant deliberate selection and contextual framing to preserve meaning, while entrustment implied a transfer of custodial responsibility—the recognition that preservation itself was a sacred, systemic act.

Its survival offers more than historical curiosity. It reveals an early example of architectural memory: a designed mechanism by which a system could preserve complexity, intention, and context across centuries.

In contemporary terms, it resembles digital architectures that preserve lineage and integrity—such as distributed ledgers or version-control systems that retain every change while safeguarding meaning over time.

Oral traditions had long carried knowledge. Stone had offered durability. But printing introduced a portable, duplicable, structured form of memory—capable of outliving its creators and adapting to new discoverers. The distinct ways these media preserve, transmit, and transform knowledge become clearer when viewed comparatively in Table 10-1.

Table 10-1. *Comparative Memory Attributes Across Media*

Medium	Primary strength	Limitation	Memory attribute supported
Oral Tradition	Flexible transmission through repetition and adaptation	Vulnerable to loss or distortion over generations	Continuity through reinterpretation
Stone inscription	High durability and physical permanence	Limited portability and context; costly to revise	Longevity and authority
Printed text	Portable, replicable, and standardized	Dependent on materials and distribution networks	Scalability and structured preservation

The *Diamond Sutra* did not simply preserve content; it preserved meaning. Its illustrations and layout shaped how the message was interpreted. Its format invited transmission. Its very survival through upheaval speaks not just to its construction, but to the cultural architecture that recognized it as something worth preserving.

In contrast, modern systems treat memory casually. Logs are purged. Histories fragment. Systems are rebuilt with little regard for what came before.

One large enterprise, after decommissioning an outdated analytics platform, discovered that years of historical transaction data had been lost—making it impossible to audit regulatory breaches that occurred during the transition. Similar patterns appear when teams develop vendor fatigue, repeatedly migrating to new tools of comparable quality and unintentionally abandoning valuable records each time. In both cases, the failure is not technical but architectural: an absence of institutional memory.

Yet the challenge remains: how to retain what matters—faithfully, meaningfully, and accessibly—so that a system can know itself over time. Memory is not a technical trace or archival habit. It is the architectural means by which a system acquires coherence, continuity, and explainability. It transforms experience into usable context. It stabilizes identity. It enables long-range accountability. Without it, systems drift—ungrounded in their past, reactive in the present, and incoherent in the future. Designed well, memory allows a system not only to persist but to evolve. It holds the shape of change. It recalls purpose even when conditions shift. When memory is neglected, systems lose their capacity to learn—like organizations that repeatedly revive failed strategies or algorithms retrained without lineage, forgetting the lessons that once made them effective. Like the Diamond Sutra, memory whispers across time: I was here. I mattered. This is what we learned. The practical consequences of these memory principles become clearer when viewed through the design outcomes outlined in Table 10-2.

Table 10-2. *Core Design Outcomes of Architectural Memory*

Design outcome	Definition	Architectural function
Coherence	The ability of a system to hold together conceptually across time, agents, and components	Ensures that decisions and behaviors align with shared logic rather than fragmenting into silos
Continuity	The retention of structural identity and learning through change	Allows systems to evolve without losing their historical grounding or sense of direction
Explainability	The capacity to reconstruct and justify past actions or configurations	Supports accountability, trust, and governance by making reasoning and lineage traceable.

10.1. Foundations of Memory As a Design Construct

To design with Memory is to recognize that experience itself is insufficient unless it becomes *usable history*. Memory is not the event itself but its structured retention—it is how what happened is retained, recalled, and made meaningful within the system's structure. Its presence enables identity to endure beyond immediate activity, allows decisions to be explained, and equips a system to learn from its own trajectory. Understanding Memory as a design construct requires clarity on its architectural definition, its reason for being, the specific architectural problem it addresses, and how it has manifested across historical systems. Each of these perspectives contributes to a coherent view of Memory not as a by-product but as a first-class concern in purposeful design.

10.1.1. Defining Memory Architecturally

Memory is the structured presence of the past within the present. It allows a system to retain what it has been—not by accident, but by design. Memory is not simply data that remains; it is *that which has been retained for a reason*, whether to support interpretation, enable continuity, or preserve identity. In practice, this structured presence appears as versioned documents, annotated workflows, and decision logs or process histories—artifacts that keep context alive rather than merely archived.

Architecturally, Memory refers to the capacity of a system to record, recall, and represent its own lived trajectory. It makes prior decisions, events, interactions, and transformations available for future use—not necessarily for reuse as Value, but for recognition, explanation, and alignment. Memory is what makes recurrence meaningful and change intelligible. In software engineering, for instance, recurrence tracking—such as logging and analyzing repeated bugs or incidents—turns failure into learning, enabling refinement instead of endless repetition.

A system without Memory may function, but it cannot remember what it has learned, how it has evolved, or why certain paths were taken. After a leadership change, for example, one public agency dismantled its long-standing analytics unit only to rebuild an almost identical one three years later—having forgotten both the rationale and lessons of the original design.

A system with misaligned Memory may recall fragments that distort rather than clarify. And a system with well-designed Memory becomes self-aware in structural terms—it can explain itself.

This distinguishes Memory from both Storage and from real-time observation mechanisms often referred to as monitoring. Monitoring captures the present. Storage retains curated Value artifacts. Memory, by contrast, carries the interpretive structure of experience—what happened, in what order, with what consequence, under what context, why it mattered, and how it informs future action. It is temporal, relational, and reconstructive. Its unit of meaning is not just the record but the storyline that makes sense of the record. These distinctions become clearer when Memory, Storage, and Monitoring are compared side-by-side in Table 10-3.

Table 10-3. *Distinguishing Memory, Storage, and Monitoring in Purpose Driven Design*

Construct	Primary intent	Temporal focus	What it retains	Architectural function in PDD
Monitoring	Observe and report current system behavior	Present (real time)	Signals, metrics, operational states	Provides immediate visibility into current system behavior; supports operational awareness but is not a PDD construct and does not preserve meaning across time
Memory	Preserve experience, rationale, and interpretation	Past informing the present	Decisions, explanations, event histories, contextual traces	Preserves history; enables coherence, accountability, and learning—making change intelligible
Storage	Retain and serve artifacts of Value for future use or exchange	Future-facing (asynchronous reuse)	Digital artifacts or physical goods	Supports the continuity and reusability of Value across time and domains

In architecture, Memory manifests as event logs, audit trails, decision histories, policy versions, configuration drifts, and all mechanisms that allow the past to re-enter design and operation. But its true role is deeper: Memory turns what was into a usable substrate for what comes next. It creates the architectural thread by which a system knows what it has been and can decide what it will become.

10.1.2. Why Memory Exists

Memory exists to make continuity possible in systems that change. Without it, every transition becomes a reset, every failure a mystery, every success an accident. Memory allows a system to carry forward not just outcomes but context—enabling alignment over time, not just at a moment. In medicine, for instance, detailed patient histories ensure that treatment decisions reflect accumulated understanding rather than isolated symptoms; in aviation, flight logs and incident reports serve the same purpose, sustaining safety through remembered context.

In sociotechnical systems, agents come and go, tools are replaced, policies evolve, and processes are revised. Memory provides the connective tissue that binds these changes into a coherent whole. It makes it possible to trace the lineage of a decision, understand the origin of a structure, or explain why a process exists in its current form.

Systems that serve Purpose over time require memory to remain accountable. Systems that pursue improvement require memory to accumulate learning. Systems that operate in regulated environments require memory to demonstrate compliance. And systems that evolve across human and machine generations require memory to avoid rediscovering what was already known—or relearning what was once painful.

Memory also enables systems to resist manipulation. When the past is known and visible, narratives can be challenged. Actions can be audited. Intent can be verified. In this sense, Memory is not only functional; it is protective. It guards against revisionism, erasure, and amnesia—whether social, procedural, or computational.

At its core, Memory exists so that the consequences of past events can remain accessible to those who were not present. It creates a bridge between action and understanding, across time. Without Memory, systems fragment. With it, they gain coherence—not by freezing the past but by remaining aware of it.

10.1.3. What Problem Memory Solves

Memory solves the problem of *architectural forgetfulness*—the loss of connection between past decisions and present conditions. In dynamic systems, change is constant. When that connection is lost, organizations pay for it in duplicated effort, repeated mistakes, regulatory penalties, and the slow erosion of strategic clarity. Without Memory, that change becomes directionless. Structures evolve without explanation, functions drift from their origin, and purposes are inherited without understanding. Memory restores the missing link between cause and consequence.

The absence of Memory creates a specific architectural failure: the inability to *justify current state*. Why is this structure in place? Why was this function deprecated? Why does this exception exist? Without historical grounding, such questions become unanswerable—or worse, answered incorrectly. Memory solves this by preserving not just outcomes, but the rationale that led to them.

It also addresses the problem of *invisible cost*. When systems lack Memory, valuable lessons are buried. Rework multiplies. Failures repeat. Institutional knowledge becomes individual dependence. Memory mitigates this by making experience shareable, by allowing systems to retain and communicate what was once hard-won.

In sociotechnical terms, Memory solves the problem of *ephemeral responsibility*. Roles change, teams reorganize, software is refactored—but accountability cannot be transient. Memory ensures that actions remain attributable, not to assign blame but to uphold continuity of understanding. It allows future agents to interpret prior behavior, not merely observe its remnants.

Lastly, Memory solves the problem of *discontinuous reasoning*. Without historical threads, decisions are made in isolation. Strategy detaches from context. Feedback loops weaken. Memory re-establishes the timeline—enabling decisions to account for what has already been attempted, observed, or proven.

What Memory solves, therefore, is not a technical inconvenience but a deep architectural vulnerability: the fragility that arises when systems cannot remember what they are evolving from.

10.1.4. Historical Patterns and Structural Evolution

The architectural role of Memory predates modern computing. Before logs and databases, civilizations encoded memory into rituals, records, monuments, and stories—not merely to commemorate but to stabilize continuity through time. Across history, systems that endured developed intentional memory structures, even in the absence of digital means. Monastic scriptoria preserved theological knowledge through disciplined copying; maritime logbooks carried navigational intelligence across generations of sailors; and early bureaucracies, from the Han Empire to the Venetian Republic, maintained meticulous ledgers to ensure institutional continuity. This long arc of memory practice is outlined across eras in Table 10-4.

Table 10-4. Evolution of Architectural Memory

Era	Dominant memory form	Mechanism of preservation	Architectural lesson
Oral and ritual (prehistoric—early societies)	Human recollection and storytelling	Repetition, performance, and shared narrative	Memory is social, interpretive, and living
Written and monumental (ancient civilizations)	Inscriptions, tablets, scrolls	Physical durability, codification	Memory gains stability but loses flexibility
Printed and bureaucratic (Renaissance—Industrial Age)	Books, ledgers, archives	Standardization and mass reproduction	Memory becomes scalable and auditable
Digital and Networked (twentieth century—present)	Databases, logs, cloud archives	Automation, replication, and indexing	Memory becomes abundant but easily fragmented
Hybrid and AI-augmented (emerging)	Human–machine ecosystems	Semantic models, continuous learning	The challenge shifts from retention to interpretation

Early oral traditions relied on trained agents—priests, scribes, elders—who served as living repositories of collective knowledge. Their memory was relational and performative, embedded in roles and transmitted through practice. This made memory adaptive but fragile: it could evolve with interpretation, but also vanish with its carriers.

Written records marked a structural shift. Tablets, scrolls, ledgers, and chronicles created external surfaces for memory—allowing it to persist independently of the agents who originated it. Yet writing also introduced new dependencies: on storage, translation, and institutional custodianship. The emergence of archives, libraries, and bureaucracies reflected an increasing need to *design for memory,* not just capture it.

The invention of printing democratized memory. Redundancy became possible. Versioning and distribution became scalable. Memory was no longer the privilege of institutions; it became an architectural layer available to society. This shift transformed how systems could preserve knowledge, policy, belief, and law—no longer local, no longer ephemeral.

With digital systems, memory became programmable—but also more volatile. While storage capacity expanded, meaningful memory structures often regressed. Logs replaced records. Snapshots replaced narratives. In many systems, what is retained is incidental; what is remembered is unstructured. This volatility arises because automation favors immediacy over interpretation, storage is cheap enough to collect without curation, and rapid iteration discourages the deliberate design of continuity—conditions that make digital memory abundant but shallow. The architecture of memory became flattened, even as the volume of data exploded.

Modern sociotechnical systems must now recover the original insight: that memory is not the same as storage, and that preservation requires structure, not just retention. The evolutionary trajectory of memory—from embodied, to inscribed, to distributed, to digital—reveals both possibilities and pathologies. It is not enough to record events; systems must decide *what kind of structure* will allow history to be usable, trustworthy, and aligned with purpose.

The patterns of the past offer a quiet warning: when memory becomes fragmented or superficial, systems become reactive, manipulable, and unaccountable. But when memory is designed—when it is recognized as a core architectural construct—it becomes an enduring source of intelligence, identity, and resilience.

10.2. Role of Memory in Sociotechnical Systems

Memory operates at multiple layers of a sociotechnical system, each with distinct architectural demands. Within a unit, memory stabilizes identity, preserves rationale, and enables continuity through change. It stabilizes identity when an organization maintains a recognizable mission despite leadership turnover; it preserves rationale when decision logs explain why a past approach was chosen; and it enables continuity when teams onboard new members through documented histories that prevent the loss of hard-won learning. This internal role allows a unit to remain intelligible to itself—especially as agents, functions, and configurations evolve. At the systemic level, memory becomes integrative: it connects units across space and time, providing the narrative thread by which coordinated behavior, accountability, and adaptation can be sustained. Here, integrative units take responsibility for shaping and curating shared memory, while operational units contribute local experience. Beyond the system's boundaries, memory plays an environmental role—mediating relationships with consumers, regulators, partners, and competitors. It anchors reputation, explains behavior, and enables trust. In all three roles, memory ensures that the system is not merely reactive but grounded in its own history—capable of responding, not forgetting.

INTERNAL—Within the unit: roles, agents, and immediate interactions

SYSTEMIC—Across units of complex systems (enterprises): coordination, alignment, and governance

ENVIRONMENTAL—Beyond the system: relationships with regulators, consumers, and partner

---

10.2.1. Internal Role: Stabilizing the Unit

A unit without memory cannot hold form. Its processes become untraceable, its decisions unexplainable, and its structure unknowable—even to itself. Internal memory serves as the stabilizing layer that captures the lived experience of the unit: what it does, how it changes, and why it behaves the way it does.

Within the unit boundary, memory records events, monitors transitions, and preserves internal logic. It traces processes as they unfold over time—capturing sequences, deviations, and outcomes. It registers the execution of jobs, the flow of tasks, and the accumulation of operational knowledge. This memory is not passive; it shapes how the unit understands its own function and performance.

Without it, agents must rely on conjecture or recollection. Decisions are repeated, not refined. Failures recur with no record of prior causes. Successes are misattributed or forgotten. Memory provides the structure by which a unit becomes self-aware—able to learn from its own behavior and evolve intentionally, not accidentally.

This role becomes especially critical during transitions: new leadership, shifting mandates, rotating roles, or evolving tools. Memory holds the thread that allows one configuration of the unit to hand over to the next without loss of integrity. It ensures that continuity is not dependent on individuals but embedded in the architecture of the unit itself.

Internally, memory may take the form of event logs, runbooks, audit trails, decision journals, workflow histories, process telemetry, or embedded annotations in tools. What matters is not the medium but the function: to make the internal life of the unit observable, explainable, and accumulative. The variety of ways units internalize and preserve their own experience becomes clearer when viewed through the mechanisms outlined in Table 10-5.

Table 10-5. *Common Internal Memory Mechanisms*

Memory form	Primary function	Example application
Event logs	Record discrete system or user actions for traceability.	Security investigations or incident analysis
Runbooks	Provide step-by-step operational recall.	Standardizing response to outages or scheduled maintenance
Audit trails	Preserve authoritative evidence of decisions and actions.	Compliance, financial reporting, or governance validation
Decision journals	Capture rationale and assumptions behind key choices.	Strategic planning, design justification, or leadership reviews
Workflow histories	Document sequence and dependency of completed processes.	Process improvement and bottleneck analysis
Process telemetry	Record real-time process signals to support learning loops.	Monitoring process efficiency or detecting drift
Embedded annotations	Contextual memory within tools or interfaces	Engineers noting exceptions directly inside code or dashboards

Memory stabilizes the unit by giving it a past. And in doing so, it creates the possibility of deliberate future.

EXAMPLE—BEFORE AND AFTER MEMORY DESIGN

Before: A product-support unit relied on informal chat logs and personal recollection to resolve client issues. When staff rotated, recurring problems had to be diagnosed from scratch, eroding trust and speed.

After: The team introduced a shared incident journal and annotated runbooks. Within weeks, resolution times dropped by half, and new analysts could explain not only what to do but why prior fixes worked. The unit's "past" became an operational asset—turning continuity into capability.

10.2.2. Systemic Role: Integration into System-Wide Coordination

In large sociotechnical systems, no single unit contains the whole story. Memory must scale beyond the boundaries of any one unit to preserve coherence across many—especially as coordination, regulation, and strategy stretch across geographies, domains, and time frames. This is the systemic role of memory: to bind the many into a comprehensible whole, and to ensure that a distributed system does not become a fragmented one.

Integrative units carry this responsibility. They maintain the memory of the system itself—not just of what each unit did but how those actions combined, conflicted, or evolved over time. Like a flagship preserving the logbook of an entire fleet, integrative units curate the historical surface on which alignment, governance, and adaptation depend.

Systemic memory includes the sequence of strategic decisions, the evolution of shared processes, the history of structural reconfigurations, the outcomes of system-wide initiatives, and the lessons drawn from previous crises or transformations. It must capture not only what was done but how units interacted, how authority shifted, and how coordination succeeded—or failed. When absent, organizations repeat failed transformations, duplicate prior efforts under new names, or lose sight of why coordination models once worked. Robust systemic memory prevents these regressions by preserving institutional reasoning, making patterns of success and failure visible before new initiatives begin.

Operational units contribute to this memory by submitting records, flags, reports, and telemetry from local activity. But it is the integrative units—regional and global headquarters—that interpret, contextualize, and retain these contributions in forms that preserve meaning across time and scale. This may involve timelines, systemic incident retrospectives, multi-unit decision registries, or platform-wide architectural digests.

Systemic memory also supports change. When architectures evolve, when units are dissolved or merged, when new capabilities are introduced—the memory maintained by integrative units prevents the loss of institutional intelligence. It allows the system to remember not only what it once was but why it changed and what it has already tried.

Without systemic memory, large systems cannot course-correct. Coordination collapses into confusion, and new decisions repeat old mistakes. But when integrative units embrace their role as memory stewards, the system becomes legible to itself—capable of evolving with context, not despite it.

10.2.3. Environmental Role: Interface with External Entities

No unit exists in isolation. Each interacts with entities beyond the system's boundaries—customers, suppliers, regulators, investors, labor markets, competitors, and others. Memory enables these interactions to persist across time, personnel, and circumstance. It forms the historical surface by which relationships are understood, agreements honored, and behavior explained.

Every sociotechnical unit maintains a memory of its environment. It remembers *who* it interacted with, *what* was exchanged, *when* and *why*, and *how* the relationship evolved. This memory is not just operational—it is reputational, regulatory, contractual, and ethical. It protects reputation through documented transparency that external partners can verify, satisfies regulators through preserved audit trails, secures contractual trust by retaining evidence of obligations, and sustains ethical credibility by recording how the unit's decisions affect others. It also mitigates disputes, supports compliance, and enables personalization, negotiation, and trust-building.

This environmental memory is often specialized. A customer-facing unit may retain transaction histories, support interactions, preference profiles, and communication patterns. A regulatory unit may preserve audit trails, submission timelines, risk classifications, and correspondence with oversight bodies. A hiring unit may record interview outcomes, candidate assessments, and long-term talent performance. While each unit focuses on a different entity type, all contribute to the system's environmental intelligence.

Externally, memory allows the system to present itself coherently over time. A regulator expects continuity between audits. A returning customer expects to be recognized. An investor expects consistent reporting. Without memory, the system appears amnesiac—disconnected from its past engagements, unaccountable for its history, and unreliable in its behavior.

Memory also enables adaptation to environmental shifts. Changes in regulation, consumer behavior, market dynamics, or partner expectations must be interpreted in light of prior experience. A system that forgets its past responses is ill-prepared for new challenges. One that remembers can evolve responsibly.

Finally, environmental memory reinforces trust. When entities outside the system see that the system remembers—not just what it did but how it treated others—they can engage with greater confidence. In this way, memory is not only inward-facing but reputational: it becomes part of how the system is perceived by the world around it.

10.3. Design Principles and Heuristics

Designing Memory is not a matter of retention alone—it is a matter of structure, intent, and alignment. What is remembered, how it is remembered, and for whom it is made accessible are all architectural choices. Table 10-6 organizes these considerations into the key decision points that define effective memory design.

Table 10-6. *Key Decision Points in Memory Design*

Design decision	Architectural choice	Potential consequence (if neglected or misapplied)
What to remember	Selecting events, rationales, or artifacts relevant to Purpose and Function	Retaining noise instead of insight; memory becomes cluttered or unusable
How to remember	Choosing structure, format, and governance—logs, narratives, or annotated records	Loss of context or interpretability; information remains inert
For whom	Defining access and visibility for roles, domains, or external parties	Overexposure creates risk; underexposure erodes trust and accountability
For how long	Establishing retention and expiration aligned with Purpose and compliance	Institutional forgetfulness or liability from unnecessary retention
How to evolve	Designing feedback loops to refresh or reinterpret memory over time	Stagnant or outdated memory misguides future decisions

Poorly designed memory accumulates noise, obscures causality, and fragments over time. Well-designed memory supports explanation, accountability, learning, and coordination. The principles and heuristics that follow are not implementation rules; they are architectural considerations that help ensure memory remains usable, trustworthy, and meaningful as the system evolves.

10.3.1. Principles of Memory-Oriented Design

Memory-oriented design approaches Memory not as leftover data or transient trace, but as a structural mechanism for preserving history, enabling explanation, and supporting continuity across time. A Memory does not stand alone—it exists in context, aligned with Purpose, integrated with Function, and shaped by the unit's role in the broader system. Designing Memory well means ensuring that what is remembered is architecturally relevant, interpretable, and aligned with the system's capacity to learn, evolve, and remain accountable.

The following principles support the integrity, usability, and coherence of Memory:

1. **Memory must preserve usable history**

   Memory is not mere retention. It should capture the sequence of relevant events, decisions, and changes in a form that supports reasoning and alignment. This includes what happened, when, under what context, and with what consequence.

   Implication: Memory that stores facts without context may remain technically complete but architecturally unusable.

2. **Memory must stabilize identity over time**

   Memory enables a unit to know what it has been—structurally, functionally, and relationally. It connects past configurations, roles, and behaviors into a coherent timeline, allowing the unit to evolve without losing itself.

   Implication: Without Memory, long-lived units risk architectural amnesia, where current state cannot be explained or governed.

3. **Memory must support continuity across agents**

   In systems where people and software change frequently, Memory must carry forward the rationale, decisions, and exceptions made by prior agents—so that new agents can act with understanding, not speculation.

   Implication: If knowledge transfer relies solely on human handover, Memory has failed to function structurally.

4. **Memory must differentiate between internal, systemic, and environmental roles**

   Each unit should retain the memory appropriate to its scope—internal operations, system-level coordination, or external relationships. Blurring these roles leads to fragmentation, duplication, or omission.

   Implication: A unit that remembers everything remembers nothing well; memory must be scoped with intent.

5. **Memory must be designed for recall, not just retention**

   What is stored must also be findable, explainable, and interpretable by the agents who need it. Raw logs or opaque records undermine the purpose of memory by requiring translation or forensics to reconstruct meaning.

   Implication: Memory that cannot be recalled when needed is functionally equivalent to forgetting.

6. **Memory must preserve the shape of events, not just their content**

   The structure, order, and interrelation of events often matter more than isolated data points. Good Memory captures progression, dependency, and flow—so that meaning is not reduced to metadata alone.

   Implication: Flattened records erase the temporal and causal logic of action, weakening reasoning and accountability.

7. **Memory must reveal decision logic, not just outcomes**

   Architectural memory includes not only what was decided but why. Retaining the assumptions, constraints, and trade-offs behind decisions allows future agents to reinterpret outcomes with fidelity to their original context.

   Implication: Without decision logic, systems can only observe behavior—not understand or improve upon it.

8. **Memory must represent meaningful absences**

   Not all memory is presence. Sometimes, what was *not* done, chosen, or triggered matters. Absences—missed alerts, skipped steps, avoided paths—are part of the story a system tells itself.

   Implication: When memory records only action, the structure of omission remains invisible and unaccounted for.

9. **Memory must resist silent mutation**

   As systems evolve, memory must not silently change alongside them. If past records adapt to new schemas, roles, or interpretations without trace, the integrity of historical reference is lost.

   Implication: Memory should retain the shape it had when it was written—even as systems grow around it.

10. **Memory must support temporal layering**

    Not all memory needs to be accessed at the same frequency or fidelity. Design should accommodate layers—frequent, recent, high-resolution access coexisting with rare, deep, archival memory.

    Implication: When all memory is treated uniformly, systems either become overwhelmed or lose long-range insight.

Together, these principles ensure that Memory is not an afterthought but a first-class architectural concern. Properly designed, Memory empowers systems to act with historical awareness, learn from experience, justify decisions, and evolve with continuity. It becomes the backbone of systemic reflection, trust, and temporal coherence—foundations that the following heuristics translate into practice.

10.3.2. Architectural Heuristics for Memory

While principles clarify what Memory must achieve architecturally, heuristics offer practical guidance for embedding it effectively within real-world systems. These are not universal rules, but tested design moves that help make Memory usable, coherent, and resilient over time. They are especially valuable when working within the constraints of legacy systems, evolving schemas, regulatory pressures, or distributed teams.

Below are thirteen architectural heuristics to support the design of intelligible, trustworthy, and maintainable Memory:

1. **Capture events as first-class citizens**

   Design for event visibility. Treat events—not just states—as foundational units of Memory. Doing so enables reconstruction of system behavior and preserves temporal logic.

   Rationale: Systems that retain only snapshots lose the ability to explain how the current state was reached.

2. **Always timestamp at source**

   Assign timestamps at the moment an event occurs, not during downstream processing. This preserves temporal accuracy and prevents confusion in ordering.

   Rationale: Delayed or ambiguous timestamps degrade historical reliability and undermine traceability.

3. **Maintain versioned representations**

   Track and store versions of schemas, policies, configurations, and critical data. Don't overwrite history—layer it.

   Rationale: Unversioned memory falsely implies uniform meaning over time, eroding interpretability.

4. **Separate Memory from Storage by intent**

   Clarify the distinction: Memory retains what happened; Storage preserves reusable Value. They may use the same infrastructure but must serve different design purposes.

   Rationale: Blending Memory and Storage leads to unclear retention logic, misaligned access, and structural drift.

5. **Use append-only logs for irreversible actions**

   For auditable or safety-critical events, avoid overwrite. Preserve immutability through append-only structures or write-once logs.

   Rationale: Mutable memory invites tampering and weakens accountability.

6. **Encode rationale, not just outcome**

   Whenever feasible, store the reasoning behind a decision: input signals, chosen path, rejected alternatives, and intended goals.

   Rationale: Without rationale, Memory preserves results but loses meaning.

7. **Design query interfaces for temporal traversal**

   Make Memory navigable. Enable queries by time, event type, life cycle stage, or cause–effect chain.

   Rationale: A memory that cannot be queried structurally is equivalent to a sealed archive.

8. **Annotate with metadata for contextual alignment**

   Add role, source, location, and environment tags to memory artifacts. Metadata supports sorting, filtering, and multi-perspective interpretation.

   Rationale: Raw records are difficult to repurpose without context for orientation.

9. **Prioritize human interpretability for critical paths**

   For governance, compliance, and safety-related logs, ensure readability by human reviewers. Avoid overly compressed or opaque encodings.

   Rationale: Machine-readable alone is not enough when human accountability is required.

10. **Integrate Memory into feedback loops**

    Feed Memory into retrospectives, diagnostics, and learning systems. Don't let it accumulate without influence.

    Rationale: Unused Memory becomes sediment—not intelligence.

11. **Architect for selective forgetting**

    Implement controlled deletion, summarization, or compaction mechanisms where appropriate. Retain meaning, not volume.

    Rationale: Endless accumulation without abstraction leads to inefficiency and liability.

12. **Enforce role-based memory access**

    Design Memory access based on roles, responsibilities, and scopes. Track who reads or writes memory artifacts.

    Rationale: Memory integrity and confidentiality require architectural governance.

13. **Simulate failure scenarios using historical data**

    Enable historical replay of system states for testing, training, or resilience analysis. Use Memory as a rehearsal space for improvement.

    Rationale: If Memory can't be reused to prevent recurrence, its value remains dormant.

These heuristics help translate the abstract concept of Memory into concrete architectural structures and operational behaviors. They reinforce the idea that Memory must not only exist—it must be constructed deliberately to support the reflective, accountable, and adaptive nature of purposeful systems. Table 10-7 maps these heuristics onto the differing demands of legacy and greenfield architectures.

Table 10-7. *Applying Memory Heuristics in Legacy vs. Greenfield Systems*

Design focus	Legacy systems (retrofitting context)	Greenfield systems (designing anew)
Capture mechanisms	Integrate with existing logs and databases; extract implicit memory from past records and tools.	Define structured memory points from the outset—events, rationales, and metadata models.
Traceability	Establish lineage gradually through audits and documentation of current states.	Bake traceability into workflows and APIs by design.
Governance and access	Work within current roles and compliance rules; use incremental transparency.	Design access models and retention rules explicitly before scale introduces complexity.
Format and structure	Normalize disparate data formats; prioritize readability and interpretability over completeness.	Choose consistent schemas early to ensure future interoperability.
Cultural integration	Encourage teams to reuse and annotate existing artifacts.	Introduce memory as a core system capability tied to Purpose and Function.
Evolution and feedback	Use incident retrospectives and refactoring cycles to evolve memory practices.	Build feedback loops directly into the architecture for adaptive learning.

This comparison underscores that while the principles of Memory remain constant, their implementation must adapt to context—constraints in legacy systems versus creative freedom in greenfield design.

10.4. Realization Across Implementation Planes

Memory does not exist only in the abstract. It takes shape across the three implementation planes—*social, physical,* and *digital*—each contributing to its persistence, accessibility, and interpretation. In the social plane, Memory is preserved through human recollection, shared narrative, and documented experience. In the physical plane, it is embedded in artifacts, signage, logs, and tools. In the digital plane, it is encoded, versioned, queried, and often overproduced. But Memory is not merely stored—it is realized through intentional structures that reflect the system's Purpose, constraints, and maturity. To design Memory well, we must recognize where and how it manifests, and what each context demands of it.

10.4.1. Manifestation in the Social and Physical Planes

Memory is not a digital invention. Long before systems were virtualized, history was preserved through people, artifacts, spatial arrangements, and material routines. In many domains—such as traditional crafts, regulatory archives, low-infrastructure settings, or crisis-resilient systems—the social and physical planes remain the primary memory surfaces. Even in digitally mature systems, they continue to serve distinct architectural functions: preserving continuity, transmitting meaning, and enabling local autonomy.

In the *social plane,* Memory is maintained through human recollection, shared narratives, verbal instruction, rituals, and embodied routines. Roles often carry memory: the senior operator who recalls why a process changed, the team that remembers a workaround born of necessity, or the community that transmits lessons through storytelling and reenactment. Though volatile and subjective, social memory is immediate, adaptive, and context-sensitive. It can guide behavior even when no formal record exists.

In the *physical plane,* Memory manifests through tangible, inspectable forms. These include

- *Written records*: Ledgers, logbooks, and physical documentation serve as authoritative and auditable repositories—especially in legal, medical, and operational settings.
- *Spatial arrangements*: The physical layout of libraries, warehouses, and assembly stations often encodes history and classification logic.
- *Manual tools and record-keeping instruments*: Punch cards, index files, or annotated charts preserve process memory through structure and repetition.
- *Architectural and infrastructural traces*: Buildings and machines wear their history—via retrofits, signage, or operator markings.

- *Symbols and artifacts*: Plaques, certificates, or commemorative objects anchor collective memory in persistent, visible forms.
- *Permanent archival media*: Stone carvings, metal plates, and microfiche outlast software platforms, preserving memory across generations.

The physical and social planes emphasize *durability, interpretability, and resilience.* They support environments where *digital fragility, power constraints, or cultural norms* make physical or human memory preferable. Even where digital systems are available, physical memory may be used deliberately: to ensure trust, provide emergency access, or serve as the final fallback for mission-critical knowledge.

Architecturally, Memory in these planes deserves first-class consideration—not as legacy overhead, but as structural layers that carry meaning and enable autonomy. When designed purposefully, they complement digital systems without depending on them. When over-relied upon without reinforcement, they become brittle. The challenge is not to replace them, but to integrate them into a broader, coherent memory architecture.

10.4.2. Memory in the Digital Plane

The digital plane allows Memory to scale, persist, replicate, and analyze—but this power is not automatic, and its usefulness is not guaranteed. Memory in digital systems must be deliberately structured, explicitly governed, and thoughtfully surfaced. Without that, it risks becoming an unmanageable accumulation of data without meaning or relevance.

In the digital plane, Memory takes many forms—each with its own architectural role:

- *Event logs* capture what happened and when, often with high granularity.
- *Metadata catalogs* organize what exists, where, and under what schema.
- *Version control systems* track changes in logic, configuration, and content.
- *Change data capture (CDC) streams* encode what changed, in real time, and why.

- *Temporal databases* allow systems to query states at a given point in history.
- *Audit trails* offer regulated visibility into system activity and access.
- *Decision rationale repositories* store the "why" behind key choices.
- *Observability systems* preserve runtime memory—metrics, traces, and signals.
- *Comment threads, logs, and documentation* represent informal but influential memory surfaces.

This diversity is a strength—when aligned to Purpose and accessible to the agents that need it. But architectural misuse is common: memory that exists in low-fidelity logs no one can query, in fragmented tools no one integrates, or in formats no one interprets. Digital capacity alone does not solve the memory problem; it simply moves it into a higher bandwidth failure mode.

Well-designed digital memory systems share several qualities:

- *Semantic alignment*: What is remembered reflects the unit's Purpose and structural role—not just infrastructure defaults. *Example*: tagging datasets by business function rather than by storage location.
- *Layered retention*: Different layers retain different durations or resolutions of memory. *Example*: high-resolution data for 90 days, summarized trends for five years.
- *Time-aware design*: Historical states are preserved with unambiguous timestamps and recoverable versions. *Example*: versioned configuration files or immutable event logs.
- *Dual accessibility*: Memory remains both machine-readable for automation and human-interpretable for governance, comprehension, and trust. *Example*: logs readable by parsers and auditors alike.
- *Causal coherence*: Memory surfaces enable reconstruction of *what led to what*, not just *what occurred*. *Example*: traceable lineage between data inputs and model outputs.

But risks remain:

- *Silent drift*: Memory that changes retroactively—due to schema evolution, retrofitting, or reprocessing—erodes its own integrity. For example, a financial analytics team discovered that historical reports no longer matched prior audit data after a silent schema migration.
- *Opaque formats*: Data pipelines may store history in compressed or pre-aggregated formats with no retained causal shape. A logistics firm, for instance, lost the ability to trace shipment delays once aggregation routines replaced detailed event logs.
- *Platform dependency*: Memory tied to proprietary tools or tightly coupled stacks may become unreadable or unrecoverable over time. When a vendor-specific archive was discontinued, an insurance company lost decades of claim histories it could no longer decrypt.
- *Overcollection*: Retaining everything dilutes significance and burdens the system with undifferentiated noise. One healthcare provider's unlimited log retention created petabytes of unclassified data, masking genuine compliance alerts.

Digital memory must not be treated as a passive by-product of computation. It is an *architectural capability* that supports coordination, explanation, resilience, and governance. When designed poorly, it burdens the system and confuses its agents. When designed well, it enables the system to *know what it has done, why it behaved as it did, and how it can act more intelligently tomorrow*.

10.4.3. Implementation Surfaces and Sociotechnical Contexts

Memory is not just something a system has—it is something a system *does*. It must be made visible, addressable, and usable through specific implementation surfaces. These surfaces exist across all planes—social, physical, and digital—but their design and function must reflect the sociotechnical context in which they operate.

A *memory surface* is any structured mechanism through which the past is preserved, interpreted, or retrieved. It may be a data structure, a workflow, a process, or even a social practice. What matters is not the medium, but its ability to support clarity, continuity, and coordination.

In practice, implementation surfaces vary based on system maturity, autonomy, compliance posture, and operating constraints:

- In *regulated environments*, memory may surface through immutable audit logs, signed decisions, and versioned documentation subject to formal retention and review.
- In *operational teams*, memory appears in dashboards, incident retrospectives, changelogs, annotated codebases, and shared naming conventions.
- In *distributed sociotechnical systems*, memory is federated: Each unit retains responsibility for its own historical logic, while integrative units maintain higher-order timelines and semantic continuity.
- In *resilient environments*, memory is reinforced with offline or human-readable forms: Printouts, physical binders, laminated exception-handling steps, or visual cueing artifacts.

These implementation surfaces do more than hold data—they *encode meaning, enable orientation,* and *constrain reinterpretation.* Their usefulness depends not just on what is retained but how it is structured:

- *Surfaces must match memory type*: A causal timeline requires different representation than a metadata snapshot or a rationale annotation.
- *Surfaces must support role-appropriate access*: Not all memory is for all agents. Systems must align visibility and editability with accountability.
- *Surfaces must enable change-aware continuity*: As units evolve, memory surfaces must retain traceability across reconfigurations, migrations, and transitions.

- *Surfaces must support both proactive and forensic use*: Memory should serve real-time explanation as well as retrospective investigation.

These surfaces are also where *social legitimacy* is enacted. Who contributes to memory? Who curates it? Who controls its visibility? Memory is not neutral—it reflects power, process, and purpose. Well-designed surfaces expose those dynamics while preserving trust.

Ultimately, implementation surfaces are *where memory becomes architecture*—bridging the abstract timeline of what has occurred with the concrete realities of how agents interact, how systems coordinate, and how value is preserved over time. Without these surfaces, Memory remains a latent potential. With them, it becomes an active force in shaping future behavior.

10.5. Strategic Impact of Memory Interfaces

Memory is not valuable simply because it exists—it becomes valuable when it can be surfaced, referenced, and acted upon. The design of Memory interfaces determines what historical knowledge is accessible, who can use it, and how it influences behavior. These interfaces include not just APIs and queries, but dashboards, audit trails, naming conventions, logs, archives, and human-readable annotations. Some are embedded within operational flows; others sit above them, governing systemic oversight.

MINI-CASE—COMPLIANCE THROUGH TRANSPARENT MEMORY INTERFACES

A global financial institution facing repeated audit delays redesigned its compliance reporting dashboard into an interactive memory interface. Every regulatory submission became traceable to its data source, responsible unit, and approval timestamp. Instead of static reports, auditors could explore a live decision trail showing "why" as well as "what." Within one year, audit resolution times fell by 60%, and regulatory findings dropped to near zero. The transformation wasn't due to new policy—but to memory made visible, navigable, and explainable.

Well-designed Memory interfaces clarify system behavior across time, reduce interpretive burden, support trust, and enable adaptation. Poorly designed ones obscure causality, overload users with noise, or fragment accountability. At scale, these choices shape how organizations learn, how systems explain themselves, and how evolution proceeds.

10.5.1. Nuances of Recall and Exposure

Not all Memory is meant to be remembered equally—nor is all memory equally recallable. Even when stored persistently, historical traces vary dramatically in *how* they are surfaced, *to whom* they are exposed, and *what interpretive work* is required to make use of them. Table 10-8 organizes these variations, showing how different agents engage with memory at different cadences.

Table 10-8. *Typical Recall Frequency by Agent Role*

Agent role	Memory recall frequency	Primary use of memory	Example
Operational agent (e.g., analyst, operator, AI process)	High—continuous or daily	Task execution, anomaly detection, process correction	Accessing recent logs or alerts to adjust real-time workflows
Supervisory agent (e.g., team lead, system orchestrator)	Moderate—weekly or situational	Coordination, troubleshooting, short-term optimization	Reviewing incident summaries or weekly performance histories
Strategic agent (e.g., executive, policymaker, governance AI)	Low—periodic or event-driven	Long-term learning, accountability, evolution	Referencing decision archives or postmortem reports during planning cycles
External agent (e.g., regulator, auditor, partner)	Intermittent—triggered by external events	Verification, compliance, trust-building	Inspecting audit trails or provenance records during reviews

These differences are not cosmetic; they shape how systems learn, how agents adapt, and how accountability is distributed. Designing for recall is not merely an act of retention—it is a structural decision about *who gets to remember what,* and *why.*

Surface of Recall: Query vs. Signal

Some memories are *queried*—explicitly retrieved in response to a request. Others are *signaled*—emitted as part of a pattern, report, or side-channel. Interfaces for query recall tend to be interactive, contextual, and role-sensitive. Signal recall, by contrast, is often periodic, preformatted, and system-initiated. Each requires different affordances: querying demands indexing, lineage, and partial exposure; signaling depends on curation, timing, and trust. Overreliance on one mode can distort the perceived shape of history.

Visibility by Role: Access vs. Relevance

Recall is always filtered by *who is asking* and *what they are allowed to see*. But beyond access control lies another dimension: *relevance filtering*. Not every agent needs to recall every part of Memory. A compliance auditor may need full traces; an operator may need only anomalies. Designing recall surfaces with purpose-aligned visibility reduces overload and supports intent-driven interpretation. Flat exposure leads to misuse; tailored exposure supports clarity and decision-making.

Temporal Orientation: Snapshot vs. Sequence

Memory may be recalled as *a state at a moment* or *a timeline of events*. Snapshot recall enables comparison and regression analysis. Sequence recall enables causality, story-building, and justification. Poor interface design often flattens timelines into aggregated tables or hides transitions behind irreversible transformations. Systems that support both temporal modes—state and transition—offer richer insight and stronger forensic capability.

Granularity and Summarization

Memory can be exposed in raw, granular form—or through structured summaries and abstractions. Granularity supports low-level debugging and high-resolution insight. Summarization supports decision-making, explanation, and trust. Systems that allow navigation between levels—*drill down* and *roll up*—avoid the trap of overexposure or overcompression. A summary without a path to detail becomes dogma; a log without interpretation becomes noise.

Bias and Omission: Memory As Power

What is recallable reflects more than architecture—it reflects power. Omitted events, overwritten states, unrecorded rationale, and inaccessible logs all shape the system's perceived history. Memory interfaces encode not just what happened, but what is *treated as having happened.* Architecting for accountable recall means surfacing not only facts but absences, corrections, and uncertainties. Exposure must include the possibility of doubt.

Memory interfaces are not neutral APIs. They are *structural assertions about trust, agency, and continuity.* The subtleties of recall—its surface, scope, structure, and clarity—determine whether Memory supports alignment and learning, or reinforces fragmentation and blame. To architect Memory is to design the system's relationship with its own past—and to determine who gets to carry it forward.

10.5.2. Measurable Benefits and Architectural Payoffs

Well-designed Memory is not just a record of the past—it is a strategic enabler of clarity, trust, evolution, and resilience. Table 10-9 organizes these strategic benefits, showing how different qualities of memory translate into organizational and technical advantage.

Table 10-9. *Strategic Benefits of Well-Designed Memory*

Quality of memory	Organizational outcome	Technical outcome
Clarity	Faster onboarding, fewer misinterpretations, improved cross-team coordination	Simplified debugging, transparent lineage, reduced duplication
Trust	Stronger stakeholder confidence and compliance credibility	Reliable audit trails, verified provenance, consistent data integrity
Evolution	Continuous improvement through learning from past actions	Version control and feedback loops supporting iterative design
Resilience	Stability during turnover or disruption; ability to recover quickly	Robust backups, replayable histories, and redundancy for recovery

Its architectural payoff lies in *what it preserves, how it is accessed, and how it shapes future action*. The following benefits are not abstract; they are measurable in how systems reduce friction, accelerate learning, and improve coordination over time.

A. **Historical Clarity Reduces Explanation Overhead**

When Memory is well-structured and context-rich, agents spend less time reconstructing what happened and why. Clear timelines, annotations, and causal traces reduce the need for meetings, escalations, or guesswork.

Payoff: Decisions can be explained without reassembled artifacts. Onboarding, auditing, and retrospective analysis become simpler, faster, and more reliable.

B. **Temporal Continuity Enables Smarter Action**

Access to structured history—of decisions, changes, interactions, or failures—enables adaptive behavior. Systems with Memory learn not just from what succeeded but from how it evolved.

Payoff: Agents make better choices when they understand precedent. Processes improve when failure patterns are recognized early. Governance adapts faster when historical context is preserved.

C. **Interpretability Builds Trust Across Roles**

Memory surfaces that are human-readable, structured, and interpretable by diverse roles build confidence in the system. When Memory can be inspected and explained—not just queried or processed—it supports alignment.

Payoff: Cross-functional teams can share context. External regulators or internal reviewers can validate behavior. Trust becomes a property of design, not an act of faith.

D. **Traceability Supports Accountability and Change**

Stable, discoverable memory trails allow units to evolve safely. When the past is preserved and versioned, new implementations, policies, or workflows can be tested and justified without erasing what came before.

Payoff: Change no longer requires forgetting. Systems can shift direction without collapsing coherence. Roles remain accountable even as responsibilities adapt.

E. **Memory As a Strategic Surface**

Over time, systems that treat Memory as a design surface—not a back-end feature—develop faster reflexes, cleaner interfaces, and stronger institutional resilience. Memory becomes not just a cost of compliance, but a *source of architectural strength.*

Payoff: Memory enables insight products, training loops, policy refinement, and inter-unit cooperation. What was once a burden becomes a backbone.

When Memory is treated as a structural capability—designed for recall, exposure, interpretation, and reuse—it becomes an evolutionary asset. The architectural dividends compound: reducing friction, enabling feedback, and preserving the coherence of change. What a system remembers—and how it remembers—determines what it is capable of becoming.

10.6. Implementation Patterns

Memory may appear to be a passive by-product of system operation—but in architectural terms, it is an active decision about what a unit or system chooses to preserve, expose, and evolve with. Implementation patterns define not only the technical substrate of Memory but also its strategic stance. They reveal whether a system prioritizes trust or control, learning or compliance, integration or autonomy.

The patterns below reflect recurring Memory configurations across sociotechnical systems. Each makes different trade-offs about scale, meaning, retention, responsibility, and recoverability. These are not mutually exclusive—many systems blend them. But each brings its own architectural posture and cultural consequences.

1. **Stateless Memory**

   The system retains no memory beyond its current state.

   All events, transactions, and signals are processed and discarded unless explicitly passed downstream.

Used when: Ephemeral behavior is sufficient; audit, continuity, and learning are not required.
Design Tip: Stateless systems must be designed with strong functional boundaries and defensible failure modes—since no corrective memory exists.

2. **Partitioned Memory**

   Each sociotechnical unit retains memory relevant to its role, autonomy, and accountability.

   Boundaries reflect trust, privacy, and operational independence. Partitioning may occur across teams, domains, or organizational layers.

   Used when: Ownership, data sovereignty, or decentralized control is important—for example, in Data Mesh.

   Design Tip: Make Memory interfaces discoverable and compatible across partitions to enable meaningful reuse and coordination.

3. **Protected Memory**

   Access to memory is constrained by context, identity, or consent.

   Not all agents can recall or inspect everything. Memory becomes role-aware and time-aware—protecting sensitive histories while enabling appropriate transparency.

   Used when: Legal, ethical, or reputational concerns require selective visibility.

   Design Tip: Architect Memory as an access-aware surface—not just as a data vault. Security is insufficient without semantic protection.

4. **Federated Memory**

   Memory is collectively maintained by multiple units—coordinated but not centralized.

   Each participant contributes its own view of history; shared meaning emerges through negotiation, standards, and interlinking.

Used when: Global traceability is needed across distributed actors, but control must remain local.

Design Tip: Agree on minimal lineage and metadata contracts. Without federated memory design, platform coherence decays over time.

5. **Reflective Loop Memory**

   Memory feeds back into system behavior.

   Outcomes, feedback, and exceptions are retained and used to adjust models, rules, or priorities. The system evolves by learning.

   Used when: Improvement, personalization, or adaptive control is needed.

   Design Tip: Make feedback mechanisms observable and testable. Uncontrolled reflection can destabilize behavior or amplify error.

6. **Semantic Memory**

   The system retains meaning—not just data.

   Events are stored with context, intent, and interpretation. Semantic alignment is maintained across time so that past behavior remains intelligible.

   Used when: Explanation, audit, or cross-domain understanding is critical.

   Design Tip: Preserve structure, not just content. Meaning degrades when schemas drift or process context is lost.

7. **Retrieval-Oriented Memory**

   Memory is designed to be queried, reconstructed, and reused.

   Rather than merely retained, memory is indexed, contextualized, and exposed for future consumption—by systems or humans.

   Used when: Historical insights are reused across cycles, domains, or roles.

   Design Tip: Design for narrative, not just record. If memory cannot be traversed, it cannot support understanding.

8. **Governed Memory**

    Retention, transformation, and exposure are subject to policy and oversight.

    Memory exists within a larger logic of stewardship: what may be remembered, how long, in what form, and with what controls.

    Used when: Accountability, compliance, or institutional trust are at stake.

    Design Tip: Make memory governance observable and testable—part of the system's surface, not just its documentation.

9. **Evolving Memory**

    Memory is not static; it is maintained through cultural and technical care.

    Important memories must be updated, reinterpreted, or refactored as contexts shift. Agents—human or artificial—play roles as memory stewards.

    Used when: Organizational learning, product evolution, or systemic renewal require continuity with change.

    Design Tip: Treat memory like a codebase. Refactor for clarity, prune for relevance, annotate for future readers.

10. **Meaningful Forgetting**

    Some memories must expire, decay, or be intentionally removed.

    Forgetting is not failure—it is an architectural choice that protects systems from clutter, trauma, drift, and irrelevance.

    Used when: Renewal, privacy, or realignment with Purpose requires letting go.

    Design Tip: Design forgetting as a life cycle. Avoid passive decay; prefer intentional, explainable deletion with architectural awareness.

In practice, systems rarely implement just one of these patterns. A compliance-oriented system may blend *Governed, Protected,* and *Retrieval-Oriented* memory. A learning platform may emphasize *Reflective, Evolving,* and *Partitioned* memory. A modern data ecosystem—such as a Data Mesh—may depend on the interplay of *Federated, Semantic,* and *Governed* memory to remain coherent at scale. Table 10-10 organizes these implementation patterns, highlighting their primary use cases, trade-offs, and risks.

Table 10-10. *Summary of Memory Implementation Patterns*

Pattern	Primary use case	Key trade-offs	Typical context/ maturity level	Risk profile if misapplied
Stateless memory	Lightweight systems where continuity is unnecessary	Low overhead vs. no learning or traceability	Early prototypes, ephemeral services	Loss of accountability or replayability
Partitioned memory	Federated environments needing local autonomy	Local control vs. cross-system visibility	Mature organizations balancing independence and compliance	Fragmentation, duplication
Protected memory	Sensitive or regulated data requiring controlled recall	Security vs. accessibility	Financial, healthcare, or defense sectors	Inaccessible memory causing operational blind spots
Federated memory	Shared understanding across distributed units	Coherence vs. governance complexity	Global enterprises, consortiums, data meshes	Conflicting interpretations across nodes
Reflective loop memory	Systems that learn from outcomes and feedback	Adaptability vs. stability	Adaptive AI pipelines, continuous improvement loops	Feedback amplification or instability

(*continued*)

Table 10-10. *(continued)*

Pattern	Primary use case	Key trade-offs	Typical context/ maturity level	Risk profile if misapplied
Semantic memory	Retaining meaning, not just data	Interpretability vs. modeling overhead	Knowledge graphs, explainable-AI systems	Schema drift or semantic inconsistency
Retrieval-oriented memory	Fast, role-specific reconstruction of history	Accessibility vs. context depth	Investigations, analytics, customer support	Misinterpretation due to partial context
Governed memory	Memory subject to policy and oversight	Accountability vs. agility	Regulated enterprises, safety-critical domains	Bureaucratic lag, outdated controls
Evolving memory	Continuous adaptation through maintenance and curation	Responsiveness vs. historical consistency	Learning organizations, long-lived systems	Loss of lineage or revision fatigue
Meaningful Forgetting	Systems needing deliberate data expiry	Renewal vs. knowledge loss	Privacy-sensitive or rapidly changing environments	Over-deletion or compliance breaches

The choice is not merely technical. Each pattern encodes a theory of history—what is worth remembering, who owns the memory, and how it should shape what comes next.

10.7. Interlocking Dependencies and Alignment

Memory is not a detached utility. It is structurally entangled with every other design construct—shaped by what the unit exists to do, how it expresses that purpose, what processes it runs, and what structures sustain them. As a result, architectural misalignment around Memory often signals deeper dysfunction: a disconnect between history and intent, between what is done and what is remembered, or between what is observed and what is acted upon.

Several key dependencies shape the design and behavior of Memory:

10.7.1. Alignment with Purpose: Selectivity and Significance

Purpose defines what the system exists to serve. Memory must align by preserving the right kind of history—relevant, interpretable, and context-aware. If Memory stores too little, it becomes fragile. If it stores indiscriminately, it becomes meaningless. When Purpose is clear, it guides what is worth remembering and what should be forgotten.

Misalignment occurs when Memory preserves events or entities that no longer support the current strategic intent—or ignores histories that are essential to it.

10.7.2. Alignment with Value: Traceability and Provenance

Value creation is cumulative. Memory enables that accumulation—by recording how Value was produced, delivered, reused, or lost. This includes lineage, feedback, exceptions, and learning.

A Memory surface that cannot support Value tracing undermines both transparency and trust. Without proper provenance, Value becomes unaccountable.

Misalignment occurs when systems emit Value but fail to capture the usage patterns, outcomes, or dependencies that shape its future relevance.

10.7.3. Alignment with Function: Generation and Recall

Function is one of the primary sources of memory. Every invocation, response, or exception leaves a trail. Memory design must match the structure and intent of Function—capturing enough to reconstruct meaning, but not so much that the trace overwhelms the signal.

Misalignment occurs when Functions emit opaque or inconsistent traces—or when Memory lacks interfaces to retrieve them in meaningful ways.

10.7.4. Alignment with Process: Continuity and Correction

Process depends on memory to operate across time—resuming, branching, retrying, or learning from experience. Memory ensures that tasks don't vanish, that decisions have context, and that exceptions have traceable causes.

Misalignment occurs when processes mutate without updating memory models—or when memory fails to reflect the actual execution path, making diagnostics and reuse unreliable.

10.7.5. Alignment with Structure: Ownership and Boundaries

Structure defines where Memory resides and who stewards it. Integrative units often maintain systemic memory; operational units contribute localized histories.

Misalignment occurs when Memory lacks a structural home—leading to orphaned records, fragmented narratives, or duplicated traces without a clear source of truth.

10.7.6. Alignment with Order: Governability and Life Cycle

Memory is not timeless. Order defines retention periods, redaction policies, access rules, and deletion logic. A memory that ignores Order invites drift or liability.

Misalignment occurs when Memory persists longer than it should, violates compliance boundaries, or lacks clarity on who governs it.

10.7.7. Alignment with Culture: Interpretation and Stewardship

Even with perfect structure, Memory is useless without cultural interpretation. Culture determines whether Memory is used for learning, blame, renewal, or justification.

Misalignment occurs when Memory is hoarded but not examined, accessed but not understood, preserved but not updated—turning systems into museums of obsolete history.

These dependencies are mutual. Misalignments accumulate slowly but corrode quickly. A Process that evolves without updating its Memory interface may confuse future observers. A Purpose that changes without reshaping what is remembered may paralyze reflection. A Culture that hoards memory without examining it may drift into unaccountable repetition.

Strong Memory design does not merely preserve the past—it creates surfaces of alignment across time, role, and domain. When Memory and the rest of the architecture evolve together, systems become both coherent and adaptive: able to reflect on what has been, without being trapped by it.

10.8. Failure Modes and Antipatterns

Poorly structured Memory erodes a system's ability to learn, coordinate, or evolve. It leads to architectural confusion, distorted meaning, and reactive behaviors instead of reflective capabilities. Because Memory touches all design constructs, its failures can propagate silently—until accountability breaks down, alignment collapses, or systems begin repeating their past mistakes at scale.

The following antipatterns identify common Memory design failures and the architectural misalignments behind them. Each reveals a different way in which Memory may fail to support continuity, understanding, and coherence over time.

1. **Misuse of Storage as Memory and Memory as Storage**

    - *Definition*: Memory is confused with Storage, resulting in blurred responsibilities between what is remembered and what is preserved as Value.

    - *Symptoms*: Analytical dashboards or model outputs are treated as long-term Memory; logs or transient histories are reused as stable Value artifacts.

    - *Consequence*: Memory is placed under production pressure, losing temporal nuance; Storage becomes bloated with non-authoritative, ephemeral traces.

    - *Architectural Fault*: Misalignment between Purpose, Value, and Memory—undermining clarity and contaminating reuse boundaries.

2. **Orphan Memory**

    - *Definition*: Memory is collected but has no accountable owner or structural home.

    - *Symptoms*: Retained histories without maintenance; unclear retention policies; records persist by inertia.

- *Consequence*: Liability exposure, data rot, compliance risk, and eventual inaccessibility.
- *Architectural Fault*: Misalignment with Structure (no responsible steward) and Order (no life cycle governance).

3. **Hollow Trace**
   - *Definition*: Memory exists but lacks semantic or relational meaning.
   - *Symptoms*: Flat logs with no business context; missing identifiers or triggers; history that cannot be interpreted.
   - *Consequence*: Diagnosis fails; postmortems become guesswork; learning is superficial or misleading.
   - *Architectural Fault*: Misalignment with Function and Culture—trace is technically correct but narratively meaningless.
4. **Compliant but Non-Reflective Memory**
   - *Definition*: Memory is maintained for compliance, not for learning or feedback.
   - *Symptoms*: Perfect audit trails, but no change in behavior; exception patterns recur unexamined.
   - *Consequence*: Organizational stagnation; missed improvement opportunities; brittle evolution.
   - *Architectural Fault*: Misalignment with Process (no feedback loop) and Culture (low interpretive usage).
5. **Siloed Memory**
   - *Definition*: Each unit retains isolated history with no systemic integration.
   - *Symptoms*: Conflicting records; manual reconciliation; duplicated insight generation.
   - *Consequence*: Fractured coordination, inconsistent truth, and slow strategic alignment.
   - *Architectural Fault*: Misalignment with Structure (no federated model) and Purpose (no shared coherence).

6. **Fragile Learning**

    - *Definition*: Learning systems operate on unstable or unverified memory foundations.
    - *Symptoms*: Models overfit; edge cases dominate; adaptation becomes volatility.
    - *Consequence*: System behavior degrades over time, especially under changing inputs.
    - *Architectural Fault*: Misalignment with Governance and Order—no stabilization phase or human review of reflective loops.

7. **Role-Free Logging**

    - *Definition*: Memory is written generically, with no alignment to agent, role, or responsibility.
    - *Symptoms*: Cannot determine who did what; trace lacks actor-context or role-awareness.
    - *Consequence*: Weak accountability; poor debugging; limited insight into human-agent interaction patterns.
    - *Architectural Fault*: Misalignment with Structure and Culture—no encoding of social or operational boundaries.

8. **Volatile Memory**

    - *Definition*: Memory is lost at reboot or reset—whether by design or oversight.
    - *Symptoms*: No persistent trace of critical events; diagnosis requires reenactment or guesswork.
    - *Consequence*: Debugging failure, recurrence of preventable issues, loss of trust.
    - *Architectural Fault*: Misalignment with Order (no persistence layer) and Process (no archival boundary).

9. **Conflated Memory**

    - *Definition*: Operational state and long-term memory are stored together.
    - *Symptoms*: Performance bottlenecks; temporal confusion; memory interpreted as real-time state.
    - *Consequence*: Bugs, ambiguity, and system instability under concurrent use.
    - *Architectural Fault*: Misalignment with Process and Order—no separation of present and past.

10. **Decoupled Memory**

    - *Definition*: Memory exists but is inaccessible or undiscoverable.
    - *Symptoms*: Data hoarded in cold storage; no indexing, search, or interface exposure.
    - *Consequence*: Memory is ignored; duplication occurs; platforms lose introspection.
    - *Architectural Fault*: Misalignment with Function (no access pattern) and Structure (no discoverability).

11. **Memory Without Narrative**

    - *Definition*: Even complete memory fails to support understanding or meaning-making.
    - *Symptoms*: Teams have full access to trace but derive no insight; onboarding is difficult.
    - *Consequence*: Loss of institutional knowledge; decision friction; weak architectural maturity.
    - *Architectural Fault*: Misalignment with Culture and Value—retention without interpretation or curation.

Memory failures are not just technical glitches—they are architectural misstatements about what matters, what is knowable, and what the system is capable of becoming. A well-structured Memory surface enables not only operational continuity but organizational maturity: it remembers with purpose, evolves with insight, and connects past decisions with future coherence.

KEY TAKEAWAYS FOR MEMORY DESIGN

1. *Define what must endure*: Clarify which decisions, contexts, and events are essential for continuity and accountability.
2. *Structure memory for meaning, not volume*: Avoid indiscriminate retention; design for interpretability and reuse.
3. *Align memory with Purpose and Function*: Ensure what is remembered reinforces the system's reason for being and operational goals.
4. *Balance transparency and protection*: Design access and retention rules that preserve trust without compromising safety or compliance.
5. *Continuously review for drift*: Treat Memory as a living construct—periodically audit relevance, integrity, and interpretive value.

10.9. Summary: Memory As a Design Construct

Within Purpose Driven Design, *Memory is the system's reflective core*—the internal mechanism by which experience is preserved, meaning is accumulated, and coherence is maintained over time. Unlike runtime execution or static configuration, Memory operates across moments and beyond contexts. It allows systems to retain what happened, interpret why, and anticipate what might recur.

Memory resolves the fundamental tension between stateless performance and stateful continuity. It stabilizes identity, supports feedback, and informs adaptation—without requiring constant recalibration or rejustification. Effective Memory is not merely data retention; it is structured recall, filtered significance, and purposeful traceability.

Realized effectively, Memory functions as

- *A continuity substrate*, stabilizing behavior across time and enabling cross-process coherence
- *An interpretive resource*, supporting diagnostics, learning, and reflective decision-making
- *An accountability surface*, anchoring traceability, auditability, and historical reconstruction
- *An integration mechanism*, enabling alignment across units and coordination across time zones, shifts, or system boundaries
- *A bridge between constructs*, connecting Purpose, Value, Function, Process, Structure, Order, Culture, and Storage through the accumulation and curation of trace

The architectural treatment of Memory reveals that it is not simply a historical artifact or technical utility. It is a compositional element of intelligence—designed to reinforce coherence, enable learning, and preserve the integrity of interaction over time. When neglected, Memory becomes fragmented, volatile, or misleading—causing systems to forget what matters and repeat avoidable failures. When deliberately structured and aligned, Memory transforms into strategic infrastructure: a scaffold for integrity, clarity, resilience, and insight.

In the age of AI and complex sociotechnical systems, Memory is no longer just about scale or latency. It is *Big Meaning*, not just Big Data—a design surface for ensuring that past experience informs future orientation.

CHAPTER 11

Storage

In a traditional warehouse, products are arranged by category. Books go on one shelf, electronics on another, clothing in a separate section. The logic is straightforward: organize items the way people expect to find them. But when Amazon began scaling its fulfillment operations in the early 2000s, it faced a different challenge. With tens of thousands of unique items arriving and departing daily, fixed locations and rigid categories created bottlenecks, misplacements, and wasted space.

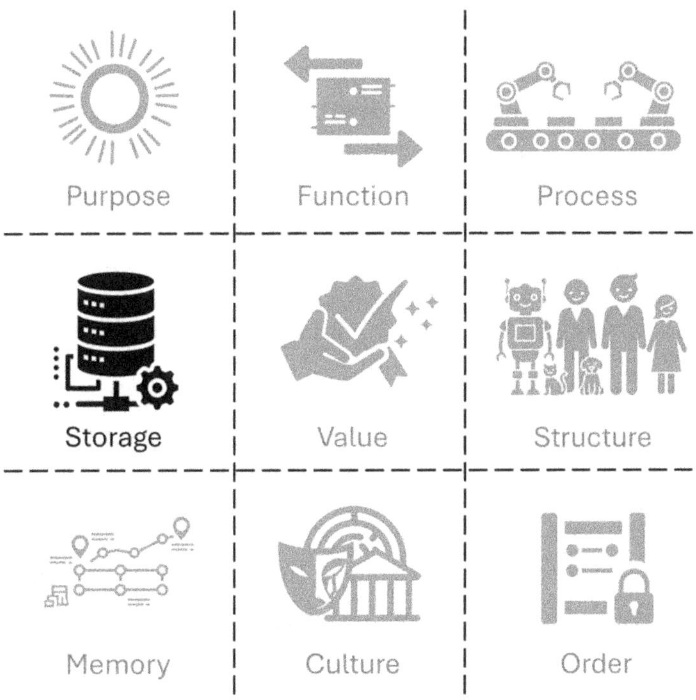

A. Milchman, *The Purpose Driven Design Framework for Modern Enterprises*,
https://doi.org/10.1007/979-8-8688-2223-0_11

The solution was counterintuitive: abandon order.

Instead of organizing items by type, Amazon's fulfillment centers adopted *chaotic storage*. Items are placed wherever space is available—a system that appears random but is in fact algorithmically organized and meticulously tracked through metadata. A copy of a best-selling novel might sit next to a screwdriver, a pair of socks, and a blender. The shelves look random—like a puzzle poured onto a warehouse floor. But behind this apparent disorder is a system of meticulously tracked inventory, barcode scanning, and real-time location mapping.

Each item's position is known *not because it's logical* but because it's recorded. When a worker places an item on a shelf, the system logs the exact bin and location. When the item is needed, the system retrieves it—efficiently, accurately, and without requiring the worker to remember categories or layouts. The chaos works because it's *computationally structured* (each item is tagged with a unique barcode linked to a dynamic index that knows its exact bin location, enabling the system to optimize retrieval routes and balance inventory automatically).

This model enabled Amazon to achieve unprecedented storage density, rapid intake of diverse products, and fast fulfillment. More importantly, it separated the logic of *retrieval* from the logic of *human intuition, delivering clear operational benefits:*

- *Faster and more consistent retrieval times*
- *Scalable inventory management independent of human memory*
- *Reduced training time for new workers or automated agents*
- *Improved spatial utilization and dynamic reallocation of stock*

What looked like disorder to people became hyper-efficiency when managed by software. The warehouse no longer needed to "make sense" to humans. It only needed to support *retrieval at speed and scale.*

The lesson extends far beyond warehouses. In digital systems, storage is often assumed to follow visible hierarchies—folders, tags, categories. But at scale, those structures become brittle. Chaotic storage demonstrates that *retrievability does not require predictability.* With the right indexing, validation, and location tracking, systems can store enormous volumes of data, objects, or assets in ways that are opaque to users but operationally optimal. In sociotechnical systems, this principle scales: storage need not be self-explanatory to every agent, but it must be trustworthy, addressable, and responsive to systemic needs.

Importantly, chaotic storage is not *unstructured*. It is *deliberately decoupled* from the need for intuitive arrangement. The structure exists, but it lives in metadata, not physical layout. What matters is not where something is placed, but that the system *knows where it is and how to get it.*

Amazon's fulfillment architecture redefined how organizations think about space, access, and optimization. It revealed that *storage is not about what things look like—it's about what things are for, and how reliably they can be found when needed.*

In this sense, storage is not only about preservation but also about *functional readiness*. What you store, how you store it, and what retrieval mechanisms support it all determine whether your system is adaptable, responsive, and usable under pressure. In cloud computing, for instance, the distinction between cold and hot data storage illustrates the same principle—speed and accessibility, not just retention, define whether stored information remains operationally useful.

Storage provides the physical or digital substrate through which a unit retains what it creates, acquires, or receives. It differs from Memory, which captures history and context—Storage is about the preservation of artifacts: data, products, components, and outputs. In this chapter, Storage is treated not as a passive repository, but as an architectural choice that influences access, security, performance, and continuity. It is never neutral. Designed well, Storage becomes a foundation for scale, modularity, and resilience. Designed poorly, it quietly strangles progress—hidden until systems stall or break. What storage means in practice—order by appearance versus order by function—comes into focus when the models are set side-by-side in Table 11-1.

Table 11-1. *Traditional vs. Chaotic Warehouse Storage Models*

Feature	Traditional storage	Chaotic storage (e.g., Amazon fulfillment model)
Organization principle	Grouped by product category or type	Placed wherever space is available; order managed through metadata and software
Human/robot involvement	Predominantly human; navigation and retrieval depend on visual cues and manual sorting	Predominantly robotic or human-assisted automation; retrieval directed by real-time algorithms
Human intuition	Layout is predictable and designed for easy manual navigation	Layout appears random to humans but is computationally optimized for efficiency
Dependence on metadata	Limited; physical location and signage guide access	Complete; every item indexed, tracked, and retrieved through digital identifiers
Operational efficiency	Straightforward but less flexible; performance declines as inventory grows	Highly scalable; adaptive to fluctuating inventory and demand
Environmental and economic considerations	Requires consistent lighting, climate control, and wide aisles for human operators	Reduced need for lighting or heating in automated zones; optimized space utilization reduces cost per item
Architectural lesson	Visible order aids comprehension but limits scalability	Hidden, metadata-driven order trades visibility for automation and adaptability

The sections that follow examine how Storage operates as a design construct, outline its guiding principles, and conclude with patterns and pitfalls that reveal its systemic impact.

11.1. Foundations of Storage As a Design Construct

Storage is not merely a place where things are kept—it is a purposeful architectural construct. Its role begins with a clear definition: to preserve what the system produces, receives, or must retain for future use. Storage exists not to accumulate but to enable continuity, reuse, replication, and operational readiness. It solves a fundamental problem in sociotechnical design: how to retain and access valuable artifacts—data, products, components, and externally received Value—without relying on immediate availability or repeated creation. The Value a unit receives from suppliers, regulators, or external partners—such as certifications, inputs, obligations, or privileges—must be stored as reliably as internally generated assets. Across time, storage patterns have evolved from local and static to distributed and dynamic, reflecting shifts in scale, coordination models, and technological possibilities. Understanding this evolution helps clarify what Storage *is for*, and how it has shaped the responsiveness and resilience of systems. Table 11-2 maps this evolution, highlighting how shifts in storage patterns have shaped architectural scalability and resilience.

Table 11-2. *Evolution of Storage Patterns*

Era/pattern	Dominant form of storage	Defining characteristic	Architectural implication
Proximity/Local (early systems)	On-premises archives, file cabinets, local disks	Physically close to the user or process it served	Tight coupling of storage and use; low scalability
Indexed/automated (industrial and early digital)	Cataloged repositories, mainframes, early databases	Introduction of indexing and automation for retrieval	Improved accessibility; early separation of content and location
Centralized/ distributed (modern digital)	Data centers, networked file systems, cloud storage	Global reach with shared but managed access	Enabled scale and coordination across units
Static/dynamic (contemporary and emerging)	Cloud-native object stores, edge caching, self-optimizing systems	Storage adapts in real time to demand and context.	Decouples persistence from place; storage becomes an active system function.

11.1.1. Defining Storage Architecturally

Architecturally, *Storage* is the construct that preserves Value in artifact form—durable, discoverable, and ready for reuse. Unlike *Memory*, which captures the evolving trajectory of the unit by recording what happened and why, Storage maintains *what is needed* for continued Function and Process. Memory preserves history; Storage preserves artifacts.

As established in Chapter 10, Memory retains the interpretive history of what occurred and why, while Storage preserves tangible artifacts of Value for reuse. Process, in turn, consumes what Storage provides—transforming stored Value into renewed Function. Together, they form a continuous loop: Memory explains, Storage retains, and Process regenerates.

Storage is not passive containment. It is a structural decision to retain what the system produces, receives, or requires—in ways that make it usable *later, elsewhere, or again.* Artifacts may include data sets, trained models, configuration states, permits, licenses, documents, components, or any other form of Value that the unit may need to deploy, transform, validate, or transfer. Whether physical or digital, the architectural function of Storage is to decouple effort from repetition, production from consumption, and creation from immediate availability. In manufacturing, this may mean retaining validated design components for reuse across product lines; in data architecture, it could mean preserving standardized datasets so analytics or machine-learning processes can build on trusted sources rather than starting from scratch.

Storage plays a bidirectional role in Value flow:

- In one direction: *Process → Value → Storage → Function*

  A Process produces Value that is stored as an artifact and later retrieved by a Function—enabling decision-making, fulfillment, presentation, or delivery.

- In the other: *Function → Value → Storage → Process*

  A Function initiates or retrieves Value—such as parameters, assets, or entitlements—which is stored and passed to a Process that executes based on that Value.

This dual pattern positions Storage as an architectural bridge between the producers and consumers of Value. It allows a system to operate asynchronously, coordinate across time, and maintain continuity across internal and external boundaries. This bidirectional relationship is illustrated in Figure 11-1.

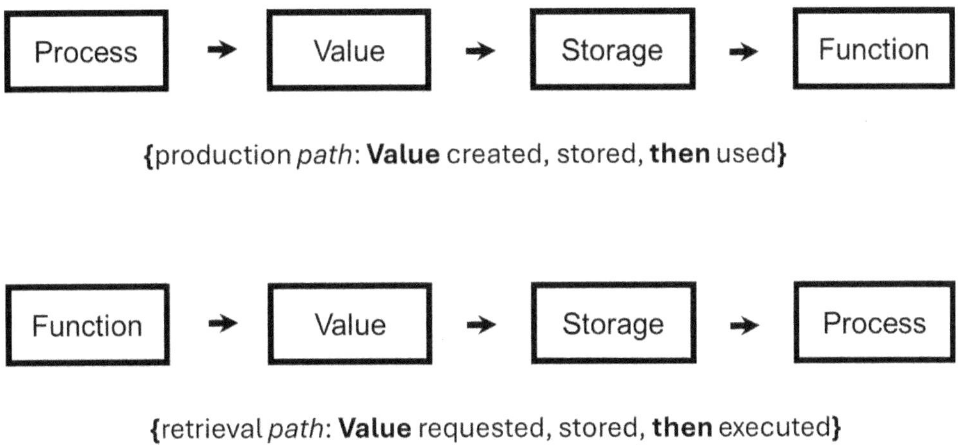

Figure 11-1. *Bidirectional flow of Value through Storage*

Importantly, Storage does not require the system to remember *why* an artifact exists—only *what* it is, *where* it is, and *how* to retrieve it. That logic lives in metadata, governance, and version control—not in location or visual organization.

Architecturally, Storage is defined by the guarantees it provides:

- *Persistence*: Value remains present and intact across time.
- *Usability*: Artifacts remain structurally valid for future Functions or Processes.
- *Accessibility*: Authorized agents or systems can retrieve Value when needed.
- *Decoupling*: Stored Value can be reused without re-executing the original effort.

Storage transforms outputs into resources. It turns flow into form. Designed well, it creates structural readiness—the ability to activate retained capability when the moment arrives.

11.1.2. Why Storage Exists

Storage exists to solve a specific architectural problem: how to preserve Value across time, context, and use—without requiring continuous presence, re-execution, or re-creation.

In sociotechnical systems, both *Function* and *Process* depend on inputs that may not be available at the moment of need. Without Storage, every action would require immediate production. Every decision would hinge on real-time generation of facts. Every output would need to be recreated from scratch. This tight coupling renders systems fragile, inefficient, and unscalable.

Storage introduces *temporal decoupling*. It allows Value to be produced once and used many times. It enables a Process to output a dataset today that a Function can consume tomorrow. It supports a Function retrieving a license granted last year and passing it to a Process executing now. It allows the system to remember what it has received, produced, or transformed—not as narrative history (as in Memory), but as usable, durable artifacts.

ASYNCHRONOUS PROCESSES ENABLED BY STORAGE

Physical systems	Digital systems
Manufacturing: Components are produced, stored, and later assembled on demand—decoupling production from fulfillment.	*Data Pipelines*: Datasets are ingested, validated, and stored once, then reused asynchronously by analytics, ML models, or APIs.
Pharmaceutical Supply Chains: Vaccines are manufactured and stored under controlled conditions until distributed to specific regions.	*Cloud Queues and Caches*: Transactions or messages are stored temporarily, allowing producers and consumers to operate independently.
Publishing: Printed materials or physical media are warehoused and released gradually to align with market demand.	*Versioned Repositories*: Code, configurations, or trained models are stored and re-deployed as needed without repeating the creation process

Storage exists so that a unit can

- Reuse what it has already invested effort into creating.
- Retain what it has received from external stakeholders.
- Redeploy what it offers to others on demand.

- Recover quickly from disruption without starting over.
- Enable asynchronous coordination across time, teams, and technologies.
- Operate at scale without redundant effort or manual rework.

It also supports value management: allowing stored artifacts to be indexed, versioned, audited, governed, and distributed—internally and externally. In complex systems, this is essential for compliance, traceability, controlled evolution, and safe collaboration across roles and boundaries.

Functionally, Storage extends the reach of both Process and Function. In one direction, a Process produces Value that is stored and later retrieved by Function; in the other, Function requests or configures Value that is stored and then used by Process. This bidirectional logic enables a system to operate with architectural rhythm rather than chaos.

In summary, Storage exists to make Value persist beyond the moment of execution. It is the infrastructure of reuse, the architecture of continuity, and the substrate through which purposeful systems scale.

11.1.3. What Problem Storage Solves

Storage solves the problem of *transience*. It ensures that valuable artifacts—once created or received—do not vanish, degrade, or become inaccessible. In purposeful systems, it is not enough to produce Value; that Value must be preserved in a form that can be reused, delivered, or activated at a later time. When NASA engineers restored Apollo-era tapes in the 2000s to recover telemetry data nearly lost to format decay, the effort underscored how fragile Value becomes without deliberate storage design—an expensive reminder that preservation must be planned, not assumed.

Function and Process generate outputs that carry utility: products, services, datasets, models, configurations, contracts, and other assets. Without Storage, these outputs are ephemeral. What is not stored becomes unusable beyond its moment of creation—forcing the system into a cycle of re-creation, real-time dependency, and operational fragility.

In the absence of Storage, systems encounter structural limitations:

- *Duplication of effort*: Without preserved artifacts, outputs must be re-created repeatedly, consuming time and resources with no compounding benefit.
- *Temporal coupling*: Producers and consumers of Value must operate concurrently, as no intermediary exists to buffer or bridge their interactions.
- *Operational fragility*: Without reliable Storage, the system cannot absorb disruptions, replay actions, or recover from loss.
- *Loss of value*: Artifacts that are not retained—whether physical, digital, or procedural—cease to be accessible, making previous investments obsolete and non-transferable.

Storage introduces a durable architectural layer that transforms *ephemeral action into persistent capability*. It allows the system to accumulate what it has achieved, to structure it for use, and to govern it over time. In doing so, Storage addresses the core challenges of immediacy, volatility, and impermanence.

Specifically, Storage solves for

- *Decoupling creation from usage*, enabling asynchronous consumption across roles, systems, and timelines
- *Enabling scale*, allowing a single Value artifact to be used repeatedly across different contexts or consumers
- *Supporting distribution*, making stored Value accessible across physical, digital, or organizational boundaries
- *Ensuring consistency and reliability*, by maintaining authoritative, governed versions of what is stored

Ultimately, Storage solves the architectural challenge of how to *keep value available, structured, and durable*—making purposeful systems resilient, repeatable, and ready to operate beyond the moment of creation.

11.1.4. Historical Patterns and Structural Evolution

Storage has always existed in some form—chests, scrolls, ledgers, warehouses, archives. But as systems grew in complexity and scale, the architecture of Storage evolved. What began as visible containment has become invisible infrastructure. While the external form of Storage has changed, its purpose remains constant: to preserve value-bearing artifacts across time, context, and use.

From Proximity to Abstraction

In early systems, Storage was physical, local, and self-evident. Items were stored near where they were used, and their presence could be confirmed by sight or touch. Books sat on shelves. Contracts lived in drawers. Coordination was shaped by physical access—if you needed something, you had to go where it was kept.

As systems grew, proximity became a constraint. Storage moved from physical proximity to conceptual abstraction. Artifacts could be located elsewhere, as long as they were findable. Catalogs, indices, and registries emerged—not to store the thing itself, but to point to it. This marked the beginning of *addressability* as a storage property. Similar abstraction occurred in knowledge management, where libraries and research archives extended beyond physical stacks into cataloged, searchable collections—turning proximity into a problem of reference rather than location.

From Manual to Automated Retrieval

Once artifacts could be indexed, systems no longer required human memory or intervention to retrieve them. Retrieval could be automated. With the advent of digital file systems and databases, Storage became programmable. Instead of looking for a file in a drawer, a system could query for an object by name, ID, or metadata. The same logic reshaped corporate knowledge systems, as manual librarianship gave way to automated indexing, search, and recommendation engines that retrieve information without human mediation. This opened the door to large-scale, high-frequency, asynchronous reuse.

From Centralized to Distributed

Initially, digital Storage followed the logic of the physical: central repositories, mainframes, and on-premises data centers. But over time, storage decoupled from location. Distributed systems—cloud object stores, content delivery networks, data lakes—allowed Storage to be physically and logically fragmented, yet logically coherent. Artifacts no longer needed to live in one place. They needed to be *reachable, consistent,* and *governed.* A parallel transformation unfolded in cloud infrastructure, where workloads and data are now dispersed across multiple regions for resilience—demanding coordination mechanisms that echo distributed storage design.

In addition, Distributed Storage brought new challenges: consistency, latency, redundancy, access control. But it also enabled architectural shifts: elastic scale, global access, and asynchronous system design.

From Static to Dynamic

Early Storage was static—write once, read many. In modern systems, Storage is often dynamic. Artifacts evolve. They are versioned, refreshed, streamed, cached, or overwritten. Storage is no longer a cold archive but a living system: logs record every change, registries track each deployment, caches accelerate access, and snapshots preserve known-good states. This dynamic principle also governs software development pipelines, where continuous integration and version control systems keep artifacts mutable yet traceable across rapid release cycles.

These patterns support operational adaptability: rollback, branching, rehydration, and recomposition. Storage became an enabler of controlled change, not just continuity.

From Private to Shared and Governed

Where Storage was once private—tied to a person, team, or machine—it is now often shared. Data catalogs, model registries, document repositories, and software artifact stores are accessible across organizational units. With shared access came the need for shared rules: retention policies, access controls, audit trails, and usage constraints. Comparable pressures shaped the rise of collaborative data platforms and open-access repositories, where governance frameworks balance transparency, intellectual ownership, and ethical reuse.

Today, Storage is both a technical capability and a governance surface. It carries not just what is stored but who can see it, use it, change it, or transfer it—and under what conditions.

As systems evolved, Storage moved

- From *where things live*
- To *how things persist*
- To *how systems recover, scale, and collaborate*

Its structural role has grown more central, more intentional, and more complex. Yet the core problem it solves remains unchanged: how to preserve what matters, so it can be used when it matters most.

11.2. Role of Storage in Sociotechnical Systems

Storage plays a structurally distinct role at each level of a sociotechnical system—but always through constructs embedded within specific units. In many cases, especially with digital artifacts, these Storage constructs are part of the unit that produces or consumes the Value. In other cases—most often involving physical artifacts—dedicated storage units are created to hold Value on behalf of others. Regardless of form, Storage always remains unit-bounded: there is no disembodied or free-floating Storage in the architecture. Within operational units, Storage preserves instances of Value they have produced or received, supporting autonomy, recovery, and reuse. For example, a finance team's shared document repository allows analysts to reuse validated forecasting models rather than rebuild them, ensuring consistency and accelerating each reporting cycle. In integrative units, it enables coordination by holding system-wide artifacts under governance. For instance, a regional headquarters maintains consolidated financial statements, cross-division performance dashboards, and strategic reports that let local units operate autonomously while staying aligned with enterprise-level objectives. At the boundary, it maintains externally sourced Value for traceability and future use. For example, a regulatory affairs team stores all submitted filings and corresponding feedback from oversight bodies, allowing the institution to track its compliance history and respond consistently to future audits. *Thus, Storage stores what will be used—preserving instances of Value so they remain accessible, reusable, and dependable across time, context, and scale.*

11.2.1. Internal Role: Stabilizing the Unit

Every sociotechnical unit—whether operational or integrative—must be capable of functioning with some degree of autonomy. Storage plays a critical internal role in enabling that autonomy by preserving instances of Value the unit has produced or received. These may include *physical components, final product batches, approved documents, regulatory filings, legal agreements, analytic reports,* or other value-bearing artifacts relevant to the unit's Function and Process.

Without internal Storage, a unit would be forced to recreate or reacquire needed Value every time it acted. This would introduce fragility, waste, and real-time dependency. With Storage, the unit gains the ability to

- *Resume operation* after interruptions or delays
- *Reuse previously produced artifacts* rather than rebuild them from scratch
- *Decouple production from immediate consumption*, allowing for buffering, scaling, and asynchronous delivery

The form Storage takes depends on the nature of the artifacts:

- For physical outputs, Storage may be temporary and spatial—bins, shelves, loading docks, staging areas—designed to hold materials until they are shipped, assembled, or transferred to a dedicated storage unit.
- For digital outputs, Storage is often embedded—file systems, object stores, internal repositories, or document vaults—accessible directly by the unit's agents or systems.

In either case, internal Storage serves as a stabilizing layer between action and dependency. It reduces the need for constant synchronization, mitigates volatility in supply or demand, and provides a reliable substrate on which the unit's Function and Process can operate.

By preserving what it needs to use—now or later—a unit becomes structurally self-supporting. It can contribute to the broader system without requiring immediate upstream or downstream availability, and it can absorb variation in pace, availability, or performance. Internal Storage thus underpins the operational coherence and resilience of the unit itself.

11.2.2. Systemic Role: Integration into System-Wide Coordination

While operational units rely on internal Storage to stabilize their own activity, system-wide coordination requires certain artifacts to be preserved and made accessible beyond the boundaries of any single unit. This role is fulfilled by Storage constructs maintained by *integrative units*—those responsible for enabling coherence, interoperability, and reuse across the broader sociotechnical system.

System-wide Storage is not an abstract layer. It is always implemented by real units: a shared data product hosted by an integrative team, a licensed document repository curated by a governance body, a secure vault managed by a compliance function, or a central inventory held by a supply and logistics group. These integrative units do not simply aggregate artifacts—they govern, version, authorize, and expose them to others in the system, enabling shared understanding and composability.

Such Storage constructs preserve instances of Value that

- Must be **reused by multiple units** (e.g., master reference data, product definitions)
- Must be **trusted as authoritative** (e.g., certified documents, regulatory records)
- Must be **shared across contexts** (e.g., shared components, configuration baselines)

Without these systemic forms of Storage, coordination becomes fragile and inefficient:

- Units may rely on out-of-date or duplicated artifacts.
- Redundant effort proliferates across silos.
- Core assets lose their integrity, traceability, or reuse potential.

Importantly, systemic Storage enables architectural decoupling. It separates the unit that produces Value from the many units that consume it—sometimes asynchronously, and often without direct coordination. A single stored instance can be used across workflows, departments, or geographies, with guarantees of consistency and accessibility.

Where physical Value is involved, dedicated storage units—such as warehouses, distribution hubs, or fulfillment centers—may play this systemic role. For digital Value, the pattern may involve centralized repositories, federated catalogs, or managed registries, depending on the system's architecture.

The outcome is not just reuse but alignment: when units draw from the same stored Value under shared rules, their actions reflect a common understanding of what the system produces, delivers, or upholds. System-wide coordination is not simply achieved through communication or control—it is anchored in *what is stored, how it is stored, and who ensures its availability and integrity.*

11.2.3. Environmental Role: Interface with External Entities

Sociotechnical systems exist in open environments. They interact continuously with customers, suppliers, regulators, partners, investors, and other external actors. Storage plays a critical role at this boundary by preserving *artifacts of Value* that move between the system and its environment—both *received* and *delivered.*

These artifacts include

- *Received Value*: Physical components from suppliers, externally sourced datasets or models, regulatory directives, licenses, or certified documents that serve as system inputs
- *Delivered Value*: Finished products, packaged digital assets, or completed service outputs that the system provides to its consumers or partners

In both directions, Storage ensures that these artifacts persist beyond the moment of transaction. A system must retain what it *receives* to make decisions, validate compliance, or reproduce behavior. It must retain what it *delivers* to confirm completion, support follow-up, or enable future reuse. Without environmental Storage, external exchanges become ephemeral and unverifiable.

This Storage may take many forms:

- Warehouses, staging areas, shipping inventories, or return facilities that manage physical goods moving into or out of the system
- Digital repositories, structured delivery folders, customer-accessible archives, *or* secure vaults that durably retain received or delivered Value artifacts in digital form

Importantly, Storage at the boundary is *bidirectional* and *unit-bound*. Every such construct is owned by a specific unit tasked with managing external interaction—whether operational or integrative in nature. This design ensures clear accountability for the persistence, structure, and accessibility of exchanged artifacts.

Architecturally, environmental Storage enables traceability without re-execution, allowing the system to refer back to what was received or delivered even after the original interaction has ended. It buffers against loss, dispute, and delay—preserving the system's integrity across exchanges. In logistics networks, for example, shipment data, digital bills of lading, and sensor logs are retained across partner systems, allowing any participant to verify delivery or provenance without repeating the transport process.

Storage at the boundary affirms not just that a system interacts—but that it remembers what has entered and what has left.

11.3. Design Principles and Heuristics

Designing Storage is not merely about provisioning capacity—it is about ensuring the *right* Value is durably retained, made accessible at the *right* level of structure and governance, and aligned with the system's Purpose, Function, and Process. Whether embedded within operational units or implemented by integrative ones, Storage must be designed deliberately, not left as an afterthought. This part distills practical principles and heuristics to help designers approach Storage as a first-class architectural construct: one that preserves what matters, supports coordination and autonomy, and reinforces the durability of Value across time, boundary, and scale.

11.3.1. Principles of Storage Design

Storage must be designed as an active architectural construct—not merely as a passive by-product of Function or Process. Its purpose is to preserve artifacts of Value in a form that enables future use, delivery, or reuse. These principles guide how Storage should be integrated into the design of sociotechnical units—ensuring it supports continuity, composability, and autonomy across time and system boundaries.

1. **Storage must preserve artifacts of Value—not just whatever happens to be generated**

   Storage is meaningful only when it retains things that can be used. Artifacts of Value—whether physical, digital, or service-based—must be intentionally stored so they remain accessible, complete, and functionally relevant.

> **WARNING: ANTIPATTERN—DATA HOARDING VS. INTENTIONAL STORAGE**
>
> Data hoarding occurs when systems retain everything "just in case," without evaluating Value, relevance, or use. This behavior burdens infrastructure, obscures meaning, and increases legal and security risks. Intentional Storage, by contrast, involves deliberate selection and structuring of artifacts that support Purpose and Function. Effective architecture curates; it does not accumulate.

   Implication: If a system stores outputs that cannot be reused, delivered, or consumed, it is storing noise, not Value.

2. **Stored Value must retain structural integrity and usability over time**

   Preservation without structure invites loss through entropy. Artifacts must remain intact, interpretable, and usable throughout their life cycle.

   Implication: Storage must include versioning, compatibility, and integrity—ensuring future agents can reuse what is stored without guesswork.

3. **Value must be accessible to authorized agents and discoverable in context**

   Stored artifacts are only useful if they can be located and retrieved when needed. Discoverability is a design concern, not a documentation problem.

Implication: If agents cannot find what has been stored, the system may need to reproduce or reacquire it—negating the benefit of Storage. In large organizations, misfiled or inaccessible artifacts can account for millions in rework and duplication costs—studies of digital asset management show that recreating lost materials can cost up to ten times more than maintaining a retrievable repository.

4. **Storage must decouple production from consumption across time and space**

   Without Storage, all interactions become immediate and brittle. With Storage, systems gain the ability to buffer, scale, and distribute.

   Implication: If every act of Value creation requires a synchronous consumer, the system lacks resilience.

5. **Storage must be governed with intent—not left to accumulate by default**

   Unbounded retention creates sprawl, liability, and confusion. Storage should follow policies aligned with Purpose and actual usage.

   Implication: If no one knows why something is being stored, or for how long, the system is architecturally incoherent.

6. **Stored Value should be optimized for reuse and composability**

   Artifacts should be designed to travel: modular, self-contained, and interoperable across contexts.

   Implication: If outputs must be manually reshaped or restructured before reuse, Storage is adding cost rather than capability.

7. **Security, privacy, and compliance must be designed into Storage constructs**

   Many Value artifacts involve obligations. Protection is not optional.

   Implication: If the system cannot guarantee secure retention, it should not retain.

8. **Storage must align with Purpose—not with infrastructure convenience**

   Stored Value must be justifiable in terms of what the unit is designed to offer or support—not simply based on ease of capture or availability of capacity. For example, a digital media team curating a library of films and images for a cultural platform may choose between two approaches: convenience-driven storage—dumping every received file into a cloud folder "because storage is cheap"—and purpose-driven storage, which keeps only finalized, culturally significant, and reusable assets. The latter remains searchable, accountable, and aligned with the platform's mission.

   Implication: Convenience-driven Storage leads to bloat; Purpose-driven Storage leads to clarity.

Storage is not an afterthought—it is an enabling mechanism. These principles ensure that what a system stores enhances its effectiveness, aligns with its intent, and supports its evolution. A system that stores with discipline preserves not only what it makes, but also its ability to make what matters.

CHECKLIST—FUTURE-PROOFING STORED ARTIFACTS OF VALUE

1. *Durability*: Can the artifact withstand degradation—physical, digital, or procedural—over its required lifetime?
2. *Accessibility*: Can authorized agents reliably locate, retrieve, and redeploy it when needed?
3. *Usability*: Does the stored artifact retain the form and structure necessary for direct reuse or delivery without conversion loss?
4. *Compatibility*: Will future systems or tools still recognize the format, interface, or packaging?
5. *Security and Protection*: Are integrity, confidentiality, and controlled access preserved across all storage layers?
6. *Life Cycle Management*: Are retention, rotation, and archival policies defined so artifacts remain relevant without unnecessary accumulation?

7. *Scalability*: Can the storage system handle growth in volume or distribution without impairing performance or reliability?
8. *Operational Readiness*: Can stored artifacts be activated or delivered without reconstruction or manual intervention?

11.3.2. Architectural Heuristics for Storage

Architectural heuristics for Storage provide practical guidance for designing, organizing, and managing storage systems that align with the principles of Purpose Driven Design. These heuristics translate high-level principles into actionable design patterns that support durability, accessibility, reuse, and governance of value-bearing artifacts.

1. **Store outcomes, not activities**

   Prioritize preserving artifacts that represent finalized Value—products, service outputs, or other deliverables. Avoid storing transient process states unless they carry independent value.

   Rationale: Storage should reflect what the system produces, not what it passes through on the way.

2. **Separate mutable from immutable artifacts**

   Design distinct storage paths for changeable drafts versus final, tamper-proof outputs. Treat immutability as a design feature for trust, traceability, and reproducibility.

   Rationale: When artifacts can't be trusted to remain stable, reuse and compliance both degrade.

3. **Design for intentional retrieval paths**

   Structure stored artifacts so they can be found and retrieved deliberately. Avoid deep nesting, ambiguous naming, or contextless dumping grounds.

   Rationale: If artifacts are hard to locate, they will be duplicated or ignored—defeating the purpose of Storage.

4. **Index for use, not just preservation**

   Metadata should reflect how artifacts will be reused—by whom, in what context, and under what constraints.

   Rationale: Unindexed Storage may retain value, but cannot deliver it.

5. **Use tiered storage based on value decay**

   Assign artifacts to different storage tiers according to how frequently they are accessed and how critical they remain over time. Not all Value ages the same; some assets require instant retrieval, while others can be archived or summarized for long-term reference. This value-based tiering—and how it applies across artifact types—is illustrated in Table 11-3.

Table 11-3. *Example Matrix: Tiered Storage by Artifact Type*

Storage tier	Typical artifacts of Value	Purpose/access pattern
Hot (active)	Frequently accessed digital media, live product catalogs, current financial reports	Immediate operational use; high availability and low latency required
Warm (reference)	Published ebooks, finalized engineering designs, recent marketing assets	Periodic reference or reuse; moderate access speed with cost efficiency
Cold (archival)	Completed film projects, historical contracts, legacy product documentation	Long-term preservation for compliance, provenance, or heritage purposes
Deep archive (preservation)	Rare physical artifacts, master negatives, long-term cultural records	Minimal access; maximum durability and environmental protection

   Rationale: Not all Value ages the same; Storage should reflect life cycle realities.

6. **Avoid silent obsolescence**

   Detect and surface stale, unused, or conflicting artifacts. Use signals such as expiry tags, access logs, or version drift.

Rationale: Without intervention, obsolete artifacts accumulate and pollute decision-making.

7. **Bind ownership to retention**

   Ensure every artifact has a responsible role or agent attached—someone accountable for its accuracy, relevance, and security. Ownership ensures accountability for what is kept, why, and for how long. Organizations that neglect this principle often face severe consequences: several high-profile GDPR and financial audit cases have resulted in multimillion-dollar fines after "orphaned" datasets—retained without stewards—were found to contain outdated or unprotected personal information.

   Rationale: Orphaned Storage leads to architectural drift and operational risk.

8. **Leverage naming and versioning conventions**

   Establish clear, consistent formats for naming, structuring, and evolving stored artifacts.

   Rationale: Consistency enables automation, auditability, and shared understanding.

9. **Use Storage as a boundary between roles or systems**

   In distributed architectures, treat Storage as a contractual interface between producers and consumers. Prefer neutral formats and avoid hidden dependencies.

   Rationale: Storage can decouple systems safely—if it's treated as an interface, not an implementation detail.

10. **Automate life cycle and compliance enforcement**

    Automate retention, encryption, access control, deletion, and logging wherever possible. Manual governance doesn't scale and rarely holds up under audit pressure. In a manual asset-storage workflow, archived documents, digital media, or product files often remain beyond their retention window—taking up costly storage space and exposing the organization to compliance risks. In an automated system, each artifact of Value is tagged with retention metadata at ingestion, automatically flagged for renewal or disposal when its term expires, and logged for audit verification. The result is leaner storage, lower cost, and continuous audit readiness.

Rationale: Manual governance doesn't scale and rarely holds up under audit.

These heuristics are not exhaustive rules but design shortcuts that reduce architectural error and support maintainability, efficiency, and systemic coherence. They help transform Storage from a passive holding mechanism into an active enabler of purpose-aligned value continuity.

11.4. Realization Across Implementation Planes

Storage is not an abstract intention—it must be made real. Every sociotechnical unit realizes Storage through a combination of social practices, physical constructs, and digital systems. These implementation planes operate in parallel: physical repositories preserve tangible artifacts, digital systems manage and protect virtual ones, and social agreements assign meaning, responsibility, and control. Effective Storage design requires coherence across these planes—ensuring that what is stored can be found, trusted, reused, and governed, no matter its form. This part explores how Storage is manifested, distributed, and maintained across the full spectrum of implementation surfaces.

11.4.1. Manifestation in the Social and Physical Planes

Storage begins in the physical world and is legitimized through social practice. Before any digital construct can exist, a sociotechnical unit must first determine *what* is worth preserving, *who* is responsible for it, and *how* it is to be maintained. This determination is social: it is embedded in roles, agreements, norms, and governance expectations.

In the physical plane, Storage takes the form of tangible repositories—warehouses, lockers, shelves, containers, or safes. These are not passive spaces. Their design and upkeep reflect deliberate decisions about access, duration, categorization, and accountability. Physical Storage may be centralized (as in dedicated facilities) or distributed (as in local staging zones), but in both cases, it imposes constraints on space, availability, and environmental resilience. These constraints directly shape the architecture of the unit.

Socially, Storage is reinforced by practices that define what is retained, why it matters, and how long it should remain. These include inventory protocols, check-in/check-out procedures, archival norms, chain-of-custody requirements, and even unwritten customs. Storage is not simply a technical act—it is a collective stance about what is valuable enough to keep.

These commitments take many forms. Below, we outline how physical Storage is realized in practice, and what design considerations govern its structure and reliability.

Forms of Physical Storage

- *Warehouses and Shelving Systems*: Used to store inventory, raw materials, finished goods, or replacement parts. These spaces are typically optimized for accessibility (e.g., FIFO/LIFO), spatial efficiency, and environmental control.
- *Containers and Packaging*: Enable the safe transfer and temporary storage of value artifacts, such as equipment kits, medical supplies, or hardware components. Packaging design affects not only protection but also reusability, traceability, and identity.
- *Libraries and Archives*: Preserve intellectual value in physical formats—books, blueprints, contracts, design specifications. These require indexing systems, environmental safeguards, and access protocols to maintain long-term integrity.
- *Tool Rooms and Asset Lockers*: Store instruments, equipment, and specialized tools for reuse. These often include permission controls, sign-out logs, and readiness indicators to ensure accountability and operational readiness.

Design Considerations

- *Protection*: Physical storage must guard against degradation caused by moisture, heat, dust, pests, or physical shock. Environmental controls, insulation, and robust infrastructure are critical design concerns.

- *Organization and Labeling*: Effective storage relies on physical indexing—barcodes, RFID tags, labeled bins, shelving maps—that reduce retrieval time, support inventory control, and minimize operational disruption.
- *Security and Access Control*: Sensitive or valuable artifacts require controlled access—locks, surveillance, permission hierarchies, and physical authentication mechanisms to prevent unauthorized retrieval or loss.
- *Redundancy and Distribution*: For systems requiring high availability or failover capacity, physical Storage may include geographically distributed backups, duplicate stockpiles, or offsite facilities to ensure resilience.

Operational Integration

Physical Storage is not isolated—it is deeply intertwined with process flow and operational patterns:

- *In manufacturing*, storage buffers are integrated into production lines, supplying just-in-time components or tooling.
- *In service delivery*, materials and kits are pre-staged near the point of use, enabling fast, reliable execution.
- *In archival practices*, active working libraries are often paired with deep storage or offsite archives to balance access and preservation.

Risks and Constraints

- *Space and Scalability*: Physical Storage is always constrained by real-world limits—floor space, weight, layout—and may require expansion planning or structural upgrades.
- *Decay and Obsolescence*: Artifacts stored too long may lose utility, degrade physically, or become incompatible with current systems or processes.

- *Inventory Inaccuracy*: Without monitoring and reconciliation mechanisms, physical Storage systems risk miscounts, misplacements, or ghost inventory—undermining reliability and decision-making.

In the social and physical planes, Storage is realized as a durable and accountable commitment: to preserve, protect, and manage the Value that gives a system its capability to function, recover, and deliver over time.

11.4.2. Storage in the Digital Plane

In the digital plane, Storage becomes radically scalable, dynamic, and abstractable. Unlike physical Storage—bound by space and material—digital Storage can be replicated, versioned, distributed, and accessed across time zones and organizational boundaries. Cloud object stores such as Amazon S3 or Google Cloud Storage exemplify this scale, holding petabytes of data replicated across regions for durability. Version-control systems like Git preserve every change to codebases, enabling teams in different time zones to work asynchronously on the same assets. Immutable logs in blockchain and ledger-based platforms extend the idea further—ensuring that once recorded, stored Value cannot be altered without trace. Yet this power introduces architectural complexity: digital Storage must be intentionally structured to avoid becoming fragmented, misaligned, or obsolete.

At its core, digital Storage preserves *digital artifacts of Value*. These may include culturally familiar forms such as eBooks, movies, songs, paintings, articles, games, or even digital tickets and NFTs—alongside more technical artifacts like service outputs, packaged apps, icons, or recorded sessions. Whether produced by the unit or received from external entities, these artifacts are retained because they can be used, reused, or delivered—supporting the continuity of Function and Process.

Digital Storage constructs are typically realized through cloud storage systems, databases, file repositories, content delivery networks (CDNs), object stores, and archival layers. But these technologies are only meaningful when paired with design logic: naming conventions, schemas, metadata, access policies, and life cycle controls that preserve structure, retrievability, and governance.

Digital Storage Constructs

- *Object Storage Systems*: Used for storing and retrieving unstructured digital artifacts—documents, files, binaries, or models—with high scalability and metadata tagging
- *Relational and Non-relational Databases*: Hold structured or semi-structured digital artifacts of Value, such as product catalogs, reference datasets, or transaction records
- *Package Repositories and Version Control Systems*: Retain modular, versioned Value artifacts—code libraries, container images, infrastructure modules—supporting composability and traceability
- *Digital Archives and Cold Storage*: Used for long-term retention of low-access Value artifacts—contracts, audit logs, legacy deliverables—under compliance or audit requirements
- *Shared Drives and Delivery Folders*: Provide controlled spaces where agents or customers can access, download, or upload digital artifacts intended for exchange

Design Considerations

- *Durability and Integrity*: Digital artifacts must be stored with mechanisms that detect and prevent corruption, ensure checksum validation, and preserve consistency across versions.
- *Searchability and Metadata*: Storage without metadata is digital oblivion. Well-designed systems index artifacts with contextual tags, descriptions, and linked references to support discoverability.
- *Access Control and Authorization*: Stored artifacts must be governed through role-based or attribute-based access control, ensuring that only the right agents can retrieve, edit, or delete.

- *Versioning and Immutability*: Many digital artifacts of Value benefit from immutability or explicit versioning to ensure reproducibility, rollback, and auditability.
- *Life cycle Automation*: Storage should integrate automated rules for expiration, archival, retention extension, or deletion—driven by compliance, usage patterns, or system goals.

Risks and Constraints

- *Unbounded Accumulation*: Digital systems make it easy to store everything, but without governance, this leads to storage bloat, rising costs, and degraded usability.
- *Silent Decay*: Digital artifacts may appear intact but become functionally obsolete—due to format drift, broken dependencies, or outdated schemas.
- *Permission Drift and Exposure*: Misconfigured access controls can lead to sensitive Value artifacts being exposed externally or misused internally.
- *Loss of Context*: If artifacts are stored without metadata, links, or structural cues, their original meaning may be lost—turning reusable Value into digital debris.

Table 11-4 organizes these digital-storage risks, detailing their impacts and the strategies required to contain them.

Table 11-4. *Digital Storage Risks and Mitigations*

Risk	Description	Real-world example	Architectural consequence	Mitigation strategy
Unbounded accumulation	Indiscriminate retention causes storage bloat, higher cost, and poor retrieval efficiency.	A large media company stored every video variant "just in case," driving costs up 40% and slowing asset searches.	Reduced system responsiveness and unmanageable audit surface	Implement life cycle policies and tiered retention based on Value decay.
Silent decay	Artifacts appear intact but become unreadable or unusable due to outdated formats or dependencies.	A financial regulator discovered old filings in obsolete file formats that could no longer be opened during a compliance review.	Loss of usable Value and credibility of institutional continuity	Schedule periodic format validation and conversion audits; maintain migration plans.
Permission drift and exposure	Evolving access rules and role changes leave sensitive artifacts overexposed or orphaned.	A healthcare vendor's cloud share exposed private records after personnel changes left open permissions.	Security breaches and regulatory violations	Automate access revocation; use role-based policies and regular permission audits.
Loss of context	Artifacts lose meaning when stripped of metadata, lineage, or structural cues.	A publishing platform archived manuscripts without version tags, causing confusion about which edition was authoritative.	Stored artifacts become inert or misused; rework required	Mandate metadata schemas, linkage standards, and contextual documentation.

In the digital plane, Storage must be designed not only for capacity but for clarity. What matters is not what can be stored—but what can be found, trusted, and used. Architecture must ensure that digital Storage supports reuse, distribution, and governance without losing sight of the Value it was meant to preserve.

11.4.3. Implementation Surfaces and Sociotechnical Contexts

Storage is not a monolith—it manifests through varied surfaces shaped by the sociotechnical context in which a unit operates. Whether physical or digital, centralized or distributed, permanent or ephemeral, every Storage implementation reflects deliberate architectural trade-offs based on what is being stored, how it is used, and who is accountable for it.

Some Storage surfaces are embedded directly within a unit's operational flow—an equipment locker in a workshop, a shared folder for marketing assets, a service endpoint that publishes downloadable reports. Others are more infrastructural or cross-cutting—archival systems maintained by integrative units, compliance vaults, or shared object stores used across teams. What matters is not the underlying technology but the *design logic* that governs what is preserved, where it lives, and how it can be used.

The context of use deeply influences the Storage form:

- In a logistics-heavy unit, Storage might be deeply physical—organized around racking systems, buffer zones, and packaging flow.
- In a knowledge-driven unit, Storage might prioritize access, searchability, and curation—enabling fast retrieval of articles, design elements, or past deliverables.
- In a software-producing unit, Storage may focus on modularity and reproducibility—version-controlled packages, signed binaries, release archives.

Implementation must reflect not only *the nature of the artifacts* being stored but also *the rhythms of the system*—how frequently artifacts are produced or consumed, how long they need to persist, and who interacts with them.

In large systems, Storage may be delegated to dedicated Storage units or platforms, especially when physical volume, digital sensitivity, or regulatory oversight demands specialization. In such cases, operational units interface with these Storage providers through contracts, APIs, protocols, or fulfillment mechanisms—but always remain responsible for what is stored and why.

Ultimately, Storage is realized through design decisions that bind architecture to context. It is not enough to retain Value; systems must do so *intentionally*—on surfaces that make sense to the humans, tools, and processes involved. Architecture must ask not only *what* is stored but *where, how, and with what understanding*.

11.5. Strategic Impact of Storage

Storage creates strategic leverage by making value durable, discoverable, and reusable. It reduces dependency on immediate availability, enables asynchronous coordination, and strengthens a unit's capacity to deliver, adapt, and recover. Beyond its operational utility, Storage supports architectural goals such as modularity, scalability, and continuity—allowing systems to function reliably across time and boundary. When embedded thoughtfully, it transforms isolated actions into lasting contributions, enabling units to compound their own output and interoperate more effectively with others.

11.5.1. Nuances of Fit and Readiness

Storage is not merely a passive container—it is a mechanism that ensures Value remains *usable, discoverable, and deliverable* after its initial creation. It allows Value artifacts—whether physical or digital—to remain structurally intact and *ready for transfer, reuse, or activation*, regardless of when or where they are consumed. *Fit* refers to structural and contextual compatibility: whether the stored artifact can be interpreted, accessed, and used without modification. *Readiness* refers to its availability and clarity at the moment it is needed. These characteristics enable sociotechnical systems to scale, coordinate, and serve across time, space, and actors—without requiring perfect simultaneity or shared presence.

Persistence As Readiness

Storage preserves artifacts not as static leftovers, but as *Value-in-waiting*—prepared for use beyond the moment of creation. What is stored is not a memory trace or internal reference, but an actionable, bounded output: a product, service artifact, or deliverable, complete and structured for consumption.

- In physical systems, this may be a finished product packaged for shipment, a set of pre-assembled parts staged for delivery, or a service kit awaiting deployment.
- In digital systems, it might be an ebook, a movie, a completed report, a concert ticket, a mobile app, a certified document, or a downloadable service result—each prepared for handoff or activation.

Readiness ensures that recipients—whether other units or entities in the environment—can access the preserved artifact without needing the producer to be present or active.

Temporal Mediation

Storage decouples creation from consumption. In its absence, producers and consumers must act in sync: the service must be performed as it's used; the product must be handed over as it's made. With Storage, the system gains temporal freedom: what's created today can be delivered or reused tomorrow—by another unit, or by a customer, client, regulator, or partner.

- A shared services unit can store onboarding kits for other teams to use on demand.
- A mobile ticket purchased asynchronously can be presented hours or days later.
- An information product can be produced in batch and served via a digital archive.

This decoupling introduces robustness, elasticity, and flexibility into the system—breaking the brittle dependence on simultaneity.

Structured Accumulation

Storage does not imply hoarding. It enables selective accumulation of *Value-bearing artifacts* that may serve future purposes—whether as reusable components, standardized deliverables, or compliance-ready outputs.

- A standardized API contract can be stored and accessed by multiple consuming units.
- A validated clinical result can be packaged and transmitted to a regulator.
- A bundle of finished design assets can be reused across marketing channels.

Accumulation allows systems to scale without redoing effort. It permits the construction of higher-order capabilities from previously created components—provided those components remain fit and ready.

Architected Reuse

Stored artifacts can only be reused if they are *visible, understandable, and composable.* Storage design must enable this: what is stored must not only be preserved but also be *made findable, clear, and usable by others*—within or outside the system.

- Digital asset libraries enable multiple units to draw from a shared pool of finished materials.
- Metadata and classification systems make stored artifacts discoverable.
- Stable packaging and versioning ensure artifacts are safe to reuse without revalidation.

Without structure and clarity, stored artifacts become clutter or a liability. With it, they become infrastructure for composability and systemic efficiency.

Optimization Surface

Storage introduces opportunities to optimize across time and flow—reducing cost, delay, and effort while increasing quality and control.

- Pre-cached results reduce response time in interactive systems.
- Pre-staged service kits streamline delivery in field operations.
- Archived outputs provide auditability and reduce legal or compliance risk.

Here, Storage acts not just as a utility but as a strategic layer—enabling the system to improve through coordination, buffering, and selective reactivation of past effort.

Storage is not memory. It is the architectural solution to *when Value cannot—or should not—be consumed at the moment it is created.* It allows systems to serve across asynchronous timelines, across organizational units, and across boundaries with their environments. Fit ensures that what is stored can be interpreted and used; readiness ensures it can be retrieved and trusted. When both are designed well, Storage becomes an invisible engine of continuity, composability, and delivery. What each construct does in the Value life cycle—remembering, preserving, or transforming—comes into focus in Table 11-5.

Table 11-5. *Differentiating Memory, Storage, and Process in the Value Life Cycle*

Construct	Primary focus	Temporal role	Typical artifacts or outputs	Architectural Purpose
Memory	Context, rationale, and historical awareness	Past → present	Decision logs, audit trails, policy rationales, version histories	Explains why things occurred and enables learning or accountability
Storage	Retention and reuse of artifacts of Value	Present → future	Finished goods, digital assets, books, datasets, blueprints	Ensures what is created remains usable, deliverable, and durable
Process	Transformation and production of new Value	Continuous/ real-time	Manufacturing steps, data processing, service execution	Converts inputs (from Storage) into new outputs (for Memory or Storage)

11.5.2. Measurable Benefits and Architectural Payoffs

Storage extends the impact of effort beyond its moment of creation. It allows Value to be preserved, transferred, reused, or delivered—without relying on synchronous coordination or shared context. While often invisible, Storage quietly powers some of the most strategic capabilities of a sociotechnical system. When thoughtfully implemented, it contributes to resilience, continuity, scalability, and accountability. The benefits below are not abstract ideals—they are directly observable and architecturally significant.

A. **Resilience Through Redundancy and Recovery**

   Storage provides a buffer against volatility, enabling systems to recover from failure, interruption, or loss.

   - Artifacts can be restored after outages or errors.
   - Historical data enables rollback, audit, and forensic analysis.
   - Redundant copies support fault tolerance and disaster recovery.

   Payoff: Storage increases system durability and trustworthiness—especially under stress or uncertainty.

B. **Continuity Across Time and Roles**

   Storage preserves institutional memory, allowing systems to maintain coherence even as agents change.

   - New agents can access past decisions, designs, and rationale.
   - Long-term projects retain context across handoffs.
   - Legacy knowledge remains usable and interpretable.

   Payoff: Storage enables systems to evolve without losing identity, purpose, or accumulated insight.

C. **Scalability Through Modularization**

   Stored artifacts can be modularized and composed into larger systems, supporting scale without chaos.

   - Components reused across units and contexts
   - Interfaces standardized for plug-and-play integration
   - Complexity managed through layered abstraction

   Payoff: Storage supports growth by enabling composability, standardization, and architectural clarity.

D. **Accountability Through Traceability**

Storage creates a trail of actions, decisions, and changes—supporting transparency and governance.

- Version histories and changelogs document evolution.
- Audit trails enable compliance and ethical oversight.
- Attribution systems clarify ownership and responsibility.

Payoff: Storage reinforces integrity and accountability—especially in regulated or high-stakes environments.

E. **Delivery Decoupling Enables Elastic Coordination**

Storage uncouples the timing between production and consumption, allowing Value to be created and delivered asynchronously.

- Artifacts can be generated in advance and distributed on demand.
- Producers and consumers do not need to coordinate in real time.
- Systems gain elasticity by buffering Value across time.

Payoff: Storage allows sociotechnical systems to operate with flexibility, absorbing variation in timing, load, and availability.

F. **Systemic Reuse Across Boundaries**

Storage makes Value artifacts available to other units, domains, or external partners—without rework.

- Deliverables are stored once and used many times across teams.
- Templates, toolkits, or packages circulate through ecosystems.
- Digital or physical products flow across organizational or regional boundaries.

Payoff: Storage amplifies the system's productive capacity by enabling each unit's output to benefit many others.

Without Storage, units are constrained to the present—forced to act synchronously, unable to retain what was created or received, and unequipped to deliver, reuse, or respond over time. With it, they retain what matters—what was produced, what was delivered, and what is still usable. Whether physical or digital, internal or external, past or future-facing, Storage ensures that Value can persist, circulate, and accumulate. Its payoff is not only operational—it is architectural.

11.6. Implementation Patterns

Storage is not merely a place where things are kept—it is a set of design choices that govern *how* artifacts of Value persist, evolve, and circulate. Whether the context is digital, physical, or hybrid, certain implementation patterns repeatedly emerge across well-architected systems. These patterns are not technology-specific. They describe stable solutions to recurring architectural challenges—revealing how Storage becomes modular, usable, traceable, and governable.

A. **Structural Patterns**

How Value is made persistable and composable

Modularization

Break complex artifacts into self-contained components that can be stored, reused, and recombined.

- Enables selective updates, partial delivery, or substitution
- Supports composability across systems and contexts

Packaging

Bundle related artifacts into coherent units—for example, deployment kits, product SKUs, service bundles.

- Simplifies distribution and retrieval
- Embeds dependencies and usage assumptions

Symbolic Compression

Store abstracted representations—such as pointers, hashes, manifests, or descriptions—instead of full content.

- Reduces duplication and storage cost
- Preserves identity without requiring material replication

B. **Life Cycle and Identity Patterns**

How artifacts evolve and remain traceable

Versioning

Track the evolution of stored artifacts through explicit versions.

- Maintains history while allowing improvement
- Supports rollback, comparison, and branching

Attribution

Associate artifacts with their creators, contributors, and stakeholders.

- Enables accountability, recognition, and ethical traceability
- Strengthens trust in provenance and intent

Preservation

Safeguard artifacts for the long term—even if they are no longer active.

- Protects against loss due to obsolescence or inattention
- Supports audit, reuse, and institutional memory

C. **Distribution and Availability Patterns**

How stored Value reaches agents or systems

Replication and Caching

Store redundant or transient copies closer to where they're needed.

- Improves speed, resilience, and fault tolerance
- Reduces load on authoritative Storage

Localization

Adapt artifacts to specific languages, jurisdictions, or use contexts.

- Enhances relevance and accessibility
- Enables legal, cultural, or technical compliance

Broadcast and Subscription

Expose stored Value to other units or agents via push/pull mechanisms.

- Reduces friction in distribution
- Enables system-wide reuse without direct coordination

D. **Governance and Control Patterns**

How access and modification are managed

Access Control

Enforce permissions on stored artifacts—read, write, delete, share.

- Protects Value from misuse or loss
- Supports roles, contexts, and boundaries

Audit Logging

Capture every interaction with Storage—what changed, who changed it, and when.

- Enables transparency and accountability
- Facilitates compliance and forensic review

Retention and Expiry

Define policies for how long artifacts are kept—and when they are removed.

- Prevents clutter and reduces liability
- Aligns Storage behavior with life cycle needs

These patterns are not mutually exclusive. A single artifact may be modular, versioned, cached, and access-controlled simultaneously. But understanding them as discrete design strategies helps architects make Storage explicit, deliberate, and aligned with system intent.

11.7. Interlocking Dependencies and Alignment

Storage does not operate in isolation. Its relevance, structure, and behavior are shaped by the needs of other design constructs—and by the Purpose the system exists to fulfill. The artifacts a system stores, the way it organizes them, and how long they persist all reflect decisions about what matters, what must be reused, and what must be delivered. Properly aligned Storage reinforces architectural intent; misaligned Storage introduces ambiguity, waste, and fragility.

11.7.1. Alignment with Purpose: Justification and Selectivity

Purpose defines what the system exists to serve. Storage reflects that Purpose by preserving what is worth keeping—what the system may need again, what others may need, or what must be retained for evidence or accountability. If Storage accumulates what no longer aligns with Purpose, it becomes a burden, not an asset. Selectivity is essential: not all that is produced should be stored; only that which contributes to Purpose across time and context.

11.7.2. Alignment with Value: Boundaries and Semantics

Storage draws boundaries around Value—marking what has been completed, received, or made available. It transforms ephemeral interaction into tangible consequence. For this reason, Storage must respect the semantics of Value. Storing a report that no longer reflects updated data, or archiving a design that was never approved, introduces semantic confusion. Storage is not just about preservation—it is about preserving *meaningful* artifacts of Value.

11.7.3. Alignment with Function: Exposure and Deliverability

Function depends on Storage to access previously created or received Value. If Storage fails to make Value discoverable, interpretable, or accessible, Function stalls—or worse, misfires. Conversely, storing artifacts with no corresponding Function leads to orphaned assets and unclear responsibilities. The link between Storage and Function must be tight: what is stored should support what is offered, and what is offered should point clearly to what is stored.

11.7.4. Alignment with Process: Flow and Reusability

Process creates Value. Storage holds it. If Process is not designed with Storage in mind, artifacts may be transient, inconsistent, or lost. And if Storage does not reflect process patterns—such as checkpoints, staging areas, or intermediate outputs—it disrupts flow. Alignment here ensures that what Process produces can be reliably handed off, paused, resumed, or reused.

11.7.5. Alignment with Structure: Ownership and Stewardship

Storage implies responsibility. Someone must know what is stored, why, and for how long. This is a structural concern: units must be accountable for their stored artifacts. Misalignment occurs when Storage spans boundaries without clear stewardship, or when artifacts outlive the unit that produced them. Structural clarity ensures that every piece of stored Value has a custodian—and that handoffs are formalized when ownership shifts.

11.7.6. Alignment with Order: Policy and Control

Order governs how long things are kept, who can access them, and under what conditions they must be deleted, archived, or reviewed. Without clear alignment, Storage becomes either a liability (hoarding unnecessary or sensitive information) or a vulnerability (lacking necessary traceability or retention). Storage aligns with Order when retention policies, access controls, and classification schemes reinforce the system's internal rules and external obligations.

11.7.7. Alignment with Culture: Practices and Expectations

Storage reflects what a system *chooses* to remember and make available. Culture determines whether artifacts are diligently captured or casually discarded; whether naming and organization follow norms or emerge ad hoc; whether access is open, selective, or guarded. Misalignment between Storage and Culture creates friction: agents may avoid Storage systems they don't trust, or misuse them in ways that conflict with broader expectations.

11.7.8. Alignment with Memory: Transition from Trace to Artifact

Memory preserves history to support interpretation and learning. But when patterns, records, or observations acquire practical value beyond reflection—becoming usable by other units or entities in the environment—they can be treated as by-products, packaged, and transitioned into Storage for delivery and reuse.

Storage acts as connective tissue across the design space. It makes Purpose material, preserves Value, supports Function, buffers Process, depends on Structure, enforces Order, reflects Culture, and monetizes (or externalizes) Memory. Alignment is not optional—it is what transforms Storage from inert accumulation into active infrastructure. A well-aligned Storage construct amplifies continuity, coherence, and clarity across the entire system.

11.8. Failure Modes and Antipatterns

Storage may seem inert—merely a container—but its design has far-reaching effects on system behavior, clarity, resilience, and flow. Poorly architected Storage silently erodes coordination, increases risk, and disconnects effort from impact. These antipatterns expose recurring failures that stem not from missing features but from architectural misalignment across the system.

1. **Storage Without Ownership**

   Definition: Artifacts are retained without a responsible unit curating, managing, or governing their life cycle.

   Symptoms: Abandoned folders, unclear permissions, lack of stewardship or version control.

   Consequence: Accumulated risk, duplicated work, inconsistent reuse, security exposure.

   Architectural Fault: Misalignment with Structure (no accountable surface), or with Order (no governance logic or life cycle definition).

2. **Shadow Storage**

   Definition: Unofficial and unmanaged repositories are created to bypass slow or restricted Storage systems.

   Symptoms: Use of local drives, hidden folders, email chains, or chat uploads to share artifacts.

   Consequence: Fragmented knowledge, lost artifacts, governance breakdown.

   Architectural Fault: Misalignment with Culture (norms permit workarounds), or with Order (official pathways are too rigid or obstructive).

3. **Unstructured or Unindexed Storage**

   Definition: Artifacts are stored without naming conventions, metadata, or search-friendly organization.

   Symptoms: Impossible-to-navigate folders, duplication of artifacts, confusion over versions.

   Consequence: Time loss, failed retrieval, risky reuse, user frustration.

   Architectural Fault: Misalignment with Value (stored but not usable), or with Process (retrieval not embedded in flow).

4. **Improper Format or Interface**

   Definition: Artifacts are stored in formats or containers incompatible with downstream needs.

   Symptoms: Misuse, transformation errors, inaccessible deliverables, wasted work.

   Consequence: Rework, failure to integrate, reduced adoption.

   Architectural Fault: Misalignment with Function (form does not match interface), or with Structure (unclear or incompatible delivery surface).

5. **Overaccumulation Without Purpose**

   Definition: Artifacts are stored indefinitely without any clear intention for reuse or delivery.

   Symptoms: Bloating repositories, escalating costs, degraded performance.

   Consequence: System becomes harder to navigate and maintain; clutter obscures what matters.

   Architectural Fault: Misalignment with Purpose (storage is symbolic rather than functional), or with Order (no expiration, retention, or cleanup policy).

6. **Premature Disposal**

   Definition: Artifacts are deleted or overwritten before downstream usage is complete or compliance conditions are met.

   Symptoms: Missing deliverables, irreproducible outputs, broken integration paths.

   Consequence: Failed service delivery, legal risk, repeated effort.

   Architectural Fault: Misalignment with Process (timing assumptions ignored), or with Value (persistence needs misjudged).

7. **Overreliance on Ephemeral Storage**

   Definition: Critical artifacts are stored in transient environments—developer laptops, short-lived containers, temporary folders.

   Symptoms: Artifacts disappear without notice; systems behave differently across environments.

   Consequence: Lost work, debugging delays, inconsistent behavior.

   Architectural Fault: Misalignment with Structure (no durable surface), or with Order (no escalation from temporary to persistent Storage).

8. **Disconnected from Function**

   Definition: Stored artifacts are not surfaced or accessible via the Functions that use or expose them.

   Symptoms: Manual retrieval, dependency on intermediaries, duplicated uploads.

   Consequence: Inefficiency, user frustration, broken handoffs.

   Architectural Fault: Misalignment with Function (Storage not wired into the interface), or with Process (handoff not embedded).

9. **Misused as Memory**

   Definition: Storage is overloaded with artifacts meant for reflection, not value transfer.

   Symptoms: Logs, observations, or raw traces are treated as deliverables; narrative context is missing.

   Consequence: Misinterpretation, audit failure, failure to learn or reuse effectively.

   Architectural Fault: Misalignment with Memory (trace data stored without interpretation), or with Value (unclear utility).

These failure modes are not merely operational gaps—they are symptoms of conceptual misalignment. Storage succeeds not when it's full, but when what's retained remains *findable, relevant, deliverable,* and *governed.* Clear architectural purpose, ownership, and structural integration are the antidotes to these silent breakdowns.

SELF-AUDIT: DIAGNOSING STORAGE FAILURE RISKS

1. *Ownership*: Does every stored artifact have an accountable custodian?
2. *Structure*: Are repositories organized with clear naming, metadata, and retrieval logic?
3. *Governance*: Are retention, access, and disposal rules enforced consistently?
4. *Alignment*: Is what's stored directly tied to Purpose and intended reuse?
5. *Life cycle*: Are artifacts reviewed, refreshed, or retired on a defined schedule?
6. *Security*: Are permissions current and scoped appropriately across units?
7. *Integration*: Are Storage systems interoperable, avoiding duplication or shadow repositories?
8. *Resilience*: Could the system recover quickly if a repository failed or a vendor changed?

11.9. Summary: Storage As a Design Construct

Within Purpose Driven Design, storage is the system's durable capability for preserving and delivering value-bearing artifacts—across time, roles, and organizational boundaries. It enables purposeful systems to function continuously and cohesively, even when creators and consumers are asynchronous, distributed, or structurally disjointed. Without Storage, output vanishes, reuse becomes impossible, and continuity erodes into repetition.

Unlike Memory, which supports learning and interpretation, Storage supports transfer, distribution, reuse, and resilience. It ensures that what matters—products, service outcomes, documents, kits, deliverables—can be retained, structured, and made available. Whether material or digital, ephemeral or enduring, Storage turns effort into reusable capability.

Realized effectively, Storage functions as

- *A continuity anchor,* preserving produced or received Value across time and enabling handoffs between agents, roles, or units
- *A delivery buffer,* decoupling production from consumption to support asynchronous operation and scalable interaction

- *A reuse surface,* allowing components, templates, or past outputs to be applied in new contexts with minimal friction
- *A resilience mechanism,* enabling recovery from failure, reconstruction after interruption, and auditability under scrutiny.
- *A structural interface,* facilitating interaction not just between units, but with external stakeholders—consumers, suppliers, regulators—through stable, governed artifacts

Storage is not a passive container. It is an active architectural surface—accountable, structured, governed, and aligned. Its design must be intentional: defining what should persist, who owns it, how it is retrieved, and when it expires. When Storage is absent or misaligned, systems fracture; when well-designed, it enables scalable, resilient, and meaningful participation in coordinated action.

In the age of modular systems and asynchronous exchange, Storage is no longer an operational afterthought. It is the backbone of architectural coherence—the logic of how systems retain and deliver what they produce, long after the process ends or the actor departs.

FIVE-MINUTE STORAGE HEALTH CHECK

Use these five quick questions to assess whether your organization's Storage design aligns with architectural principles of Purpose Driven Design.

1. *Purpose Alignment*: Can every major repository or storage system explain how its contents serve the organization's Purpose and Function?
2. *Value Focus*: Are only artifacts of demonstrable Value (deliverables, data products, documents, or goods) retained, not incidental outputs?
3. *Life Cycle Management*: Are there clear policies and automated mechanisms for rotation, archiving, and disposal?

4. *Accessibility and Readiness*: Can authorized agents easily retrieve what they need in a usable, validated form?
5. *Governance and Stewardship*: Are ownership, compliance, and auditability maintained without relying on ad hoc intervention?

If you can answer "yes" to all five, your Storage architecture likely supports continuity and scalability. If not, begin by reviewing alignment with Purpose (Principle 8) and ownership and life cycle practices (Heuristics 7 and 10).

CHAPTER 12

Governance, Life Cycle, and Evolution of PDD

While each PDD construct is structurally distinct and instantiated within the boundaries of a sociotechnical unit, none are exempt from the demands of ongoing governance, controlled evolution, and life cycle management.

In this context:

- Governance ensures accountability, coherence, and compliance across constructs and units—defining who decides, how oversight occurs, and what standards apply.
- Evolution manages intentional adaptation over time—allowing constructs to change purposefully without losing identity or coherence.
- Life cycle management defines how constructs are created, matured, reviewed, and eventually retired—preserving clarity, traceability, and architectural integrity throughout their lifespan.

PDD constructs are not static definitions; they are actively maintained components of the architecture. For example, as a technology enterprise scales globally, its Order construct—initially defined by simple approval chains—must evolve into a modular governance model with delegated authority and automated compliance. This evolution allows the organization to remain agile without losing accountability, preserving viability through structured adaptability.

The constructs' implementation surfaces—interfaces, processes, policies, and behaviors—must adapt to environmental changes, systemic feedback, and internal refinements without losing coherence. For instance, a healthcare platform may update

A. Milchman, *The Purpose Driven Design Framework for Modern Enterprises*,
https://doi.org/10.1007/979-8-8688-2223-0_12

its data-sharing interfaces in response to new privacy regulations and patient feedback, yet its underlying governance model—access tiers, auditability, and role boundaries—remains intact. Feedback informs improvement without dissolving architectural intent.

Rather than repeating governance mechanisms within each construct chapter, this chapter consolidates the common strategies, patterns, and life cycle practices that apply across all PDD constructs. It offers a unified architectural view of how constructs persist, adapt, and remain governable within a dynamic system of units.

12.1. Overview: Why Governance and Life Cycle Matter

Purpose Driven Design offers a composable architecture of constructs that stabilize and differentiate sociotechnical units. However, the presence of constructs alone does not guarantee coherence over time. Without governance, even well-designed constructs drift—losing alignment with their originating Purpose, diverging from one another, and accumulating inconsistencies that degrade both internal intelligibility and external viability. Governance provides the structural mechanisms through which constructs remain fit for purpose, and life cycle management ensures that constructs evolve in response to meaningful feedback rather than arbitrary change. When governance becomes too rigid, constructs stagnate; when it is too lax, coherence erodes; when it is reactive, change loses intent. Life cycle management mitigates these pitfalls by enforcing review cycles, aligning evolution with Purpose, and retiring constructs that no longer serve the system's design logic.

Governance is not synonymous with control. It is the means by which units preserve integrity while remaining adaptable. When control becomes excessive, every change demands approval and systems slow to a crawl, sacrificing responsiveness for predictability. When governance is neglected, units drift—duplicating effort, diverging from shared standards, and weakening collective coherence. Effective governance occupies the middle ground: it defines boundaries that enable adaptation without chaos. Robust governance assigns accountability, establishes interfaces for oversight, enforces consistency across roles and domains, and enables distributed decision-making without collapsing into chaos.

Life cycle management, in turn, introduces temporal structure: constructs are not created and forgotten; they move through clear stages—**Definition ➤ Instantiation ➤ Maturity ➤ Revision ➤ Retirement**—each requiring intentionality, visibility, and architectural justification. These transitions must be intentional, visible, and architecturally justified.

Together, governance and life cycle considerations form the meta-architecture that makes Purpose Driven Design sustainable in practice. They allow constructs to remain responsive without becoming reactive, modular without becoming fragmented, and stable without becoming stagnant. This chapter outlines the general strategies that apply across all constructs—versioning, stewardship, compositional contracts, adaptation mechanisms, and maturity patterns—providing a unified foundation for construct-level and system-level coherence.

12.2. Core Governance Mechanisms

While each PDD construct plays a distinct role in unit composition, they all require governance mechanisms that preserve structural integrity, ensure traceability, and enable adaptation over time. These mechanisms operate uniformly across constructs, allowing units to be governed without reinventing oversight for each design surface. What follows are governance patterns that can be applied consistently, regardless of whether the construct is Purpose, Culture, Process, or Storage.

12.2.1. Versioning and Structural Traceability

Every construct should be versioned—not just in its interface definition, but in its internal configuration, policies, and implications. In one financial system audit, multiple "current" process definitions coexisted without version identifiers, making it impossible to determine which logic governed transactions at a given time—delaying reconciliation and exposing the institution to regulatory penalties. Versioning makes changes explicit and enables rollback, impact analysis, and historical comparison. Semantic or life cycle-based versioning (e.g., Draft ➤ Stable ➤ Deprecated) helps distinguish experimental changes from structural revisions. Without versioning, constructs evolve silently, creating fragmentation and making auditability or reproducibility impossible.

12.2.2. Role-Based Stewardship

Each construct should have a designated steward—an individual or team responsible for maintaining its clarity, quality, and contextual fit.

Typical responsibilities include

- Ensuring the construct remains aligned with Purpose and other constructs
- Reviewing updates and version changes for coherence and traceability
- Managing documentation, interfaces, and points of dependency
- Acting as the escalation contact when conflicts or ambiguities arise

Without stewardship, constructs drift. In one enterprise modernization effort, the absence of a steward for the Process construct led to each business unit redefining workflows independently, creating incompatible automation patterns and collapsing systemic coherence. Stewards do not own the construct exclusively, but they are accountable for its interpretability, evolution, and governance within the unit or system.

12.2.3. Compositional Contracts

Constructs rarely operate in isolation. Governance must define the boundaries and expectations of interaction—what is required, permitted, or disallowed. These "compositional contracts" clarify how constructs integrate, expose interfaces, and conform to global norms (e.g., Value contracts specify what constitutes acceptable offerings; Storage contracts define retention policies and access boundaries). Contracts reduce ambiguity and allow modular composition across teams, domains, and time. Table 12-1 organizes these contract types, showing how each establishes boundaries, expectations, and integration rules.

Table 12-1. Sample Compositional Contracts Across PDD Constructs

Construct	Example contract definition	Purpose of the contract
Purpose	Defines the stakeholder need or mission that justifies the unit's existence	Aligns all constructs around a shared reason for being
Value	Specifies deliverables, quality standards, and exchange conditions	Ensures measurable outcomes consistent with Purpose
Function	Defines interaction interfaces and service-level expectations	Coordinates how Value is exposed and consumed
Process	Sets boundaries for sequence, dependencies, and timing	Prevents process drift and clarifies accountability for execution
Structure	Establishes role definitions, authority spans, and escalation paths	Ensures that responsibilities are explicit and interoperable
Order	Describes rules, permissions, and constraint layers	Maintains legitimacy and consistency across systems
Culture	Codifies behavioral norms and shared interpretive frameworks	Promotes cohesion and predictable collaboration
Memory	Outlines what rationale, events, or lessons must be recorded and for how long	Preserves contextual learning for future decisions
Storage	Defines retention scope, access conditions, and reusability criteria for artifacts of Value	Guarantees that what is stored remains usable and aligned with Purpose

12.2.4. Feedback-Informed Adjustment

Governance must embed listening surfaces into constructs. Memory, metrics, and usage patterns should feed back into stewards' decision-making processes. Constructs are adjusted not by whim or compliance mandates but through informed, purposeful evolution based on performance, relevance, and systemic impact. This pattern is especially critical for constructs like Process and Structure, which are prone to drift if not anchored to performance or Purpose feedback.

In one financial services platform, recurring delays in regulatory reporting were first treated as a tooling issue. However, feedback revealed that the underlying Process construct—its sequence of data validation steps—needed to be restructured, not merely accelerated. Once governance addressed the structure itself, timeliness and audit reliability improved simultaneously.

12.2.5. Alignment Audits and Dependency Reviews

Periodically, constructs must be examined in relation to their interdependencies. Are Processes still aligned with Purpose? Is Culture reinforcing or eroding Structure? Are Memory and Storage synchronized? Governance must include mechanisms to identify misalignments across constructs and initiate corrective action. These reviews help preserve the integrity of the unit as a composed system, not just a collection of individually managed parts.

ALIGNMENT AUDIT AND DEPENDENCY REVIEW CHECKLIST

1. *Identify construct dependencies*: Map the relationships between the eight active constructs and note their directionality (e.g., Purpose ➤ Value ➤ Function).
2. *Verify structural consistency*: Confirm that each construct's definitions, interfaces, and policies still reflect the current version of its dependencies.
3. *Assess directional alignment*: Ensure that Purpose continues to justify Process, that Value is still traceable to Purpose, and that Order and Culture reinforce rather than contradict each other.
4. *Document deviations*: Record any misalignments, ambiguities, or duplications detected during the review.
5. *Assign remediation ownership*: Allocate corrective actions to the appropriate construct stewards.

6. *Validate and version outcomes*: Reassess after corrections and update construct versions, dependencies, and audit logs.
7. *Archive and share findings*: Store results in the governance registry to inform future life cycle reviews.

12.2.6. Transparency and Accessibility

Construct governance must be observable. Stakeholders—internal or external—should be able to understand what a construct does, who governs it, how decisions are made, and how changes are tracked. Documentation, dashboards, and open interfaces ensure constructs are not black boxes. Transparency builds trust and enables accountability, especially when constructs are federated across domains.

Best practices for observability include maintaining construct registries that expose ownership, version, and dependency data; instrumenting dashboards that display governance metrics (change frequency, alignment status, policy exceptions); and embedding audit logs directly into decision workflows so that governance events are traceable. Common architectural enablers include version-controlled repositories for governance artifacts, lightweight workflow tools for stewardship approvals, and federated catalogs or APIs that surface construct metadata system-wide. These mechanisms make governance visible, queryable, and accountable without forcing centralization.

The core governance mechanisms are not optional—they form the enabling substrate for scalable, interpretable, and resilient units. Without them, constructs may remain elegant in theory but quickly degrade in practice. Governance, in this sense, is the meta-architecture that governs construct evolution: the structural grammar that allows constructs to persist as meaningful components within dynamic, evolving systems.

12.3. Life Cycle States of Constructs

Just as systems and units evolve, so too must the constructs that compose them. Each PDD construct—whether it governs behavior, produces value, retains memory, or stores artifacts—undergoes distinct life cycle states. These states are not abstract; they shape how constructs are introduced, managed, adapted, and eventually retired. Life cycle awareness enables responsible evolution by ensuring that changes are intentional, visible, and structurally justified. Without explicit life cycle staging, constructs tend to drift—accreting inconsistency, losing interpretability, or persisting long after their relevance has expired.

The following five states represent a generalized life cycle model applicable across all constructs:

12.3.1. Defined

A construct enters the Defined state when it is formally specified but not yet implemented. It has architectural intent and clarity of scope, but no runtime expression. For instance, a newly articulated Purpose may be documented in design artifacts or interfaces but has not yet been reflected in Function, Process, or Structure. This state ensures that constructs are not improvised—they are designed before being deployed.

12.3.2. Instantiated

The construct is now realized within the system or unit. A Process is implemented, a Memory system is activated, or a Culture program is launched. Instantiation does not guarantee completeness or maturity—it simply marks the transition from theoretical definition to operational presence. Monitoring, traceability, and stewardship become critical at this stage.

12.3.3. Active

The construct is fully operational, governed, and aligned with the unit's Purpose. It is referenced, invoked, or depended upon by other constructs or external actors. Governance mechanisms are functioning, feedback is being gathered, and adaptation cycles are underway. In this state, constructs are not only in use—they are being used well.

12.3.4. Misaligned

At this point, the construct is still active but no longer fit for purpose. Perhaps it no longer aligns with environmental demands, or it has drifted from systemic dependencies, or its usage patterns reveal obsolescence. Misalignment may be gradual (e.g., Culture eroding through informal divergence) or sudden (e.g., Order invalidated by regulatory change). This state signals the need for reassessment, revision, or replacement—not termination by default, but intentional structural inquiry.

For instance, one global analytics program discovered its Process construct had drifted—teams in different regions had reinterpreted workflow logic to fit local tools. Instead of retiring the entire framework, the steward initiated a cross-region audit, identified the viable local adaptations, and codified them into a new baseline process. Within a single quarter, the construct was re-aligned and re-activated, preserving both institutional learning and continuity of governance.

12.3.5. Deprecated

The construct is no longer recommended for use and is being phased out. This may follow a misalignment diagnosis or result from systemic redesign. A deprecated construct remains visible and traceable for audit or transitional purposes, but it is no longer treated as authoritative. For example, a Storage policy may be retained for compliance but marked as superseded by a more modern, interoperable approach. Deprecation requires clear communication to prevent ghost dependencies or legacy entanglement.

DEPRECATION COMMUNICATION AND TRANSITION CHECKLIST

- *Announce intent early*: Notify affected stewards and dependent units of planned deprecation and its rationale.
- *Define transition window*: Establish clear timelines for migration, replacement, or archival.
- *Document successor or alternative*: Provide references to the construct or policy that supersedes the deprecated one.
- *Preserve traceability*: Archive prior versions with metadata linking them to audit logs and governance records.
- *Confirm dependency closure*: Validate that no active functions or processes still rely on the deprecated construct.
- *Communicate completion*: Issue a final notice confirming that deprecation is closed and stewardship has ended.

Table 12-2 summarizes each life cycle state, outlining its defining characteristics, governance focus, and primary stewardship responsibility.

Table 12-2. *Construct Life Cycle States*

State	Description	Key governance actions	Primary steward role
Defined	Construct is formally specified but not implemented.	Confirm scope, dependencies, and intended steward.	Architect/domain steward
Instantiated	Construct is implemented in system or process.	Establish monitoring, initialize version control.	Implementation lead
Active	Construct is fully governed and used.	Collect feedback, track versions, adjust minor parameters.	Steward/governance council
Misaligned	Construct no longer fits context or Purpose.	Initiate alignment audit, propose revision or replacement.	Steward with meta-steward oversight
Deprecated	Construct is retained for reference but phased out.	Archive, document rationale, remove dependencies.	Governance archivist/ compliance lead

The choreography of state change—what moves, why it moves, and who must act—comes into focus in the transitions outlined in Table 12-3.

Table 12-3. *Typical Life Cycle Transitions*

From → to	Trigger	Governance action
Defined → instantiated	Implementation approval	Deployment and registration
Instantiated → active	Validation and adoption	Activate monitoring and feedback loops.
Active → misaligned	Performance drift or contextual change	Alignment review and remediation plan
Misaligned → active	Successful correction	Re-version and re-release
Misaligned → deprecated	Structural obsolescence	Controlled retirement
Deprecated → defined	Replacement initiated	Design of successor construct

Treating constructs as life cycle-aware objects supports more than governance—it fosters architectural maturity. By moving beyond a binary view of "active or inactive," units can intentionally curate their internal architecture, retire obsolete logic, and evolve constructs without destabilizing the whole. Life cycle states offer the scaffolding for continuity: a way to change responsibly, adapt visibly, and preserve the coherence that Purpose Driven Design demands.

12.4. Evolution Patterns Across Constructs

Not all constructs evolve at the same pace or in the same manner. Some require stability and continuity; others must adapt rapidly to shifting demands. Understanding the typical evolution patterns of each PDD construct enables designers and stewards to apply appropriate governance rhythms, resource allocations, and change thresholds. A one-size-fits-all approach to change management leads to either rigidity or chaos—depending on whether it underreacts or overreacts to the pressures of context.

Constructs can be roughly grouped into three evolutionary categories: *slow-evolving*, *moderate-cycle*, and *fast-adapting*. Each group demands a different stance on governance, tooling, and feedback responsiveness.

12.4.1. Slow-Evolving Constructs: Anchors of Continuity

Purpose, *Order*, and *Culture* form the structural spine of a sociotechnical unit. Their evolution is possible but must be deliberate, justified, and well-orchestrated.

- *Purpose* shifts only when the existential orientation of the unit changes—such as pivoting from regulatory support to market-facing innovation.
- *Order* evolves slowly, reflecting deep governance shifts, legal mandates, or formal coordination adjustments.
- *Culture* adapts incrementally through social reinforcement, making abrupt reprogramming infeasible and often counterproductive.

These constructs benefit from long feedback loops, structural audits, and change thresholds that emphasize stability over responsiveness. Their evolution is less frequent, but highly consequential.

12.4.2. Moderate-Cycle Constructs: Agents of Reconfiguration

Function and *Structure* are moderately paced in their evolution. They respond to shifts in Purpose, feedback from use, and changes in Process or Value design.

- *Function* may evolve as new user needs, channel requirements, or technologies emerge—for example, exposing APIs, changing interaction models, or redefining service contracts.
- *Structure* adapts with organizational change: roles, team compositions, automation layers, or agency blending.

These constructs require ongoing governance and versioning but with tolerable rates of change. Change must be visible, tractable, and backward-compatible when possible.

12.4.3. Fast-Adapting Constructs: Surfaces of Responsiveness

Process, Memory, and *Storage* evolve most rapidly, often on short cycles tied to operational feedback, system load, or data requirements.

- *Process* may change weekly or monthly in agile environments, reflecting continuous improvement, automation, or simplification.
- *Memory* evolves as new traceability needs emerge, data structures change, or retrospective practices mature.
- *Storage* adapts based on consumption patterns, cost constraints, life cycle policies, or technology shifts.

These constructs benefit from high observability, responsive governance mechanisms, and tight coupling with feedback loops. Their fast evolution enables the system to adapt without constantly renegotiating foundational principles.

Table 12-4 organizes these constructs by evolutionary category and suggests indicative governance rhythms and the feedback mechanisms that inform change.

***Table 12-4.** Relationship Between Construct Evolution Categories, Governance Rhythms, and Feedback Mechanisms*

Evolution category	Constructs	Governance rhythm	Typical feedback sources	Change triggers
Slow-evolving (anchors of continuity)	Purpose, Order, Culture	Annual or semi-annual structural review; high-change threshold	Strategic audits, external regulation, long-term trend analysis	Environmental shifts, mission pivots, regulatory updates
Moderate-cycle (agents of reconfiguration)	Function, Structure	Quarterly governance and dependency reviews; visible versioning	Usage metrics, dependency mapping, stakeholder feedback	New capabilities, workflow realignment, technology upgrades
Fast-adapting (surfaces of responsiveness)	Process, Memory, Storage	Monthly or continuous improvement cycles; automated feedback loops	Operational telemetry, performance metrics, incident retrospectives	Performance drift, system load changes, new data or tools

12.4.4. Evolution Misalignment As a Structural Risk

Constructs must evolve *in proportion to their role*. When slow-evolving constructs change too frequently (e.g., redefining Purpose or Culture arbitrarily), units lose coherence and legitimacy. Conversely, when fast-adapting constructs resist change (e.g., outdated Processes or frozen Storage strategies), operational effectiveness suffers.

Additionally, *temporal misalignment between constructs* can create architectural stress. For example:

- Process evolves, but Function does not adapt—users face incoherent interfaces.
- Structure is updated, but Order is not—roles exceed permissions or violate constraints.
- Memory systems are refreshed, but Culture fails to absorb lessons—feedback loops collapse.

Designers must therefore consider not only whether constructs evolve, but how they *co-evolve*. Governance should enforce both autonomy and synchronization across evolutionary tempos.

For example, a large financial platform modernized its data ingestion Process on a bi-weekly release cadence, while its Function interfaces remained quarterly versioned. Within weeks, consumers reported broken integrations and inconsistent schemas. Rather than freezing updates, governance introduced an evolution-sync protocol—aligning Process release reviews with Function dependency checks. The result restored stability without slowing adaptation, demonstrating how temporal alignment can be engineered rather than improvised.

Recognizing these evolution patterns allows units to remain resilient without becoming erratic. When each construct evolves at the right rhythm, in the right way, and with the right checks, the unit matures sustainably—retaining purpose, delivering value, and remaining intelligible in a dynamic environment.

12.5. Meta-Governance Considerations

Governance does not end at the construct level. In systems composed of multiple sociotechnical units—each with its own set of PDD constructs—*meta-governance* becomes essential. It ensures that units do not evolve in isolation, diverge beyond interoperability, or undermine system-wide coherence. Meta-governance is the architecture behind architecture: it governs how governance itself is structured, delegated, and integrated across units and constructs.

While individual construct stewards maintain local alignment, meta-governance introduces the *cross-cutting rules, tools, and roles* needed to coordinate evolution, resolve conflicts, and preserve the integrity of the larger system.

12.5.1. Federated Oversight with Local Autonomy

Meta-governance must balance standardization with autonomy. Each unit should be free to adapt its constructs based on local purpose and context, but within a shared governance frame. This often involves

- *Shared governance principles* (e.g., versioning standards, stewardship responsibilities)
- *Minimum viable contracts* for interoperability (e.g., Function schemas, Memory auditability)
- *Domain-level governance councils* that align structural changes across adjacent units

This federated model enables diversity without chaos—supporting local adaptation while avoiding fragmentation.

EXAMPLE SCENARIO—FEDERATED OVERSIGHT IN PRACTICE

A financial organization with five regional data platforms established a Global Governance Council. Each region retained autonomy for local data processes but adhered to shared policies on metadata standards and lineage. When one region introduced a new data-sharing construct, the council coordinated impact analysis and version updates across all others. This avoided divergence while preserving local flexibility—illustrating how meta-governance can synchronize evolution without centralizing control.

12.5.2. Construct Registries and Shared Catalogs

Constructs should be discoverable, auditable, and referable across units. Meta-governance relies on

- *Construct registries*: Machine-readable records of constructs, versions, dependencies, and stewards
- *Shared catalogs*: Interfaces for reusing Value definitions, Function endpoints, or Process templates
- *Tagging and classification standards*: To support filtering, comparison, and governance automation

Without registries and catalogs, constructs become opaque, duplicated, or untraceable across domains.

12.5.3. Cross-Construct Synchronization

Changes to one construct often require coordinated updates elsewhere. Meta-governance introduces

- *Change coordination workflows* (e.g., updating Process and Function when Value changes)
- *Dependency awareness graphs* to visualize inter-construct links
- *Propagation policies* to determine which changes must ripple across units

This ensures that evolution remains coherent across layers and boundaries, not just locally consistent.

12.5.4. System-Level Feedback and Governance Analytics

Meta-governance must be evidence-based. This requires

- *Aggregated telemetry* across units (e.g., usage of Functions, Process throughput, Memory access rates)
- *Health and maturity indicators* for constructs (e.g., stale Storage, undefined Purpose, excessive Culture overrides)
- *Governance dashboards* to monitor compliance, alignment, and drift

These analytics allow governance to evolve responsively, not bureaucratically.

12.5.5. Meta-Stewardship and Escalation Pathways

In complex environments, issues may transcend the authority or perspective of individual construct stewards. Meta-governance defines

- *Meta-stewards* or system architects responsible for coherence across units
- *Escalation protocols* for construct conflicts, structural misalignments, or compliance breaches
- *Cross-unit review boards* to approve major changes to foundational constructs (e.g., Purpose, Order)

This ensures that high-impact decisions are made with system-wide awareness, not in local silos.

Meta-governance is not about imposing control from above—it is about designing the conditions under which autonomy can thrive without undermining unity. It enables Purpose Driven Design to scale responsibly, ensuring that constructs evolve not just within units, but across systems, with alignment, intelligibility, and structural integrity intact.

META-GOVERNANCE READINESS CHECKLIST

1. *Shared Principles Defined*: Have enterprise-wide governance principles (e.g., versioning, stewardship, auditability) been documented and adopted by all units?
2. *Federated Councils Formed*: Are domain-level or cross-unit councils established to align structural decisions and resolve conflicts?
3. *Construct Registries Operational*: Do shared registries record versions, dependencies, and stewards across units?
4. *Change Coordination Workflow*: Is there a standard process for propagating construct changes and notifying dependent domains?
5. *System-Level Telemetry*: Are governance dashboards or analytics used to monitor alignment and drift across units?
6. *Escalation Protocols*: Are escalation pathways and meta-steward roles defined for issues that exceed local authority?
7. *Feedback Integration*: Do results from audits and analytics loop back into federated councils for continuous refinement?

12.6. Summary and Governance Checklist

Governance and life cycle management are not peripheral concerns in Purpose Driven Design—they are the structural guarantees that allow constructs to persist, adapt, and contribute meaningfully over time. Without them, constructs become brittle, incoherent, or invisible, undermining the very coherence they were designed to support. Governance provides the architecture for decision rights, oversight, and adaptation; life cycle awareness ensures that constructs do not stagnate or drift without intent. Together, they enable units to evolve without unraveling.

This chapter has outlined the foundational mechanisms—versioning, stewardship, contracts, feedback, transparency—that apply uniformly across all constructs. It has introduced life cycle stages to manage change with clarity, categorized typical evolution tempos to inform governance rhythms, and addressed meta-governance as a way to maintain alignment across multiple units and systems. These practices ensure that constructs are not only well-designed at inception but remain intelligible, functional, and structurally aligned over time.

Table 12-5 consolidates these considerations into a governance checklist, offering criteria for evaluating both construct-level readiness and systemic alignment.

Table 12-5. *Governance and Life Cycle Checklist*

Category	Checklist item
Versioning	Is the construct versioned and are changes traceable over time?
Stewardship	Is there a designated steward responsible for the construct's coherence and relevance?
Compositional contracts	Are the construct's dependencies and external expectations clearly defined and enforced?
Feedback mechanisms	Is feedback from usage, performance, or alignment actively informing improvements?
Life cycle stage	Is the construct's current life cycle stage (e.g., Defined, Active, Deprecated) explicitly declared and understood?
Change management	Are changes to the construct coordinated with dependent constructs and communicated effectively?
Inter-construct alignment	Are dependencies between this construct and others reviewed for coherence?
Transparency	Is the construct discoverable, documented, and interpretable by stakeholders?
Meta-governance integration	Does the construct participate in shared governance structures (registries, councils, dashboards)?
Evolution tempo fit	Is the rate of change appropriate for the construct's systemic role (e.g., stable Purpose, adaptive Process)?

This checklist can be embedded into design playbooks, maturity models, or governance dashboards. When applied consistently, it prevents constructs from becoming isolated artifacts or unmanaged liabilities. Instead, it helps position them as durable, accountable, and evolvable components of a coherent sociotechnical system—aligned with purpose, informed by practice, and governed by design.

Table 12-6 operationalizes these governance principles into a self-audit template, providing criteria for assessing construct-level maturity and next steps.

Use this quick self-assessment to evaluate the governance maturity of a unit or construct. Mark each item ✓ if complete, △ if in progress, or X if absent.

Table 12-6. *Governance Readiness Self-Audit (Practitioner Template)*

Category	**Audit question**	**Status (✓ / △ / X)**	**Next action/owner**
Versioning	Are all constructs version-controlled with visible change history?		
Stewardship	Is a responsible steward or team formally assigned?		
Contracts	Are inter-construct dependencies defined and maintained?		
Feedback	Do metrics or user signals inform construct updates?		
Life cycle	Is each construct's current state (Defined, Active, Deprecated) documented?		
Change mgmt	Are cross-construct changes coordinated and communicated?		
Alignment	Are dependencies reviewed for coherence at least quarterly?		

(*continued*)

Table 12-6. *(continued)*

Category	Audit question	Status (✓ / △ / X)	Next action/owner
Transparency	Is governance documentation accessible to all stakeholders?		
Meta-gov integration	Does the unit participate in shared councils, registries, and dashboards?		
Evolution tempo	Is the rate of change appropriate to the construct's role?		

A completed audit highlights both strengths and areas for governance reinforcement.

CHAPTER 13

Conclusion—Architecting with Purpose, Designing for Coherence

A system cannot govern what it cannot see, and it cannot endure what it cannot explain. Just as a pilot cannot navigate through fog without instruments, organizations lose control and coherence when their structures and decisions become opaque to themselves. This book has offered a deliberate architecture of constructs—Purpose, Value, Function, Process, Structure, Order, Culture, Memory, and Storage—not as abstract ideals, but as tangible design surfaces for building coherent, adaptive, and meaningful sociotechnical systems. These constructs form more than a taxonomy; they represent the essential scaffolding of systems capable of self-awareness, continuity, and transformation. How these constructs reinforce and shape one another comes into focus in the cyclical representation illustrated in Figure 13-1.

A. Milchman, *The Purpose Driven Design Framework for Modern Enterprises*,
https://doi.org/10.1007/979-8-8688-2223-0_13

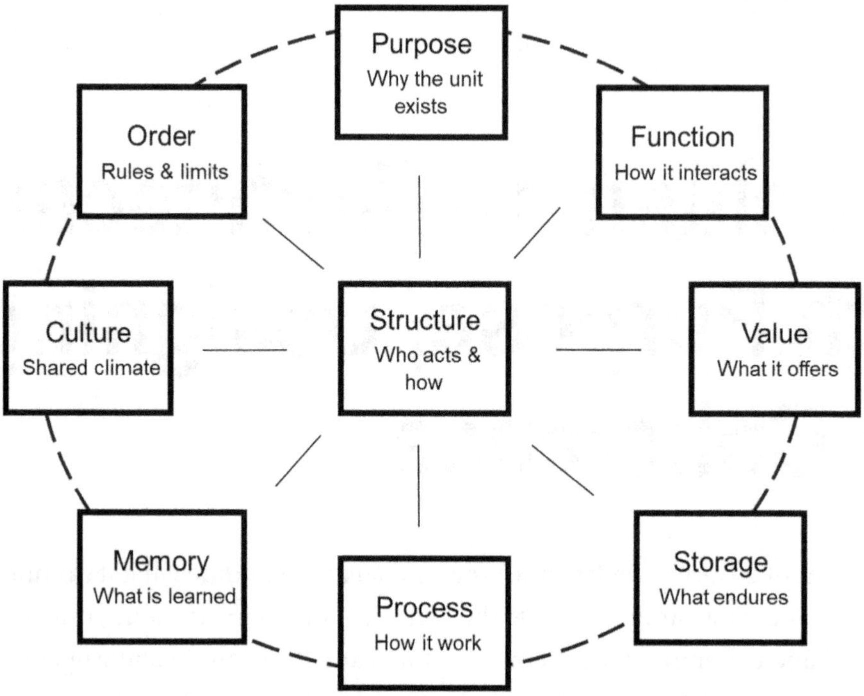

Figure 13-1. *One possible representation of the nine constructs of Purpose Driven Design. The diagram illustrates their cyclical continuity and interdependence within a sociotechnical unit but does not attempt to depict all possible relationships or feedback loops among constructs*

In this conclusion, we pause not to summarize but to step back—reflecting on what this architectural perspective enables, why coherence must be intentional, and how designing with Purpose opens new possibilities not only for system performance but for human dignity, trust, and agency.

As you design or steward systems of your own, ask

- Where is Purpose visible, and where has it become assumed?
- Do the constructs that sustain your system—Value, Function, Process, Structure—still align with that Purpose?
- Which boundaries of Order or habits of Culture support coherence, and which now inhibit it?
- What should your system choose to remember, and what can it safely forget?

- How does your system distinguish between what must be remembered (Memory) and what must be preserved for reuse (Storage)?

13.1. Recapitulating the Journey

This book has traced a path from structural confusion to architectural coherence, offering not just a new framework but a way of seeing. It began with a simple observation: that modern systems—digital, institutional, and social—are too often composed without internal differentiation, evolving not through design but through accumulation. In response, *Purpose Driven Design* introduced a set of nine composable constructs that enable sociotechnical units to operate intelligibly, align structurally, and evolve without disintegration.

Each construct addresses a distinct dimension of coherence:

- *Purpose* orients the unit toward meaning.
- *Value* externalizes that meaning through offerings.
- *Function* exposes value through usable interfaces.
- *Process* performs the work that generates value.
- *Structure* configures the agents and roles responsible for action.
- *Order* formalizes the rules and constraints under which the unit operates.
- *Culture* shapes the informal climate in which those rules and actions unfold.
- *Memory* preserves systemic recall, supporting learning and accountability.
- *Storage* retains the durable artifacts of value and activity.

During growth phases, outer constructs such as Purpose, Value, and Function expand outward to engage new contexts, while inner constructs—Process, Structure, and Order—stabilize the system; during crisis, alignment flows inward, with Process and Structure reasserting coherence around Purpose.

Rather than layering new complexity, these constructs clarify what must be present in every coherent system. They serve as building blocks—not to be applied uniformly but to be composed thoughtfully, with full awareness of their interdependencies. From Purpose to Storage, the book has shown how these constructs can be designed, governed, and evolved—not in isolation, but as a mutually reinforcing system. Table 13-1 consolidates these principles into a set of practical dos and don'ts, offering direction for maintaining coherence while managing complexity.

Table 13-1. *Practical Guidance for Composing and Evolving PDD Constructs—Common Dos and Don'ts to Preserve Coherence Without Adding Unnecessary Complexity*

Do	Don't
Treat constructs as composable surfaces, not sequential steps.	Force a uniform template across every context.
Revisit interdependencies regularly—changes in one affect all.	Evolve constructs in isolation.
Let Purpose and Order anchor change; adapt Process and Structure more freely.	Redefine Purpose or Culture impulsively.
Distinguish Memory (history) from Storage (artifacts).	Confuse traceability with hoarding.
Use Culture and Order to balance autonomy and constraint.	Substitute culture slogans for governance.

What began as a structural diagnosis has become a compositional grammar. Across contexts—social, physical, digital—the grammar of constructs has revealed the same pattern: coherence emerges not from control but from composition; resilience depends on clarity of purpose more than on uniformity; and systems mature when their constructs evolve in rhythm rather than in isolation. And what once appeared as a failure of ambition (the collapse of Babel) is now revealed as a failure of design. With this journey complete, the imperative becomes clear: to architect with intention, govern with clarity, and compose systems that carry purpose without unraveling under their own scale.

13.2. From Systems to Worlds

We do not merely build systems—we construct worlds. Every architecture reflects more than function; it embeds assumptions, priorities, values, and beliefs. The structures we design—through code, roles, rules, and interfaces—become lived environments. They shape not only how work is done but how meaning is formed, how responsibility is distributed, and how coherence is either preserved or eroded over time. Facilitating open dialogue about values and meaning—through joint design sessions, reflective reviews, or stakeholder mapping—helps ensure that what is built remains interpretable and legitimate to all who inhabit it.

When designed with dignity in mind, a health platform that exposes transparent decision paths can empower patients and clinicians to act with trust. By contrast, a system that hides its reasoning—automating eligibility or risk scores without explanation—erodes agency and turns participation into dependence.

Purpose Driven Design reframes architecture not as technical assembly, but as world-making. A system without Purpose is inert; one without Culture is unlivable; one without Memory cannot learn. These constructs are not abstractions—they are the internal conditions that make systems inhabitable by humans and extensible by machines. When composed deliberately, they allow units to express clarity, direction, and interdependence. When neglected, they produce alienation, opacity, and collapse.

When the Bauhaus school in early-twentieth-century Germany united art, craft, and industrial design, it did more than produce modern furniture—it reimagined how everyday environments could express clarity, functionality, and social progress. In that sense, it turned architecture and design into instruments of world-making.

Table 13-2 contrasts how each construct of Purpose Driven Design manifests in a coherent, well-governed system versus a fragmented or failing one.

Table 13-2. *Healthy System vs. Failing System—contrasting how each PDD construct manifests when coherence is sustained versus when it deteriorates*

Construct	Healthy system	Failing system
Purpose	Shared, lived, and continuously reinterpreted	Drifts or fragments; loses legitimacy
Value	Clear contribution recognized by stakeholders	Outputs detached from real relevance
Function	Accessible, coherent, and trustworthy	Confusing, unreliable, or overexposed
Process	Transparent, adaptive, and traceable	Hidden, brittle, or improvisational
Structure	Roles aligned and accountable	Ambiguous, overlapping, or absent
Order	Constraints enable autonomy and safety.	Rules suppress or contradict intent.
Culture	Trust-based, reflective, and evolving	Defensive, stagnant, or performative
Memory	Learning embedded in decision-making	Forgetfulness, repetition, and blame
Storage	Artifacts preserved and reusable	Cluttered, inaccessible, or decayed

In this light, design carries moral consequences. To architect is to encode a vision of what kind of world is possible—and permissible. Constructs like Order and Culture are not neutral; they influence who is included, what is prioritized, and how change is absorbed or resisted. Even the decision to store or forget is a structural stance on what matters.

Designers might ask: Whose perspective does this construct serve? Whose agency does it enable or constrain? Which priorities become invisible once the design solidifies?

This book has offered not only a framework for coherence but a reminder: every system we build participates in the making of a world. And the responsibility of the designer is not just to make it work—but to make it make sense.

13.3. The Real Stakes of Design

Design is never neutral. Every omission, every misalignment, every invisible decision accumulates into systems that either enable coherence or produce fragmentation. What fails in complex systems is not always intention—it is often structure. And what breaks first is rarely the code or the interface, but the shared understanding of what the system is for, how it works, and whom it serves.

Before moving to implementation, ask

- Is the Purpose of this design still clear and shared?
- Which constructs have been overemphasized or neglected?
- Where might users experience opacity, exclusion, or loss of agency?
- What assumptions have gone untested or unspoken?
- How will the system remain coherent as it evolves?

The stakes of design extend beyond the technical—they are human. When Purpose is unclear, decisions drift.

COMMON HUMAN CONSEQUENCES OF MISALIGNED DESIGN—AND WHAT CAN HELP REPAIR THEM

- *Loss of trust*: Restore transparency and explainability.
- *Exclusion*: Broaden participation and representation.
- *Fatigue and disengagement*: Simplify interfaces and clarify purpose.
- *Anxiety or confusion*: Increase legibility and feedback loops.
- *Dependence without agency*: Reintroduce human oversight and choice.

When Purpose is unclear, decisions drift—as when a civic platform launches without a shared goal, leaving teams to pursue conflicting metrics. When Value is ill-defined, outputs lose relevance—as when systems produce artifacts that no stakeholder recognizes as useful or meaningful. When Function is inconsistent, users disengage—like a digital service whose updates break familiar workflows each release. When Process is incoherent, execution fragments—as when work depends on improvisation, leading to delays, duplication, and failure at handoffs. When Structure is ambiguous, responsibility diffuses—as when no one knows who acts, who decides, or who owns outcomes. When Order is rigid or Culture is toxic, coordination breaks down—as in organizations where rules silence initiative and trust evaporates. When Memory fades and Storage fragments, responsibility vanishes. These are not abstract risks—they are the daily realities of misaligned systems, experienced as friction, fatigue, exclusion, and loss of trust.

The collapse of Babel was not the result of overreach—it was the result of under-architecture. The ambition was not the problem; the absence of structural differentiation was. This insight is not historical—it is diagnostic. Today's enterprises replicate the same failure patterns at scale: expanding without decomposing, integrating without coordinating, modernizing without aligning.

Design, then, is not merely a question of tools or processes—it is a question of consequence. It determines whether systems evolve or ossify, whether they empower or obscure, whether they fragment or cohere. Every organization should periodically ask: What consequences—intended or not—are our current design choices creating, and for whom? The real stakes of design are not about efficiency—they are about intelligibility, legitimacy, and the preservation of meaning at scale.

13.4. Architecting for the Future

The future will not be built from scratch—it will be composed from what we already have, restructured with greater clarity, coherence, and care. As systems grow more distributed, as automation accelerates, and as machine agency expands alongside human intent, the need for purposeful design becomes not just urgent but foundational. Architecting for the future means designing not for stability alone but for evolvability, intelligibility, and enduring alignment.

Purpose Driven Design offers a grammar for such futures. It is not a methodology to be enforced, but a set of structural constructs to be composed and recomposed as contexts shift. Whether shaping adaptive institutions, federated platforms, intelligent assistants, or civic infrastructures, these constructs provide the means to retain purpose across change, and to remain legible even as complexity scales.

It's important to note that readers need not start big; one can begin by periodically revisiting each construct—asking how it has shifted, what remains aligned, and what must be recomposed for current conditions.

The future belongs to systems that can evolve without unraveling, interoperate without collapsing into uniformity, and adapt without losing their ethical center. To design such systems is not simply a matter of technical competence—it is an act of stewardship. It requires humility to accept that coherence is not a one-time achievement but an ongoing responsibility. Like tending a garden or maintaining a craft, coherence endures only through ongoing attention and renewal.

As the boundaries between digital, physical, and social systems continue to blur, the role of the architect must also evolve—from builder of frameworks to composer of meaning. To architect for the future is to insist that structure must serve clarity, that design must encode purpose, and that coherence—across roles, functions, and constructs—is not a luxury but a precondition for systems worth sustaining.

Coherence depends on continuous attention to

- Purpose that remains explicit and shared
- Functions that reliably express that Purpose
- Processes that are transparent and adaptive
- Structures that clarify accountability and enable flow
- Culture and Order that sustain trust and alignment

13.5. Closing Reflection: The Unit As a Hopeful Construct

At the heart of Purpose Driven Design lies a quiet conviction: that the **unit**—not the enterprise, not the platform, not the system—is the most hopeful architectural form we can build. A unit is small enough to be accountable, yet structured enough to scale. It carries its own Purpose, speaks its own Value, governs its own Function, and remains legible to those who build and inhabit it. In practice, a product or sales team operating as a self-contained unit can define its own Purpose, deliver measurable Value, and adjust its internal Structure without waiting for enterprise-wide change—demonstrating how accountability and scalability can coexist.

In an age of sprawling systems and abstract platforms, the unit offers architectural dignity. It is not a container of tasks, but a vessel of meaning. Its constructs are not checklists—they are commitments: to design with clarity, to coordinate with integrity, and to evolve without losing direction. A unit does not seek to centralize; it seeks to compose—coherently, contextually, and consciously.

Designing such units is not only feasible—it is necessary. It is how we reclaim systems from ungovernable sprawl. It is how we ensure that scale does not come at the cost of sensemaking. And it is how we create architectures that are not just efficient but *livable*—capable of sustaining trust, adaptation, and shared meaning over time.

To design accountable and scalable units, ask

- Is the unit's Purpose explicit and legible to those who depend on it?
- What Value does it deliver—and to whom?
- Are Function, Process, and Structure proportionate to that Purpose?
- How does Order protect autonomy while preserving alignment?
- What mechanisms of Memory and Storage sustain its continuity over time?

The unit is our response to the fragmentation of modern complexity. It is a statement that coherence is still possible, that structure still matters, and that systems can still be designed—not as monuments to ambition but as frameworks for belonging. Treat every act of design as an invitation—to coherence, to care, and to the shared work of meaning.

Designing Toward Coherence

Purpose Driven Design offers more than a set of constructs—it provides a language for composing systems that remain coherent as they evolve. In a world where complexity accelerates and alignment decays, the ability to build with structural clarity is no longer optional. It is the foundation for adaptability, legitimacy, and meaning.

Each construct—Purpose, Value, Function, Process, Structure, Order, Culture, Memory, and Storage—does not stand alone. Together, they form the internal architecture of sociotechnical units capable of navigating change without losing direction. Their proper design and governance is what distinguishes resilient systems from those that merely accumulate.

This book began with the lesson of Babel: that collapse often follows from unstructured ambition. It ends with a simple proposition—that clarity, not control, is the future of architecture. Purpose is not just where design begins. It is how coherence is sustained.

A. Milchman, *The Purpose Driven Design Framework for Modern Enterprises*,
https://doi.org/10.1007/979-8-8688-2223-0

Index

A

A. Milchman, *The Purpose Driven Design Framework for Modern Enterprises*,
https://doi.org/10.1007/979-8-8688-2223-0

B

C

D

E

F

G

H

I, J

K

L

N

O

P, Q

T

U

V

W, X, Y, Z

www.ingramcontent.com/pod-product-compliance
Lightning Source LLC
LaVergne TN
LVHW082019200926
848481LV00010B/2889
* 9 7 9 8 8 6 8 8 2 2 2 2 3 *